Interior Designer's
Portable Handbook

D1082257

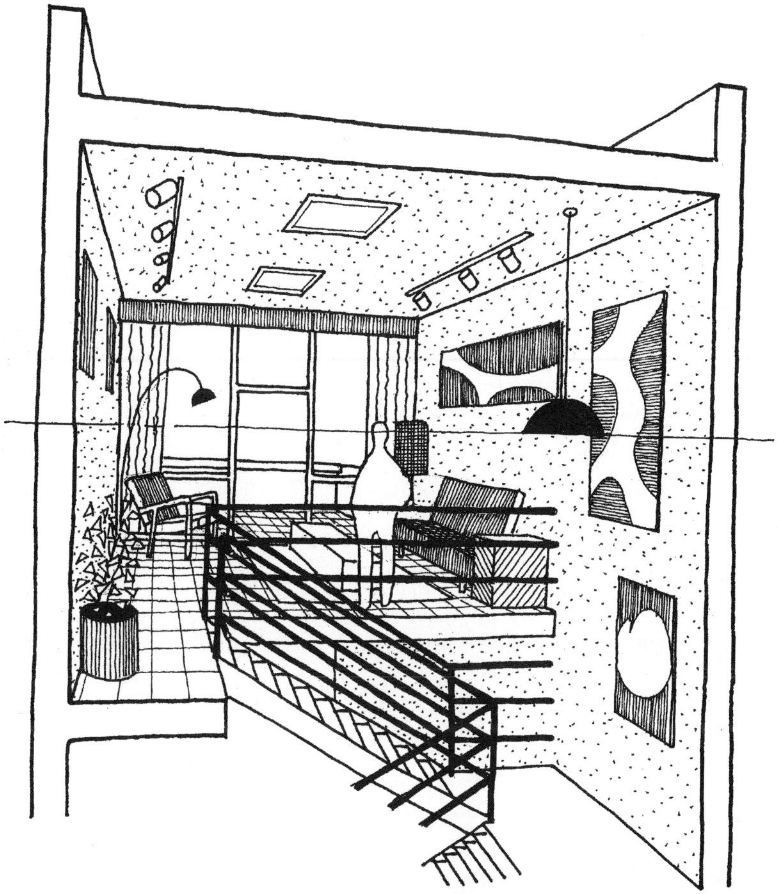

INTERIOR DESIGNER'S PORTABLE HANDBOOK

FIRST-STEP RULES OF THUMB FOR THE DESIGN OF INTERIORS

Third Edition

PAT GUTHRIE
ARCHITECT

McGraw Hill

New York Chicago San Francisco Lisbon London Madrid Mexico City
Milan New Delhi San Juan Seoul Singapore Sydney Toronto

The **McGraw·Hill** Companies

Copyright © 2012, 2004, 2000 by The McGraw-Hill Companies, Inc.
All rights reserved. Printed in the United States of America. Except as
permitted under the United States Copyright Act of 1976, no part of
this publication may be reproduced or distributed in any form or by
any means, or stored in a data base or retrieval system, without the
prior written permission of the publisher.

1 2 3 4 5 6 7 8 9 0   DOC/DOC   1 8 7 6 5 4 3 2

ISBN 978-0-07-178206-7
MHID 0-07-178206-0

*The sponsoring editor for this book was Joy Evangeline Bramble, the
editing supervisor was Stephen M. Smith, the production supervisor
was Pamela A. Pelton, and the acquisitions coordinator was Molly T.
Wyand. Virginia E. Carroll of North Market Street Graphics was the
project manager. This book was set in Times Ten by North Market
Street Graphics. The art director for the cover was Jeff Weeks.*

Printed and bound by RR Donnelley.

McGraw-Hill books are available at special quantity discounts to
use as premiums and sales promotions, or for use in corporate
training programs. To contact a representative, please e-mail us at
bulksales@mcgraw-hill.com.

This book is printed on acid-free paper.

Dedicated to:

- *Pat Anthony, who got me interested in interior design*
- *Wendy Lochner, who invited me to do this book*
- *My family (Jan, Eric, and Erin)*

ABOUT THE AUTHOR

John Patten (Pat) Guthrie, AIA, was principal of John Pat
Guthrie Architects, Inc. A now-retired resident of Scottsdale,
Arizona, Pat has been licensed as an architect in 13 states. He
is also the author of *The Architect's Portable Handbook*,
*Cross-Check: Integrating Building Systems and Working
Drawings* (both published by McGraw-Hill), and *Desert
Architecture* (self-published). Pat has taught at the FLW
School of Architecture at Taliesin West. He and his wife
Janet are the parents of two grown children, Eric and Erin.
Pat's avocations include world travel, art, sailing, and history.

Contents

Use this table as a checklist for the interior design of buildings.

___ PART 6. WOOD

___ PART 7. THERMAL AND MOISTURE PROTECTION

___ PART 8. DOORS, WINDOWS, AND GLASS

___ PART 9. FINISHES

___ PART 10. SPECIALTIES

Preface

This book is largely adapted from the author's *The Architect's Portable Handbook* (McGraw-Hill, 2010). Many things overlap between the fields of architecture and interior design. In fact, interior design could be considered a specialty in the architectural field. From *The Architect's Portable Handbook* those things dealing with engineering, with the exterior "skin," and the "bones" (structural elements) have largely been deleted. What has been added is a greater emphasis on space planning, interior finishes, equipment, and furniture.

This book is particularly oriented toward those working in the field of interior design at two levels:

1. The interior decorator who selects furnishings and room finishes.
2. The interior designer who does what the decorator does, but who also does space planning and may approach what an architect does for "tenant improvement" or remodel-type projects where whole new spaces are created within an existing building.

In either case, and everything in between, this book should be a help to anyone doing "interiors."

As with *The Architect's Portable Handbook,* this book is laid out in the C.S.I. (Construction Specifications Institute) format. The user would do well to become familiar with this, as it has become a standard in the construction industry.

This book does not deal with aesthetics. There are many books on that subject. Rather, the idea of this book is to provide the designer with fast preliminary facts to save the designer time—so more time can be devoted to the creative side of interior design.

A final word. The major reason for this third edition, besides updating costs and other information, is the new building codes.

Acknowledgments

Thanks to the following people for their professional expertise in helping with this book:

Special thanks to: Pat Anthony.

Also, thanks to: Steve Andros, Doug Collier, Ed Denham, Anthony Floyd, Rick Goolsby, Glenn Heyes, Larry Litchfield, Norm Littler, Bill Lundsford, John Messerschmidt, Marvin Nance, and Wildan Co.

How to Use This Book

The concept of this book is that of a *personal tool* that compacts the 20% of the data that is needed 80% of the time by *design professionals* in the preliminary design of *interiors* of all types and sizes.

This tool is meant to always be at one's *fingertips* (open on a drawing board or desk, carried in a briefcase, or kept in one's pocket). It is never meant to sit on a bookshelf. It is meant to be *used every day!*

Because design professionals are individualistic and their practices are so varied, the user is encouraged to *individualize this book* over time, by adding notes or changing data as experience dictates.

The addition of rough construction and furnishings **costs** throughout the book (making this type of handbook truly unique) will "date" the data. But building laws, new products, new technologies, and materials are changing just as fast. Therefore, this book should be looked on as a *starter of simple data collection* that must be updated over time. New editions *may* be published in the future. See p. 35 for more information on **costs**.

Because this book is so broad in scope, yet so compact, information can be presented only at one place and not repeated. There is little room for examples of how to use the information provided. Information is presented by simple numbers that leave the need for *common-sense judgment.*

The whole book is laid out in checklist format, to be quickly read and checked against the design problem at hand.

Where a ◯ is shown, refer to p. 519 for further explanation of references.

This book is *not a substitute* for professional expertise or other books of a more detailed and specialized nature, but it will be a continuing every-day aid that takes the more useful "cream" off the top of other sources.

Interior Designer's
Portable Handbook

NOTES

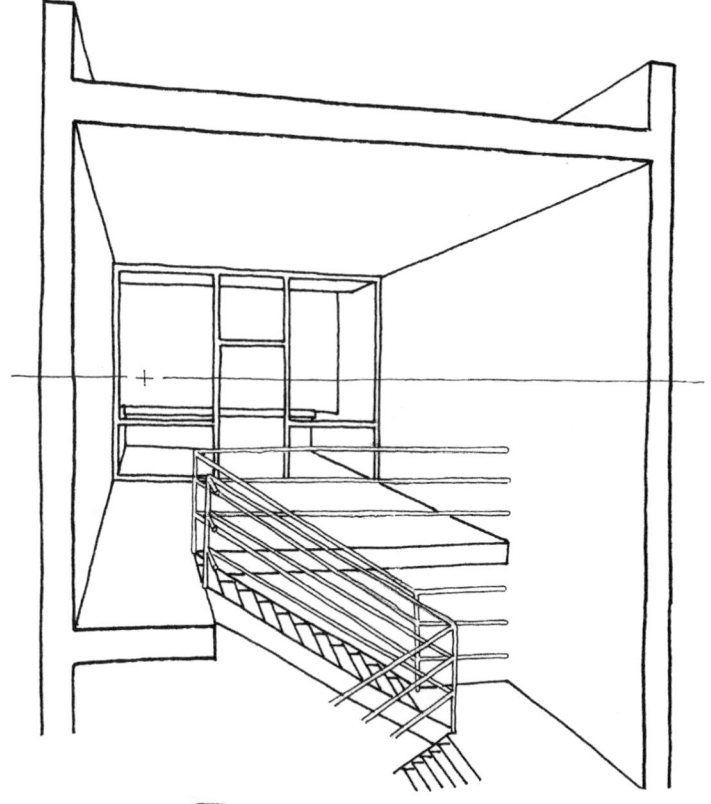

1

GENERAL

<u>NOTES</u>

__ A. PRACTICE ⑤ ㉑ ㊲ ㊴ ㊷

__ 1. Compensation (fees)

__ *a.* See Appendix for designer fees as a percentage of installation cost by building type.

__ *b.* Use the American Institute of Architects breakdown of fees as a guide:

__ Schematic design phase	15% ⎫ or	
__ Design development ment	20% ⎭	25% Preliminary design
__ Construction documents	40%	50% Construction documents
__ Bid/negotiation	5% ⎫	25% Construction
__ Construction administration	20% ⎭	administration
	100%	100%

__ 2. Rules of Thumb for Business Practice

__ *a.* *Watch cash flow:* For a small firm, balance checkbook. For a medium or large firm, use cash statements and balance and income statements. Estimate future cash flow based on past, with 15% "fudge factor," plus desired profit.

__ *b.* *Have financial reserves:* (six months' worth).

__ *c.* *Monitor time by these ratios:*

__ (1) Chargeable ratio = $\dfrac{\text{direct job labor cost}}{\text{total labor cost}}$

This tells what percent of total labor cost is being spent on paying work. The higher the percent the better. Typical range is 55 to 85%, but lower than 65% is poor. However, principals often have a 50% ratio.

__ (2) Multiplier ratio = $\dfrac{\text{dollars of revenue}}{\text{dollars of direct labor}}$

This ratio is multiplied times wages for billing rates. Usually *2.5 to 3.0.* Will vary with firm and time.

__ (3) Overhead rate: looks at total indirect expenses as they relate to total direct labor. An overhead rate of 180 means $1.80 spent for each $1.00 working on revenue-producing projects.

__ (4) Profit: measured as total revenue minus expenses. Expressed as percent of total revenue.

___ *d.* <u>*Monitor accounting reports:*</u> A financial statement consists of:

    ___ (1) Balance Sheet: Tells where you are on a given date by Assets and Liabilities.

    ___ (2) Earnings Statement (Profit and Loss): Tells you how you got there by Income less Direct (job) costs, and Indirect (overhead) costs = Profit, or Loss.

___ *e.* <u>*Mark up for reimbursable expenses*</u> (travel, printing, etc.): Usually 10%.

___ *f.* <u>*Negotiating contracts*</u>

    ___ (1) Estimate scope of services.

    ___ (2) Estimate time, costs, and profit.

    ___ (3) Determine method of compensation: See 4.

___ *g.* <u>*Contract checklist*</u>

    ___ (1) Detailed scope of work, no interpretation necessary.

    ___ (2) Responsibilities of both parties.

    ___ (3) Monthly progress payments.

    ___ (4) Interest penalty on overdue payments.

    ___ (5) Limit length of construction administration phase.

    ___ (6) Cost estimating responsibilities.

    ___ (7) For cost-reimbursable contracts, specify a provisional overhead rate (changes year to year).

    ___ (8) Retainer, applied to fee but not costs.

    ___ (9) Date of agreement, and time limit on contract.

    ___ (10) Approval of work—who, when, where.

    ___ (11) Ways to terminate contract, by both parties.

    ___ (12) For changes in scope, bilateral agreement, and an equitable adjustment in fee.

    ___ (13) Court or arbitration remedies and who pays legal fees.

    ___ (14) Signature and date by both parties.

    ___ (15) Limits on liability.

    ___ (16) Time limit on offer.

    ___ (17) Put in writing!

___ 3. **Fees and Purchasing**

The interior designer's compensation is more complicated than that for other design professionals. The designer may charge a professional fee. The designer may also be a contractor of sorts by being the purchaser and installer of furnishings, with profit from the difference between wholesale and resold price to the client. The interior designer may use either of these methods or both.

___ a. *Professional fees* can be determined by:

    ___ (1) Percentage of installed cost (furnishing or interior construction cost, or both).

    ___ (2) Lump sum.

    ___ (3) Hourly rates. **(Typical rates are $75 to $145/hr.)**

    ___ (4) Hourly rates with maximum "upset."

___ b. *Resale compensation* can be determined by:

Designer purchasing furniture from supplier at a discounted amount less than the list price. This usually ranges from 10% to 50% discount. The designer then resells furniture to client at a higher price. The difference is compensation to the designer. Other issues in the resale of furniture are:

    ___ (1) Who pays for *freight* (shipping) and insurance? "F.O.B. Factory" means buyer assumes ownership and responsibility when goods are loaded on truck at factory. "F.O.B. Destination" means the manufacturer retains ownership and responsibility of goods until they reach delivery destination. Costs of freight and insurance are borne by the manufacturer in this case.

    ___ (2) Cost of *warehousing,* if required, and who is responsible for it needs to be established. As a rule of thumb, shipping is 8% to 12% of discounted price, with 10% as a good average.

    ___ (3) Sales tax for resale of furniture.

EXAMPLE

THE INTERIOR DESIGNER SELECTS A PIECE OF FURNITURE WITH MANUFACTURER'S LIST PRICE OF $1000. THE MANUFACTURER OFFERS A 50% DISCOUNT 'TO THE TRADE' AND F.O.B. DESTINATION. THE DESIGNER PURCHASES THE FURNITURE FOR 50% OF $1000, OR $500.

THE DESIGNER MAY THEN RESELL THE FURNITURE TO THE CLIENT AT :

LIST $1000 ($1000 - $500 = $500 COMPENSATION)
+ TAX

DISCOUNT, SAY 25%. ($1000 X .25 = $250 + TAX)
$500 DIFF. BETW'N MANUF. $ DESIGNER
LESS $250 RESALE (+TAX)
$250 COMPENSATION TO DESIGNER

NOTES

NOTES

___ B. "SYSTEMS" THINKING

In the planning and design of buildings, a helpful, all-inclusive tool is to think in terms of overall "systems" or "flows." For each of the checklist items on the next 2 pages, follow from the beginning or "upper end" through to the "lower end" or "outfall":

___ 1. <u>People Functions</u>
    ___ *a.* Follow flow of occupants from one space to another. This includes sources of vertical transportation (stairs, elevators, etc.) including pathways to service equipment.
    ___ *b.* Follow flow of occupants to enter building from off-site.
    ___ *c.* Follow flow of occupants to exit building as required by code.
    ___ *d.* Follow flow of accessible route as required by law.
    ___ *e.* Follow flow of materials to supply building (including furniture and off-site).
    ___ *f.* Follow flow of trash to leave building (including to off-site).
    ___ *g.* Way finding: Do graphics or other visual clues aid flow of the above six items?

___ 2. <u>Heat</u>
    ___ *a.* Follow sun paths to and into building to plan for access or blocking.
    ___ *b.* Follow excessive external (or internal) heat through building skin and block if necessary.
    ___ *c.* Follow source of internal heat loads (lights, people, equipment, etc.) to their "outfall" (natural ventilation or AC, etc.).
    ___ *d.* Follow heat flow into materials over a year, a day, etc. and allow for expansion and contraction.

___ 3. <u>Air</u>
    ___ *a.* Follow wind patterns through site to encourage or block natural ventilation through building, as required.
    ___ *b.* Follow air patterns through building. When natural ventilation is used, follow flow from inlets to outlets. When air is still, hot air rises and cold air descends.
    ___ *c.* Follow forced-air ventilation patterns through building to address heat (add or dissipate) and odors.

___ 4. <u>Light</u>
    ___ *a.* Follow paths of natural light (direct or indirect sun) to and into building. Encourage or block as needed.
    ___ *b.* Follow paths of circulation and at spaces to provide

artificial illumination where necessary. This includes both site and building.

___ 5. <u>Energy and Communications</u>

___ *a.* Follow electric or gas supply from off-site to transformer, to breakers or panels, and to each outlet or point of connection.

___ *b.* Follow telephone lines from off-site to telephone mounting board, to each phone location.

___ 6. <u>Sound</u>

___ *a.* Identify potential sound sources, potential receiver locations, and the potential sound paths between the two.

___ *b.* Follow sound through air from source to receiver. Mitigate with distance or barrier.

___ *c.* Follow sound through structure from source to receiver. Mitigate by isolation of source or receiver.

NOTES

NOTES

C. SPECIFICATIONS (15)

___ 1. **Standard outline for writing specification sections:**
    ___ *a.* General   ___ *b.* Products   ___ *c.* Execution

___ 2. **Quick checklist on products or materials:**
    ___ *a.* What is it and what does it do?
    ___ *b.* Who is it made by?
    ___ *c.* How to apply?
    ___ *d.* What does it cost?
    ___ *e.* Warranties?

___ 3. **Detailed checklist on evaluating new products or materials:**
    ___ *a.* *Structural serviceability* (resistance to natural forces such as wind and earthquake; structural adequacy and physical properties such as strength, compression, tension, shear, and behavior against impact and indentation).
    ___ *b.* *Fire safety* (resistance against the effects of fire such as flame propagation, burnthrough, smoke, toxic gases, etc.).
    ___ *c.* *Habitability* (livability relative to thermal efficiency, acoustic properties, water permeability, optical properties, hygiene, comfort, light, and ventilation, etc.).
    ___ *d.* *Durability* (ability to withstand wear, weather resistance such as ozone and ultraviolet, dimensional stability, etc.).
    ___ *e.* *Practicability* (ability to surmount field conditions such as transportation, storage, handling, tolerances, connections, site hazards, etc.).
    ___ *f.* *Compatibility* (ability to withstand reaction with adjacent materials in terms of chemical interaction, galvanic action, ability to be coated, etc.).
    ___ *g.* *Maintainability* (ease of cleaning; repairability of punctures, gouges, and tears; recoating, etc.).
    ___ *h.* *Code acceptability* (review of code and manufacturer's claims as to code compliance).
    ___ *i.* *Economics* (installation and maintenance costs).

___ 4. **CSI format**
Use this section as a checklist of everything that makes or goes into buildings, to be all-inclusive in the planning and designing of buildings, their contents, and their surroundings:

    ___ *a.* Uniformat for preliminary systems planning:

PROJECT DESCRIPTION
10—PROJECT DESCRIPTION
___ 1010 Project Summary
___ 1020 Project Program
___ 1030 Existing Conditions
___ 1040 Owner's Work
___ 1050 Funding

20—PROPOSAL, BIDDING, AND CONTRACTING
___ 2010 Delivery Method
___ 2020 Qualifications Requirements
___ 2030 Proposal Requirements
___ 2040 Bid Requirements
___ 2050 Contracting Requirements

30—COST SUMMARY
___ 3010 Elemental Cost Estimate
___ 3020 Assumptions and Qualifications
___ 3030 Allowances
___ 3040 Alternates
___ 3050 Unit Prices

CONSTRUCTION SYSTEMS AND ASSEMBLIES

ELEMENT A—*SUBSTRUCTURE*
___ A10 Foundations
___ A1010 Standard Foundations
___ A1020 Special Foundations
___ A1030 Slab on Grade

___ A20 Basement Construction
___ A2010 Basement Excavation
___ A2020 Basement Walls

ELEMENT B—*SHELL*
___ B10 Superstructure
___ B1010 Floor Construction
___ B1020 Roof Construction

___ B20 Exterior Enclosure
___ B2010 Exterior Walls
___ B2020 Exterior Windows
___ B2030 Exterior Doors

___ B30 Roofing
___ B3010 Roof Coverings
___ B3020 Roof Openings

ELEMENT C—*INTERIORS*
___ C10 Interior Construction
___ C1010 Partitions
___ C1020 Interior Doors
___ C1030 Fittings

___ C20 Stairs
___ C2010 Stair Construction
___ C2020 Stair Finishes

___ C30 Interior Finishes
___ C3010 Wall Finishes
___ C3020 Floor Finishes
___ C3030 Ceiling Finishes

ELEMENT D—*SERVICES*
___ D10 Conveying
___ D1010 Elevators and Lifts
___ D1020 Escalators and Moving Walks
___ D1090 Other Conveying Systems

___ D20 Plumbing
___ D2010 Plumbing Fixtures
___ D2020 Domestic Water Distribution
___ D2030 Sanitary Waste
___ D2040 Rain Water Drainage
___ D2090 Other Plumbing Systems

___ D30 Heating, Ventilating, and Air Conditioning (HVAC)
___ D3010 Energy Supply
___ D3020 Heat Generation
___ D3030 Refrigeration
___ D3040 HVAC Distribution
___ D3050 Terminal and Packaged Units
___ D3060 HVAC Instrumentation and Controls
___ D3070 Testing, Adjusting, and Balancing
___ D3090 Other Special HVAC Systems and Equipment

___ D40 Fire Protection
___ D4010 Sprinklers
___ D4020 Standpipes
___ D4030 Fire Protection Specialties
___ D4090 Other Fire Protection Systems

___ D50 Electrical
___ D5010 Electrical Service and Distribution

___ D5020 Lighting and Branch Wiring
___ D5030 Communications and Security
___ D5090 Other Electrical Systems

___ D60 Basic Materials and Methods

ELEMENT E—*EQUIPMENT AND FURNISHINGS*
___ E10 Equipment
___ E1010 Commercial Equipment
___ E1020 Institutional Equipment
___ E1030 Vehicular Equipment
___ E1090 Other Equipment

___ E20 Furnishings
___ E2010 Fixed Furnishings
___ E2020 Movable Furnishings

ELEMENT F—*SPECIAL CONSTRUCTION
AND DEMOLITION*
___ F10 Special Construction
___ F1010 Special Structures
___ F1020 Integrated Construction
___ F1030 Special Construction Systems
___ F1040 Special Facilities
___ F1050 Special Controls and Instrumentation

___ F20 Selective Demolition
___ F2010 Building Elements Demolition
___ F2020 Hazardous Components Abatement

ELEMENT G—*BUILDING SITEWORK*
___ G10 Site Preparation
___ G1010 Site Clearing
___ G1020 Site Demolition and Relocations
___ G1030 Site Earthwork
___ G1040 Hazardous Waste Remediation

___ G20 Site Improvements
___ G2010 Roadways
___ G2020 Parking Lots
___ G2030 Pedestrian Paving
___ G2040 Site Development
___ G2050 Landscaping

___ G30 Site Civil/Mechanical Utilities
___ G3010 Water Supply
___ G3020 Sanitary Sewer

___ G3030 Storm Sewer
___ G3040 Heating Distribution
___ G3050 Cooling Distribution
___ G3060 Fuel Distribution
___ G3090 Other Site Mechanical Utilities

___ G40 Site Electrical Utilities
___ G4010 Electrical Distribution
___ G4020 Site Lighting
___ G4030 Site Communications and Security
___ G4090 Other Site Electrical Utilities

___ G90 Other Site Construction
___ G9010 Service Tunnels
___ G9090 Other Site Systems

ELEMENT Z—*GENERAL*
___ Z10 General Requirements
___ Z1010 Administration
___ Z1020 Quality Requirements
___ Z1030 Temporary Facilities
___ Z1040 Project Closeout
___ Z1050 Permits, Insurance, and Bonds
___ Z1060 Fee

___ Z20 Contingencies
___ Z2010 Design Contingency
___ Z2020 Escalation Contingency
___ Z2030 Construction Contingency

___ *b.* Masterformat for more detailed planning:

INTRODUCTORY INFORMATION
___ 00001 Project Title Page
___ 00005 Certifications Page
___ 00007 Seals Page
___ 00010 Table of Contents
___ 00015 List of Drawings
___ 00020 List of Schedules

BIDDING REQUIREMENTS
___ 00100 Bid Solicitation
___ 00200 Instructions to Bidders
___ 00300 Information Available to Bidders
___ 00400 Bid Forms and Supplements
___ 00490 Bidding Addenda

CONTRACTING REQUIREMENTS
- ___ 00500 Agreement
- ___ 00600 Bonds and Certificates
- ___ 00700 General Conditions
- ___ 00800 Supplementary Conditions
- ___ 00900 Addenda and Modifications

FACILITIES AND SPACES

SYSTEMS AND ASSEMBLIES

CONSTRUCTION PRODUCTS AND ACTIVITIES

DIVISION 1—GENERAL REQUIREMENTS
- ___ 01100 Summary
- ___ 01200 Price and Payment Procedures
- ___ 01300 Administrative Requirements
- ___ 01400 Quality Requirements
- ___ 01500 Temporary Facilities and Controls
- ___ 01600 Product Requirements
- ___ 01700 Execution Requirements
- ___ 01800 Facility Operation
- ___ 01900 Facility Decommissioning

DIVISION 2—SITE CONSTRUCTION
- ___ 02050 Basic Site Materials and Methods
- ___ 02100 Site Remediation
- ___ 02200 Site Preparation
- ___ 02300 Earthwork
- ___ 02400 Tunneling, Boring, and Jacking
- ___ 02450 Foundation and Load-Bearing Elements
- ___ 02500 Utility Services
- ___ 02600 Drainage and Containment
- ___ 02700 Bases, Ballasts, Pavements, and Appurtenances
- ___ 02800 Site Improvements and Amenities
- ___ 02900 Planting
- ___ 02950 Site Restoration and Rehabilitation

DIVISION 3—CONCRETE
- ___ 03050 Basic Concrete Materials and Methods
- ___ 03100 Concrete Forms and Accessories
- ___ 03200 Concrete Reinforcement
- ___ 03300 Cast-In-Place Concrete
  - ___ 03310 Structural Concrete
  - ___ 03330 Architectural Concrete
  - ___ 03340 Low-Density Concrete

DIVISION 6—<u>WOOD AND PLASTICS</u>
___ 06050 Basic Wood and Plastic Materials and Methods
___ 06100 Rough Carpentry
    ___ 06110 Wood Framing
    ___ 06120 Structural Panels
    ___ 06130 Heavy Timber Construction
    ___ 06140 Treated Wood Foundations
    ___ 06150 Wood Decking
    ___ 06160 Sheathing
    ___ 06170 Prefabricated Structural Wood
    ___ 06180 Glue-Laminated Construction
___ 06200 Finish Carpentry
    ___ 06220 Millwork
    ___ 06250 Prefinished Paneling
    ___ 06260 Board Paneling
    ___ 06270 Closet and Utility Wood Shelving
___ 06400 Architectural Woodwork
    ___ 06410 Custom Cabinets
    ___ 06415 Countertops
    ___ 06420 Paneling
    ___ 06430 Wood Stairs and Railings
    ___ 06440 Wood Ornaments
    ___ 06445 Simulated Wood Ornaments
    ___ 06450 Standing and Running Trim
    ___ 06455 Simulated Wood Trim
    ___ 06460 Wood Frames
    ___ 06470 Screens, Blinds, and Shutters
___ 06500 Structural Plastics
___ 06600 Plastic Fabrications
___ 06900 Wood and Plastic Restoration and Cleaning

DIVISION 7—<u>THERMAL AND MOISTURE PROTECTION</u>
___ 07050 Basic Thermal and Moisture Protection Materials
      and Methods
___ 07100 Dampproofing and Waterproofing
    ___ 07110 Dampproofing
    ___ 07120 Built-up Bituminous Waterproofing
    ___ 07130 Sheet Waterproofing
    ___ 07140 Fluid-Applied Waterproofing
    ___ 07150 Sheet Metal Waterproofing
    ___ 07160 Cementitious and Reactive Waterproofing
    ___ 07170 Bentonite Waterproofing
    ___ 07180 Traffic Coatings
    ___ 07190 Water Repellants
___ 07200 Thermal Protection

___ 07210 Building Insulation
___ 07220 Roof and Deck Insulation
___ 07240 Exterior Insulation and Finish Systems (EIFS)
___ 07260 Vapor Retarders
___ 07270 Air Barriers
___ 07300 Shingles, Roof Tiles, and Roof Coverings
   ___ 07310 Shingles (Asphalt, Fiberglass Reinforced, Metal, Mineral Fiber Cement, Plastic Shakes, Porcelain Enamel, Slate, Wood, and Wood Shakes)
   ___ 07320 Roof Tiles (Clay, Concrete, Metal, Mineral Fiber Cement, Plastic)
   ___ 07330 Roof Coverings
___ 07400 Roofing and Siding Panels
   ___ 07410 Metal Roof and Wall Panels
   ___ 07420 Plastic Roof and Wall Panels
   ___ 07430 Composite Panels
   ___ 07440 Faced Panels
   ___ 07450 Fiber-Reinforced Cementitious Panels
   ___ 07460 Siding (Aluminum, Composition, Hardboard, Mineral Fiber Cement, Plastic, Plywood, Steel, and Wood)
   ___ 07470 Wood Roof and Wall Panels
   ___ 07480 Exterior Wall Assemblies
___ 07500 Membrane Roofing
   ___ 07510 Built-up Roofing
   ___ 07520 Cold-Applied Bituminous Roofing
   ___ 07530 Elastomeric Membrane Roofing (CPE, CSPE, CPA, EPDM, NBP, and PIB)
   ___ 07540 Thermoplastic Membrane Roofing (EIP, PVC, and TPA)
   ___ 07550 Modified Bituminous Membrane Roofing
   ___ 07560 Fluid-Applied Roofing
   ___ 07570 Coated Foamed Roofing
   ___ 07580 Roll Roofing
   ___ 07590 Roof Maintenance and Repairs
___ 07600 Flashing and Sheet Metal
   ___ 07610 Sheet Metal Roofing
   ___ 07620 Sheet Metal Flashing and Trim
   ___ 07630 Sheet Metal Roofing Specialties
   ___ 07650 Flexible Flashing
___ 07700 Roof Specialties and Accessories
   ___ 07710 Manufactured Roof Specialties (Copings, Counterflashing, Gravel Stops and Fascias, Gutters and Downspouts, Reglets, Roof Expansion Assemblies, and Scuppers)

___ 07720 Roof Accessories (Manufactured Curbs, Relief Vents, Ridge Vents, Roof Hatches, Roof Walk Boards, Roof Walkways, Smoke Vents, Snow Guards, and Waste Containment Assemblies)
___ 07760 Roof Pavers
___ 07800 Fire and Smoke Protection
___ 07810 Applied Fireproofing
___ 07820 Board Fireproofing
___ 07840 Firestopping
___ 07860 Smoke Seals
___ 07870 Smoke Containment Barriers
___ 07900 Joint Sealers

DIVISION 8—<u>DOORS AND WINDOWS</u>
___ 08050 Basic Door and Window Materials and Methods
___ 08100 Metal Doors and Frames
___ 08110 Steel
___ 08120 Aluminum
___ 08130 Stainless Steel
___ 08140 Bronze
___ 08150 Preassembled Metal Door and Frame Units
___ 08160 Sliding Metal Doors and Grilles
___ 08180 Metal Screen and Storm Doors
___ 08190 Metal Door Restoration
___ 08200 Wood and Plastic Doors
___ 08300 Specialty Doors
___ 08310 Access Doors and Panels
___ 08320 Detention Doors and Frames
___ 08330 Coiling Doors and Grilles
___ 08340 Special Function
___ 08350 Folding Doors and Grilles
___ 08360 Overhead Doors
___ 08370 Vertical Lift Doors
___ 08380 Traffic Doors
___ 08390 Pressure-Resistant Doors
___ 08400 Entrances and Storefronts
___ 08410 Metal-Framed Storefronts
___ 08450 All-Glass Entrances and Storefronts
___ 08460 Automatic Entrance Doors
___ 08470 Revolving Entrance Doors
___ 08480 Balanced Entrance Doors
___ 08490 Sliding Storefronts
___ 08500 Windows
___ 08510 Steel
___ 08520 Aluminum
___ 08530 Stainless Steel

___ 08540 Bronze
___ 08550 Wood
___ 08560 Plastic
___ 08570 Composite
___ 08580 Special Function
___ 08590 Window Restoration and Replacement
___ 08600 Skylights
___ 08700 Hardware
___ 08800 Glazing
___ 08810 Glass
___ 08830 Mirrors
___ 08840 Plastic Glazing
___ 08850 Glazing Accessories
___ 08890 Glazing Restoration
___ 08900 Glazing Curtain Wall
___ 08910 Metal-Framed Curtain Wall
___ 08950 Translucent Wall and Roof Assemblies
___ 08960 Sloped Glazing Assemblies
___ 08970 Structural Glass Curtain Walls
___ 08990 Glazed Curtain Wall Restoration

DIVISION 9—<u>FINISHES</u>
___ 09050 Basic Finish Materials and Methods
___ 09100 Metal Support Assemblies
___ 09200 Plaster and Gypsum Board
___ 09300 Tile
___ 09305 Tile Setting Materials and Accessories
___ 09310 Ceramic
___ 09330 Quarry
___ 09340 Paver
___ 09350 Glass Mosaics
___ 09360 Plastic
___ 09370 Metal
___ 09380 Cut Natural Stone Tile
___ 09390 Tile Restoration
___ 09400 Terrazzo
___ 09500 Ceilings
___ 09510 Acoustical
___ 09545 Specialty
___ 09550 Mirror Panel Ceilings
___ 09560 Textured
___ 09570 Linear Wood
___ 09580 Suspended Decorative Grids
___ 09590 Ceiling Assembly Restoration
___ 09600 Flooring
___ 09610 Floor Treatment

___ 09620 Specialty Flooring
___ 09630 Masonry
___ 09640 Wood
___ 09650 Resilient
___ 09660 Static Control
___ 09670 Fluid Applied
___ 09680 Carpet
___ 09690 Flooring Restoration
___ 09700 Wall Finishes
   ___ 09710 Acoustical Wall Treatment
   ___ 09720 Wall Covering
   ___ 09730 Wall Carpet
   ___ 09740 Flexible Wood Sheets
   ___ 09750 Stone Facing
   ___ 09760 Plastic Blocks
   ___ 09770 Special Wall Surfaces
   ___ 09790 Wall Finish Restoration
___ 09800 Acoustical Treatment
   ___ 09810 Acoustical Space Units
   ___ 09820 Acoustical Insulation and Sealants
   ___ 09830 Acoustical Barriers
   ___ 09840 Acoustical Wall Treatment
___ 09900 Paints and Coatings
   ___ 09910 Paints
   ___ 09930 Stains and Transparent Finishes
   ___ 09940 Decorative Finishes
   ___ 09960 High-Performance Coatings
   ___ 09970 Coatings for Steel
   ___ 09980 Coatings for Concrete and Masonry
   ___ 09990 Paint Restoration

DIVISION 10—SPECIALTIES
___ 10100 Visual Display Boards
___ 10150 Compartments and Cubicles
___ 10200 Louvers and Vents
___ 10240 Grilles and Screens
___ 10250 Service Walls
___ 10260 Wall and Corner Guards
___ 10270 Access Flooring
___ 10290 Pest Control
___ 10300 Fireplaces and Stoves
___ 10340 Manufactured Exterior Specialties
___ 10350 Flagpoles
___ 10400 Identification Devices
___ 10450 Pedestrian Control Devices
___ 10500 Lockers

___ 10520 Fire Protection Specialties
___ 10530 Protective Covers
___ 10550 Postal Specialties
___ 10600 Partitions
___ 10670 Storage Shelving
___ 10700 Exterior Protection
___ 10750 Telephone Specialties
___ 10800 Toilet, Bath, and Laundry Accessories
___ 10880 Scales
___ 10900 Wardrobe and Closet Specialties

DIVISION 11—EQUIPMENT
___ 11010 Maintenance Equipment
___ 11020 Security and Vault Equipment
___ 11030 Teller and Service Equipment
___ 11040 Ecclesiastical Equipment
___ 11050 Library Equipment
___ 11060 Theater and Stage Equipment
___ 11070 Instrumental Equipment
___ 11080 Registration Equipment
___ 11090 Checkroom Equipment
___ 11100 Mercantile Equipment
___ 11110 Commercial Laundry and Dry Cleaning Equipment
___ 11120 Vending Equipment
___ 11130 Audiovisual Equipment
___ 11140 Vehicle Service Equipment
___ 11150 Parking Control Equipment
___ 11160 Loading Dock Equipment
___ 11170 Solid Waste Handling Equipment
___ 11190 Detention Equipment
___ 11200 Water Supply and Treatment Equipment
___ 11280 Hydraulic Gates and Valves
___ 11300 Fluid Waste Treatment and Disposal Equipment
___ 11400 Food Service Equipment
___ 11450 Residential Equipment
___ 11460 Unit Kitchens
___ 11470 Darkroom Equipment
___ 11480 Athletic, Recreational, and Therapeutic Equipment
___ 11500 Industrial and Process Equipment
___ 11600 Laboratory Equipment
___ 11650 Planetarium Equipment
___ 11660 Observatory Equipment
___ 11680 Office Equipment
___ 11700 Medical Equipment
___ 11780 Mortuary Equipment

___ 11850 Navigation Equipment
___ 11870 Agricultural Equipment
___ 11900 Exhibit Equipment

DIVISION 12—FURNISHINGS
___ 12050 Fabrics
___ 12100 Art
___ 12300 Manufactured Casework
___ 12400 Furnishings and Accessories
___ 12500 Furniture
___ 12600 Multiple Seating
___ 12700 Systems Furniture
___ 12800 Interior Plants and Planters
___ 12900 Furnishings Restoration and Repair

DIVISION 13—SPECIAL CONSTRUCTION
___ 13010 Air-Supported Structures
___ 13020 Building Modules
___ 13030 Special-Purpose Rooms
___ 13080 Sound, Vibration, and Seismic Control
___ 13090 Radiation Protection
___ 13100 Lightning Protection
___ 13110 Cathodic Protection
___ 13120 Preengineered Structures
___ 13150 Swimming Pools
___ 13160 Aquariums
___ 13165 Aquatic Park Facilities
___ 13170 Tubs and Pools
___ 13175 Ice Rinks
___ 13185 Kennels and Animal Shelters
___ 13190 Site-Constructed Incinerators
___ 13200 Storage Tanks
___ 13220 Filter Underdrains and Media
___ 13230 Digester Covers and Appurtenances
___ 13240 Oxygenation Systems
___ 13260 Sludge Conditioning Systems
___ 13280 Hazardous Material Remediation
___ 13400 Measurement and Control Instrumentation
___ 13500 Recording Instrumentation
___ 13550 Transportation Control Instrumentation
___ 13600 Solar and Wind Energy Equipment
___ 13700 Security Access and Surveillance
___ 13800 Building Automation and Control
___ 13850 Detection and Alarm
___ 13900 Fire Suppression

DIVISION 14—CONVEYING SYSTEMS
___ 14100 Dumbwaiters
___ 14200 Elevators
___ 14300 Escalators and Moving Walks
___ 14400 Lifts
___ 14500 Material Handling
___ 14600 Hoists and Cranes
___ 14700 Turntables
___ 14800 Scaffolding
___ 14900 Transportation

DIVISION 15—MECHANICAL
___ 15050 Basic Mechanical Materials and Methods
___ 15100 Building Services Piping
___ 15200 Process Piping
___ 15300 Fire Protection Piping
___ 15400 Plumbing Fixtures and Equipment
___ 15500 Heat-Generation Equipment
___ 15600 Refrigeration Equipment
___ 15700 Heating, Ventilating, and Air Conditioning Equipment
___ 15800 Air Distribution
___ 15900 HVAC Instrumentation and Controls
___ 15950 Testing, Adjusting, and Balancing

DIVISION 16—ELECTRICAL
___ 16050 Basic Electrical Materials and Methods
___ 16100 Wiring Methods
___ 16200 Electrical Power
___ 16300 Transmission and Distribution
___ 16400 Low-Voltage Distribution
___ 16500 Lighting
___ 16700 Communications
___ 16800 Sound and Video

NOTES

___ 1. **Programming** is a process leading to the statement of an architectural problem and the requirements to be met in offering a solution. It is the search for sufficient information to clarify, to understand, to state the problem. Programming is problem seeking and design is problem solving.

___ 2. **Use the Information Index** on pp. 32–33 as a guide for creating a program for more complex projects.

___ 3. **Efficiency Ratios:** Use the following numbers to aid in planning the size of buildings and their spaces in regard to the ratio of net area to gross area:

> **assigned areas** ─┐
>
> **unassigned areas** **50/50% = 100%** ◄─── **gross building area**
> **(circulation, walls, etc.)** ─┘

Note: The gross area of a building is the total floor area based on outside dimensions. The net area is based on the interior dimensions. For office or retail space, net leasable area means the area of the primary function of the building, excluding such things as stairwells, corridors, mechanical rooms, etc.

Common Range		
Automobile analogy	For buildings	Ratios
Super Luxury	Superb	50/50
Luxury	Grand	55/45
Full	Excellent	60/40
Intermediate	Moderate	65/35
Compact	Economical	67/33
Subcompact	Austere	70/30
Uncommon Range		
	Meager	75/25
	Spare	80/20
	Minimal	85/15
	Skeletal	90/10

The following table gives common breakdowns of unassigned areas:

Circulation	16.0	20.0	22.0	24.0	25.0
Mechanical	5.0	5.5	7.5	8.0	10.0
Structure and walls	7.0	7.0	8.0	9.5	10.0
Public toilets	1.5	1.5	1.5	2.0	2.5
Janitor closets	0.2	0.5	0.5	0.5	1.0
Unassigned storage	0.3	0.5	0.5	1.0	1.5
	30.0%	35.0%	40.0%	45.0%	50.0%

NOTES

INFORMATION INDEX

	GOALS What does the client want to achieve & why?	FACTS What is it all about?
FUNCTION What's going to happen in the building? 　People 　Activities 　Relationships	Mission Maximum number Individual identity Interaction/privacy Hierarchy of values Security Progression Segregation Encounters Efficiency	Statistical data Area parameters Manpower/workloads User characteristics Community characteristics Value of loss Time-motion study Traffic analysis Behavioral patterns Space adequacy
FORM What is there now & what is to be there? 　Site 　Environment 　Quality	Site elements (Trees, water, open space, 　existing facilities, 　utilities) Efficient land use Neighbors Individuality Direction Entry Projected image Level of quality	Site analysis Climate analysis Cope survey Soils analysis F.A.R. and G.A.C. Surroundings Psychological implications Cost/SF Building efficiency Functional support
ECONOMY Concerns the initial budget & quality of construction. 　Initial budget 　Operating costs 　Lifecycle costs	Extent of funds Cost effectiveness Maximum return Return on investment Minimize oper. costs Maint. & oper. costs Reduce life cycle costs	Cost parameters Maximum budget Time-use factors Market analysis Energy source-costs Activities & climate 　factors Economic data
TIME Deals with the influences of history, the inevitability of change from the present, & projections into the future. 　Past 　Present 　Future	Historic preservation Static/dynamic Change Growth Occupancy date	Significance Space parameters Activities Projections Linear schedule

CONCEPTS How does the client want to achieve the goals?	NEEDS How much money, space, & quality (as opposed to wants)?	PROBLEM What are the significant conditions & the general directions the design of the building should take?
Service grouping People grouping Activity grouping Priority Security controls Sequential flow Separated flow Mixed flow Relationships	Space requirements Parking requirements Outdoor space req'mts. Building efficiency Functional alternatives	Unique and important performance requirements which will shape building design.

Enhancement
Climate control
Safety
Special
  foundations
Density
Interdependence
Home base
Orientation
Accessibility
Character
Quality control
	Quality (cost/SF)
	Environmental & site
	  influences on costs
		Major form considerations which will affect building design.

Cost control Efficient allocation Multifunction Merchandising Energy conservation Cost control	Cost estimate analysis Entry budget (FRAS) Operating costs Life cycle costs	Attitude toward the initial budget and its influence on the fabric and geometry of the building.
Adaptability Tailored/loose fit Convertibility Expansibility Concurrent scheduling	Phasing Escalation	Implications of change/ growth on long-range performance.

NOTES

__ E. INSTALLATION COSTS

Note: Most costs throughout this book (and this chapter) are from the following sources:

(11) (16) (17) (33) (34) (35) (42)

___ 1. **This book has rough cost data throughout.** Rough costs are **boldface.** See App. A for rough estimates of installed costs by building type.

___ 2. **Furniture, Fixtures, and Equipment (FF&E) Costs**

    ___ *a.* The interior designer is most often specifying and sometimes installing *furniture*. See Part 2, p. 251, and Part 12, p. 383 for typical furnishing costs. Also see p. 3 for furniture cost markups and discounts.

    ___ *b.* Fixtures are *built-ins* that may be priced and installed by the interior designer or maybe by a general contractor.

    ___ *c.* Equipment costs often relate to items designed and installed by another specialist designer. Examples would be kitchen equipment by a food service company or medical equipment by a medical equipment supplier. In short, these costs do not involve a general contractor and are not construction costs.

___ 3. **General Interior Construction Costs**

If the interior designer is doing more than just space planning and specifying furnishings, then tenant improvements or remodel of existing building interiors will probably involve a general contractor and possibly an architect and maybe engineers.

    ___ *a.* If anything structural is involved, retain a structural engineer.

    ___ *b.* If plumbing is modified or added, or heating, ventilation, and air-conditioning (HVAC) is modified or added, retain a mechanical engineer.

    ___ *c.* If lighting or power is modified or added, retain an electrical engineer.

    ___ *d.* Check with the local code enforcing agencies to see if an architect is required.

For general interior construction costs given, subcontractor's overhead and profit, plus tax, are included. Both material (M) and labor (L) are included, usually with a general idea of percentage of each to the total (100%). Because there is room for only one cost per "element," often an idea of possible variation (higher or lower) of cost is given. One must use judgment in this regard to come up with a reasonable but rough cost estimate. As costs change, the user will have to revise costs in this book. The easiest way to do this will be to add historical modifiers, published each year, by various sources. The costs in this book are approx. costs for 2011. For this edition, costs have been increased 20% over the last edition and rounded off. Over the last few years costs have increased about 2 to 4% per year. Be sure to compound when using this rule of thumb.

EXAMPLE:

A CONSTRUCTION ITEM IN THE BOOK GIVES THE FOLLOWING:
$5⁰⁰/SF (40% M & 60% L)(VARIATION OF +100%
$ -20%.
THIS MEANS THAT AS A GOOD AVERAGE, THE COST OF THE
ITEM INSTALLED (WITH THE SUB-CONTRACTOR'S O.H. & P.) IS
$5/SF. THE MATERIAL COST IS APPROX. $5 x .40 =
$2/SF. THE LABOR COST IS APPROX. $5 x .60 = $3/SF.
A VERY EXPENSIVE VERSION CAN BE ROUGHLY 100% HIGHER
($5 x 2 = $10/SF) OR A CHEAPER VERSION CAN BE ROUGH-
LY 20% LOWER ($5 x .80 = $4/SF). BUT THE APPROX.
AVERAGE COST IS $5⁰⁰/SF. NOTE THAT GENERAL CON-
TRACTOR'S O.H. & P. NOT INCLUDED.

____ 4. Cost Control and Estimating

Cost estimating can be time-consuming. It can also be dangerous in that wrong estimates may require time-consuming and expensive redesign. From the beginning of a project, responsibility for cost control (if any) should be clearly established. If the designer is responsible for doing estimates, the designer should consider the following points:

____ a. Apples to Apples: In discussing costs and budgets with clients and builders, the parties must be sure they are comparing "apples to apples" (i.e., what is included and excluded). Examples of misunderstandings:

____ (1) Financing costs (are usually excluded).
____ (2) Designers' and consultants' fees.
____ (3) City or government fees (are usually excluded).
____ (4) Are furniture, fixture, and equipment (FF&E) costs included?

EXAMPLE:

A DESIGNER IS WORKING FOR A "SPEC" BUILDER. THE
DESIGNER HAS IN MIND A $60/SF BUDGET. THE SPEC.
BUILDER HAS IN MIND A $50/SF BUDGET. THE DESIGNER'S
NUMBER HAS A GENERAL CONTRACTOR'S O.H.&P. OF SAY
20%, AS IF THE PROJECT WERE BID OR NEGOTIATED WITH
AN INDEPENDENT CONTRACTOR. THE SPEC BUILDER IS
THINKING ABOUT HIS DIRECT COSTS, ONLY. THEY ARE COMPAR-
ING "APPLES TO ORANGES". BUT, IF THE BUILDER'S NUMBER
IS ADJUSTED:

$$\$50/SF \times 1.20 = \$60/SF$$

OR, IF THE DESIGNER'S NUMBER IS ADJUSTED:

$$\$60/SF \times 0.80 = \$50/SF$$

THEN, THEY ARE TALKING "APPLES TO APPLES".

___ *b.* Variables:

___ (1) *Location.* Modify costs for actual location. Use modifiers often published.

___ (2) *Historical index.* If cost data is old, modify to current or future time by often-published modifiers.

___ (3) *Building size.* The $/SF costs may need to be modified due to size of the project.

___ (4) *Shape.* More complicated shapes will cause costs to go up. Where single elements are articulated (e.g., rounded corners or different types of coursing and materials in a masonry wall), add 30% to the costs involved.

___ (5) *Quality of materials, construction, and design.* Use the following rough guidelines to increase or decrease as needed:

Automobile analogy	For buildings	%
Super Luxury	Superb	+120
Luxury	Grand	+60
Full	Excellent	+20

Intermediate	Moderate	100
Compact	Economical	−10
Subcompact	Austere	−20

Note: Increasing the quality from lowest to highest can double the cost.

___ *c.* Costs (and construction scheduling) can be affected by weather, season, materials shortages, and/or labor practices.

___ *d.* Beyond a 20-mile radius of cities, extra transportation charges increase material costs slightly. This may be offset by lower wage rates. In dense urban areas, costs for locally supplied material may actually increase.

___ *e.* In doing a total estimate, an allowance for general conditions should be added. This usually ranges from 5 to 15%, with *10%* a typical average.

___ *f.* At the end of a total estimate, an allowance for the general contractor's overhead and profit should be added. This usually ranges from *10 to 20%*. Market conditions at the time of bidding will often affect this percentage as well as all items. The market can swing 10 to 20% from inactive to active times.

___ *g.* Contingencies should always be included in estimates as listed below. On alterations or repair projects, *20%* is not an unreasonable allowance to make.

___ *h.* Use rounding of numbers in all estimating items.

___ *i.* Consider using "add alternates" to projects where the demand is high but the budget tight. These alternates should be things the client would like but does not have to have and should be clearly denoted in the drawings.

___ *j.* It is often wise for the designer to give estimates in a range.

___ *k.* Because clients often change their minds or things go wrong that cannot be foreseen in the beginning, it may pay to advise the client to withhold from his budget a confidential *5 to 10% contingency*. On the other hand, clients often do this anyway, without telling the designer.

___ *l.* Costs can further be affected by other things:

Government overhead	≈+100%
Award-winning designs are often	≈+200 to 300%

___ **5. Cost Control Procedure**

___ *a.* At the *predesign phase* or beginning of a project, determine the client's *budget* and what it includes, as well as anticipated size of the project. Back out all nonconstruction costs such as cost of furniture and fixtures, design fees, etc. Verify, in a simple format (such as $/SF, $/room, etc.) that this is reasonable. See App. A for average $/SF costs as a comparison and guideline.

___ *b.* At the *schematic design phase,* establish a reasonable $/SF target. Include a *15% to 20% contingency.*

___ *c.* At the *design development phase,* as the design becomes more specific, do a "systems" estimate. See Part 13 as an aid. For small projects a "unit" estimate might be appropriate, especially if basic plans not normally done at this time can be quickly sketched up for a "take off." Include a *10% to 15% contingency.*

___ *d.* At the *construction documents phase,* do a full unit "take off." For smaller projects, the estimate in the last phase may be enough, provided nothing has changed or been added to the project. Add a *5% to 10% contingency.*

___ 6. Typical Commercial Building Cost Percentages

Division	New const.	Remodeling
1. General requirements	6 to 8%	about 30% for general
2. Sitework	4 to 6%	
3. Concrete	15 to 20%	
4. Masonry	8 to 12%	
5. Metals	5 to 7%	
6. Wood	1 to 5%	
7. Thermal and moisture protection	4 to 6%	
8. Doors, windows, and glass	5 to 7%	
9. Finishes	8 to 12%	about 30% for divisions 8–12
10. Specialties*		
11. Equipment*		
12. Furnishings*	6 to 10%*	
13. Special construction*		
14. Conveying systems*		about 40% for mech. and elect.
15. Mechanical	15 to 25%	
16. Electrical	8 to 12%	
Total	100%	

Note: FF&E (furniture, fixtures, and equipment) are often excluded from building cost budget.

   ___ 7. <u>Guidelines for Tenant Improvements (TI) in Office Build-</u>
<u>ings:</u>
     ___ *a.* **Costs for office building frames and envelopes: <u>$30</u>**
<u>to $40/SF</u>.
     ___ *b.* **TI costs range from <u>$30 to $60/SF</u> (in extreme cases**
<u>$120/SF</u>).
   ___ 8. <u>Guidelines for Demolition</u>
     ___ *a.* Total buildings: *$5 to $7/SF*
     ___ *b.* Separate elements: *10% to 60%* of in-place con-
struction cost of element.

EXAMPLE

PROBLEM: DEVELOP A LIKELY BUDGET FOR A NEW OFFICE TENANT IMPROVEMENT IN A NEW OPEN SPACE "SHELL" BUILDING. THE T.I. SPACE IS TO BE 5000 S.F.

SOLUTION:

		LOW	AVE	HIGH
1. FROM APP P. 507 ESTABLISH AVERAGE IN PLACE COSTS ($/SF)		20	40	60
2. ASSUME FURNISHINGS ARE 30% & CONST. COSTS ARE 70% OF TOTAL ($/SF)	FURN.	6	12	18
	CONST.	14	28	42
	TOTAL	20	40	60
3. TYPICAL INTERIOR DESIGN FEES ARE (%)		3	5	6
4. CONVERT COSTS TO LUMP SUMS BY MULTIPLYING X 5000 S.F. ($)	FURN.	30000	60000	90000
	CONST.	70000	140000	210000
	TOTAL INSTALLED	$100 000	$200 000	$300 000
5. ESTIMATE DESIGNER'S FEE OF % IN LINE 3 X LUMP SUMS IN 4 ($)		$3000	$10000	$18000
6. TOTAL ESTIMATE OF COSTS TO CLIENT		$103000	$210000	$318000

NOTES

NOTES

F. HUMAN DIMENSIONS

Note: All dimensions are ranges from several sources in inches and rounded to the nearest inch. Where a range is given, lower numbers tend to be for women and the higher numbers for men. The average of the two tends to be the standard average.

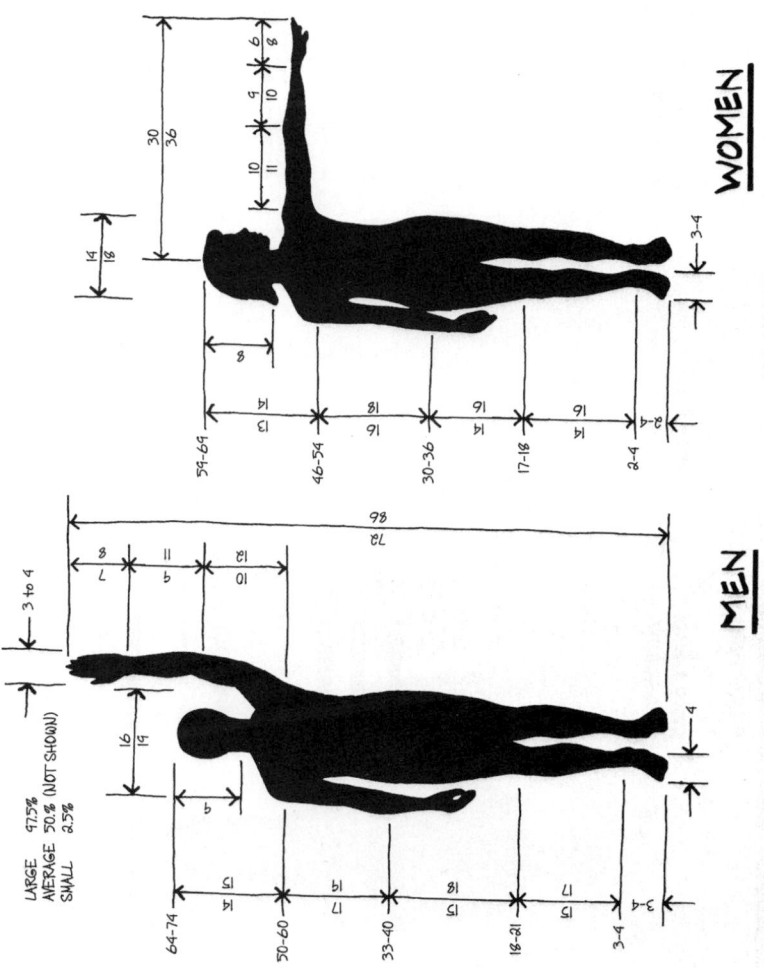

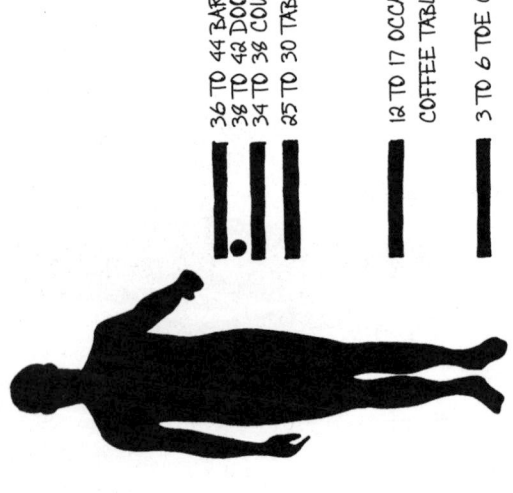

HEIGHTS

96 CEILING (TYP. RESID. CL'G)

84 EXTERIOR DOOR
80 INTERIOR DOOR

36 TO 44 BAR COUNTER
38 TO 42 DOOR KNOB
34 TO 38 COUNTER
25 TO 30 TABLE/DESK

12 TO 17 OCCASIONAL/COFFEE TABLE

3 TO 6 TOE CLEARANCE

WIDTHS

36 EXTERIOR DOOR
32 INTERIOR DOOR

36 TO 38 COUNTER TOP
24 TO 30 DESK TOP

15 TO 18 SEAT DEPTH

44

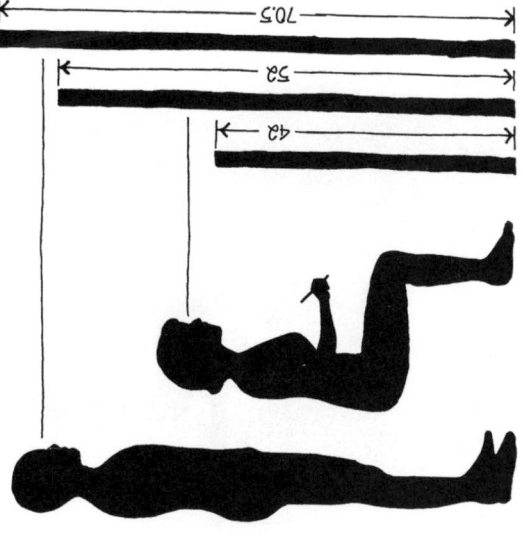

PARTITION HEIGHTS

70.5

52

43

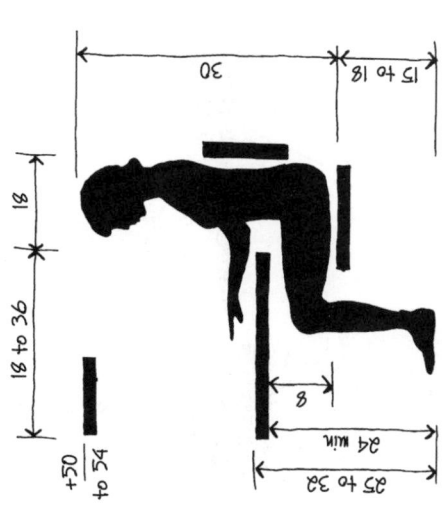

WORKSTATION

30

15 to 18

18

18 to 36

+50 to 54

8

34 min.

35 to 32

45

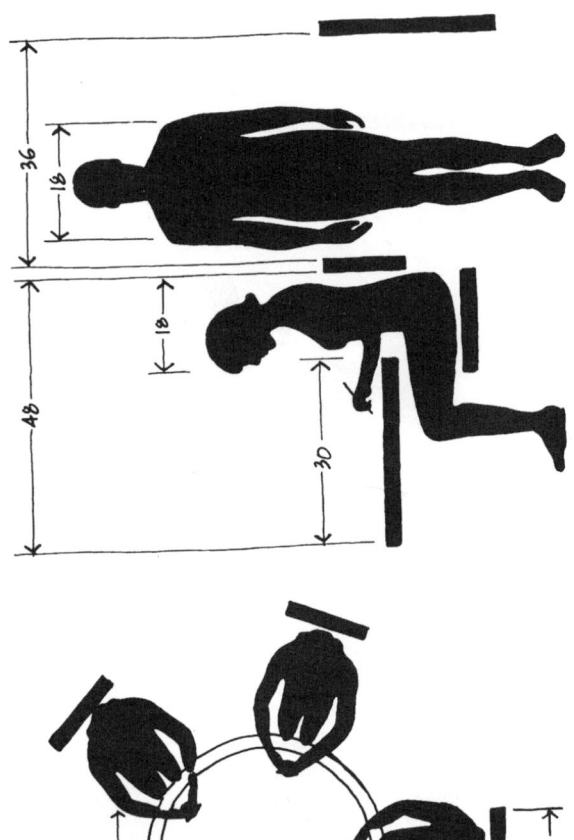

CORRIDOR/CLEARANCE

CONFERENCE

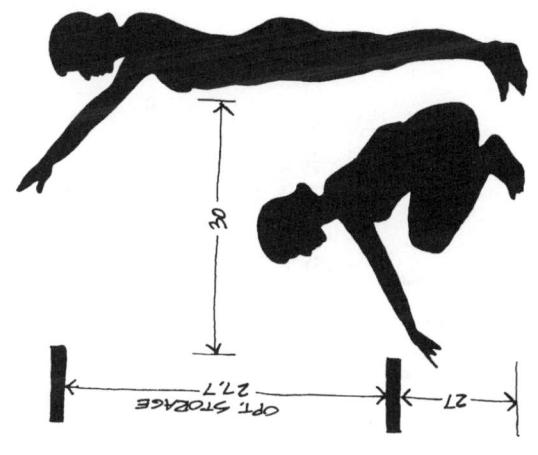

OVERHEAD / UNDER COUNTER REACH

OPT. STORAGE
27.7

27

30

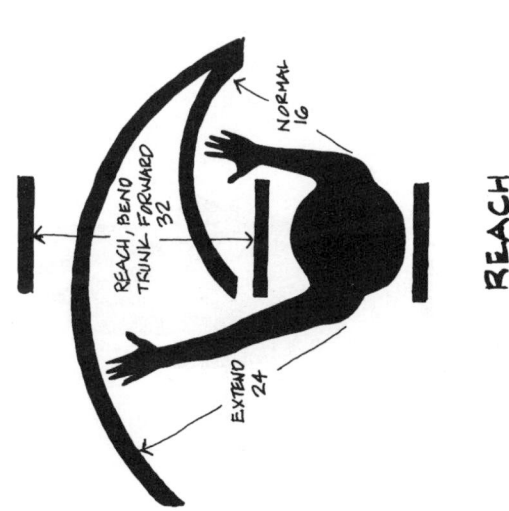

REACH, BEND
TRUNK FORWARD
32

NORMAL
16

EXTEND
24

REACH

47

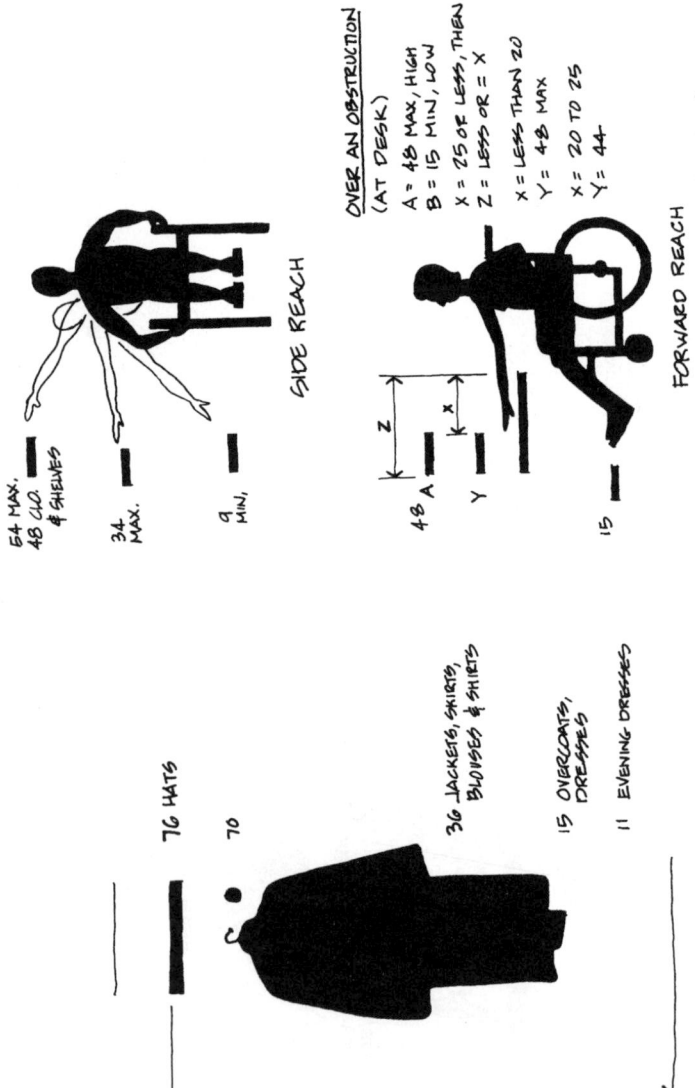

SIDE REACH

54 MAX.
48 CLO.
@ SHELVES
34 MAX.
9 MIN.

OVER AN OBSTRUCTION
(AT DESK)

A = 48 MAX, HIGH
B = 15 MIN, LOW

X = 25 OR LESS, THEN
Z = LESS OR = X

X = LESS THAN 20
Y = 48 MAX

X = 20 TO 25
Y = 44

48 A
Z
X
Y
15

FORWARD REACH

WHEELCHAIR (ADA)

76 HATS
70
36 JACKETS, SKIRTS, BLOUSES & SHIRTS
15 OVERCOATS, DRESSES
11 EVENING DRESSES
76

CLOSET

48

TYPICAL HEIGHTS

STANDING	SEATED

8'-0"	96"	TYP. RESIDENTIAL CEILING
	94	
	92	
	90	
	88	
7'-0"	84" 86	
		OFFICE DOORS
	82	
	80	RESIDENTIAL CEILING
	78	SHOWER HEADS
	76	
6'-0"	72" 74	CLOTHES LINES
		NO SEE OVER (HIGH SHELF), MEN
	70	TOP OF MIRROR
	68	HIGH SHELF, WOMEN
	66	
	64	MIN. CATWALK CLEARANCE
5'-0"	60" 62	
	58	THERMOSTATS
	56	SEE OVER — MIRROR TOP
	54	GRAB BARS, PHONE DIAL — NO SEE OVER, FLOOR
	52	HIGHEST FILE — LAMP (HIGH)
4'-0"	48" 50	PUSH PLATES — HIGH SHELF
		SHOWER VALVES, WALL SWITCH PLATES
	46	DOOR PUSH BARS — AVE. EYE LEVEL
	44	HIGH BAR, TOP — HIGH FILE, FRONT TAB
		COUNTERS, DOOR KNOBS
	42	(MAX.) HANDRAILS, BAR — SEE OVER HEIGHT
	40	TOPS, DOOR LOCKS — PHONE DIAL HEIGHT
		— FLOOR LAMP (LOW)
3'-0"	36" 38	RAILS ON STEPS (MAX.)
		HANDRAILS, COUNTERS, DOOR KNOBS — LUNCH COUNTER
	34	RAILS ON STEPS (MIN.) — HIGH FILE, TOP TAB
	32	PANIC BARS — SEWING TABLE
	30	LAVATORY RIM — STOOL FOR 42" COUNTER
	28	WORK TABLE, DESK
2'-0"	24" 26	— TYPING TABLE
		— TABLE, MIN. KNEE SPACE
	22	— CHAIR FOR 36" COUNTER
	20	
	18	WALL OUTLETS (MAX.) — BED
	16	HIGHEST STEP — WORK CHAIR
1'-0"	12" 14	RUNG SPACING — SEAT (MIN.)
		WALL OUTLET (AVE.)
	10	
	8	WALL OUTLET (MIN.) RISER (MAX.)
	6	
	4	RISER (MIN.)
0'-0"	0" 2	
		THRESHOLD (¼" to ½")

NOTES

___ G. SPACE PLANNING

___ 1. General

___ a. With so many building types, it is interesting that there really are so few "space types" that the interior designer may be involved in. The interior designer is not likely to be involved in building layout (room-to-room or space-to-space relationships, or room/space sizes, although there can be exceptions such as office T.I.). Rather, the interior designer usually inherits a room or space and must do space planning and furniture arrangement within the given conditions. Hopefully, the given conditions allow enough room.

___ b. This section deals with common spaces to most building types that the interior designer may be involved with. They run throughout all building types. Often, the primary use for the building sets the overall layout of the building in which the interior designer will not be involved. But the auxillary uses may require an interior designer. As an example, a fire station layout is mainly set by apparatus (truck) and equipment storage. But the building will usually have auxiliary spaces (reception, office, living, toilet, sleeping, dining, and cooking) that might require the assistance of an interior designer. This is usually the case. Exceptions to this general rule might be single-family residences or office space, where the designer may be involved with all the rooms and spaces.

___ c. This section has the following space types:
  ___ (1) General, p. 51.
  ___ (2) Living/Waiting, p. 52.
  ___ (3) Bedrooms/Sleeping, p. 65.
  ___ (4) Dining/Eating, p. 74.
  ___ (5) Kitchens/Cooking, p. 81.
  ___ (6) Toilets/Bathing, p. 87.
  ___ (7) Office/Work, p. 100.
  ___ (8) Retail/Shopping, p. 112.
  ___ (9) Reception/Counter/Circulation, p. 120.
  ___ (10) Group seating (performance, classrooms, etc.), p. 122.
  ___ (11) Storage, p. 124.

___ 2. <u>Living/Waiting</u>

    ___ *a.* General

Planning considerations should include adequate floor and wall space for furniture groupings, window locations, and separation of traffic ways from centers of activity. Ideally there should be no through-traffic. If such traffic is necessary, it should be at one end, with the remaining portion of the room a dead-end space. During social activities, people tend to gather or congregate in relatively small groups. A desirable conversation area is relatively small, approximately *10 feet* in diameter.

    ___ *b.* Overall layouts

        ___ (1) Residential: Use the following layouts for general planning (also see p. 388 for furnishings).

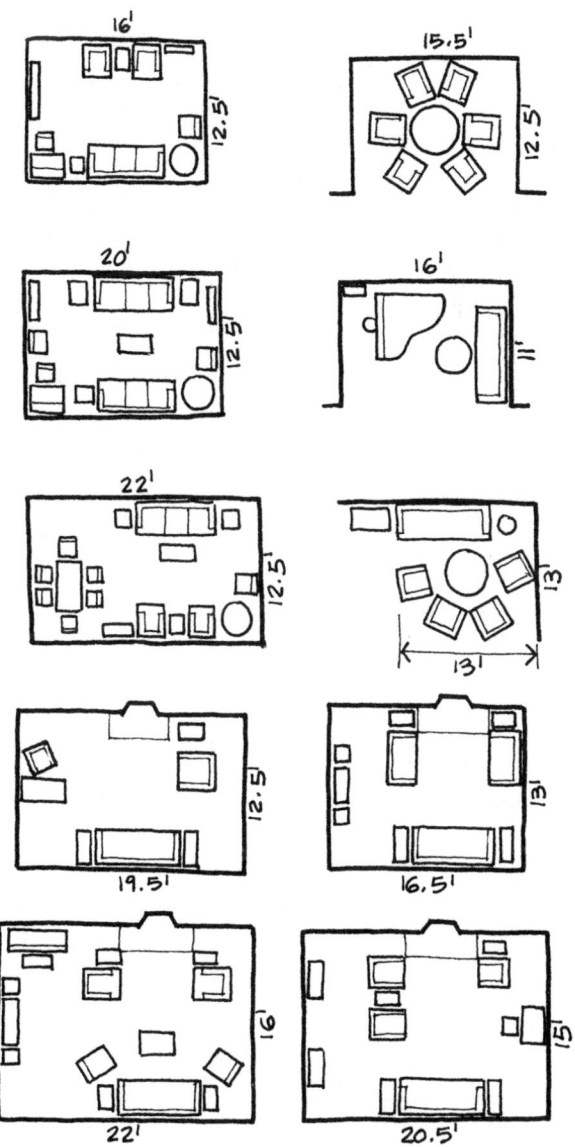

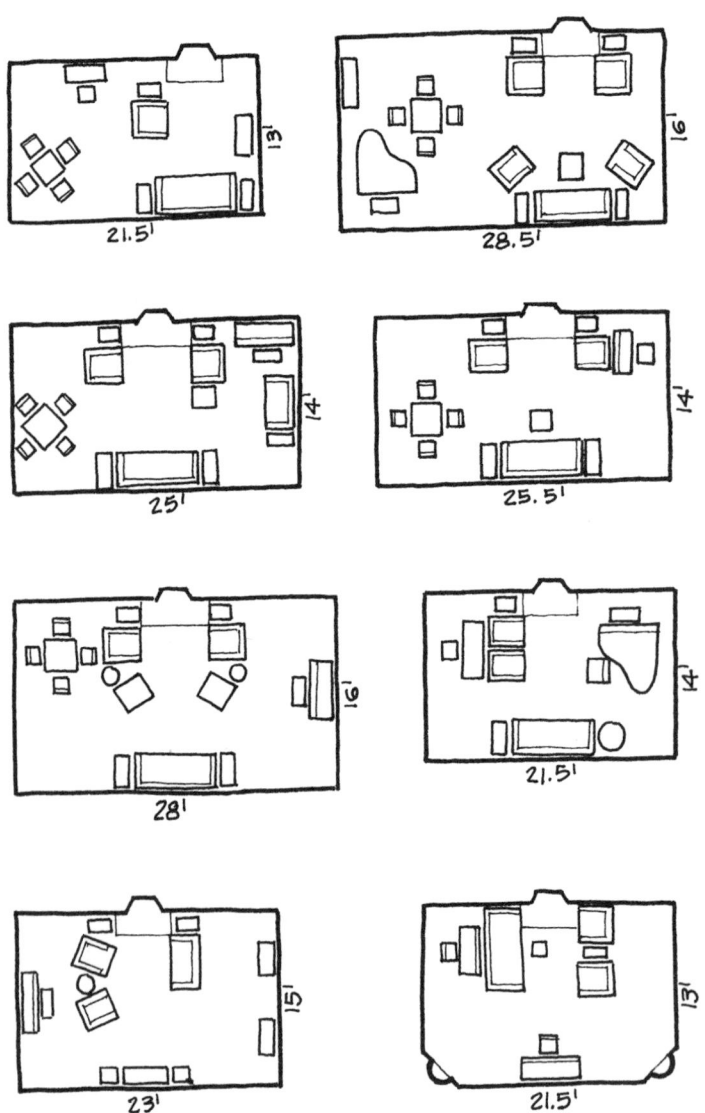

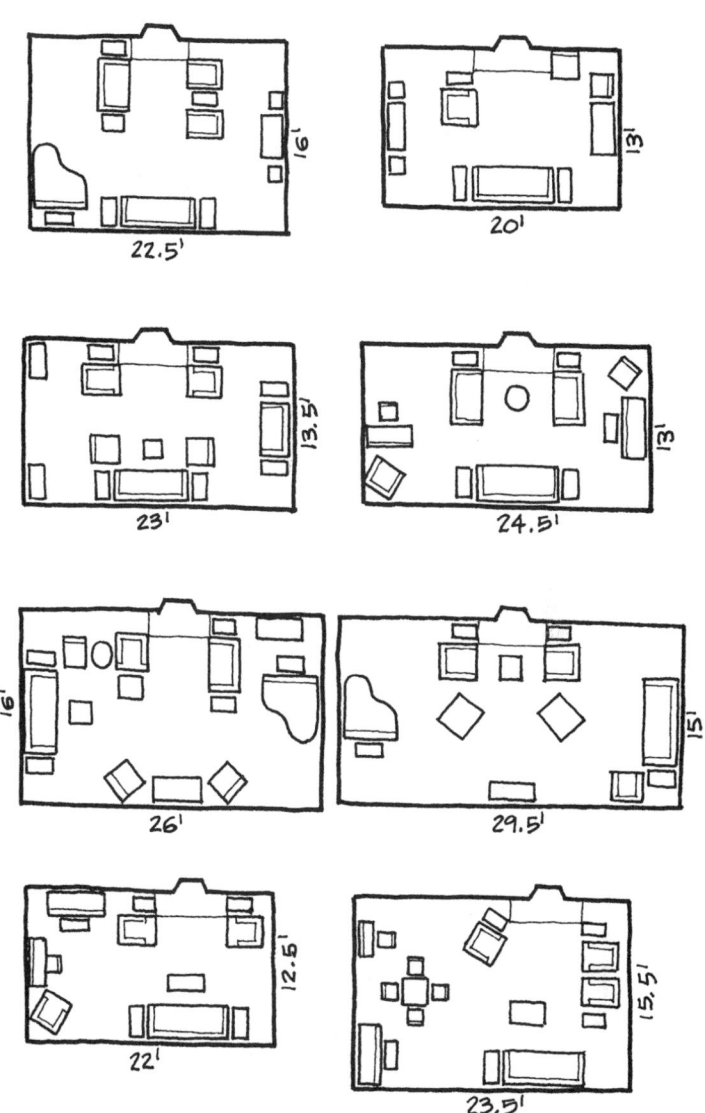

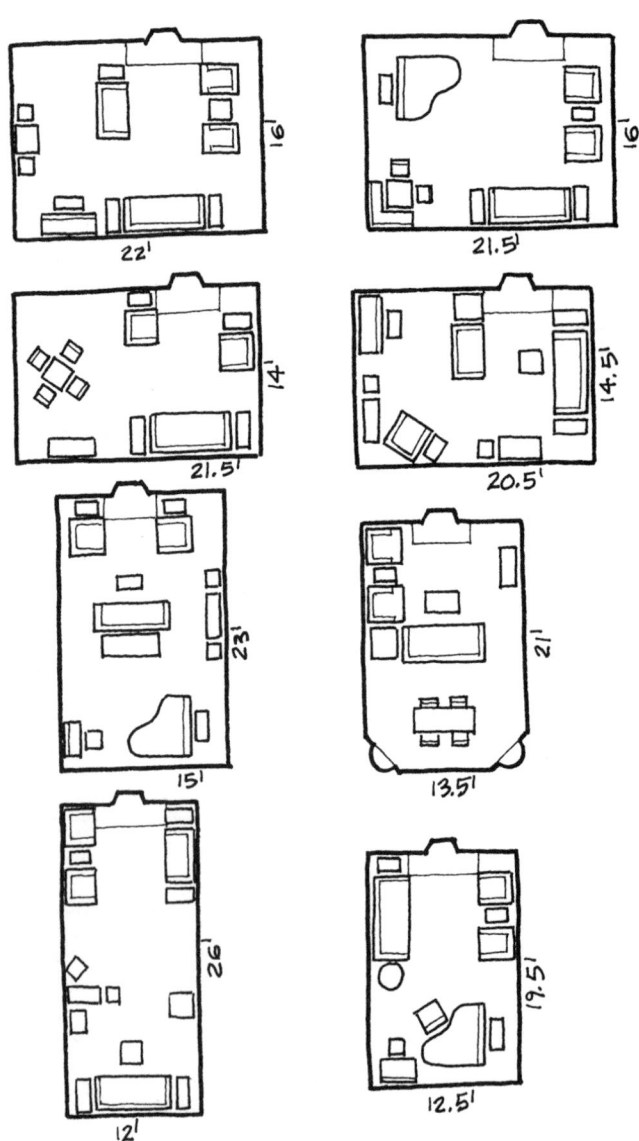

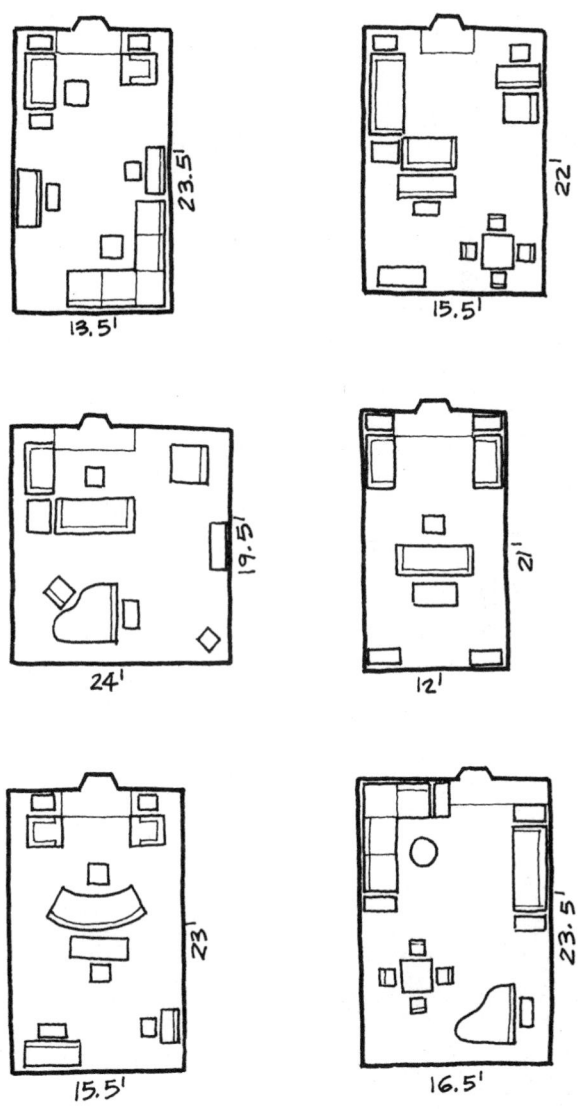

___ (2) **Public waiting:** Much of the information for residential living areas can be used for public areas. The following layouts depict typical office waiting and public spaces.

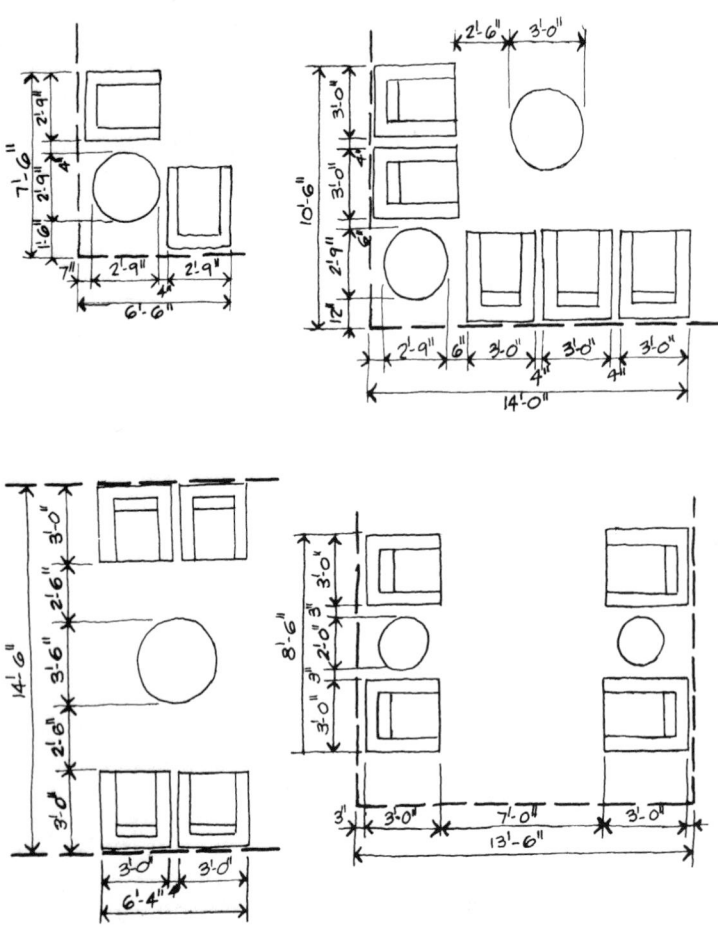

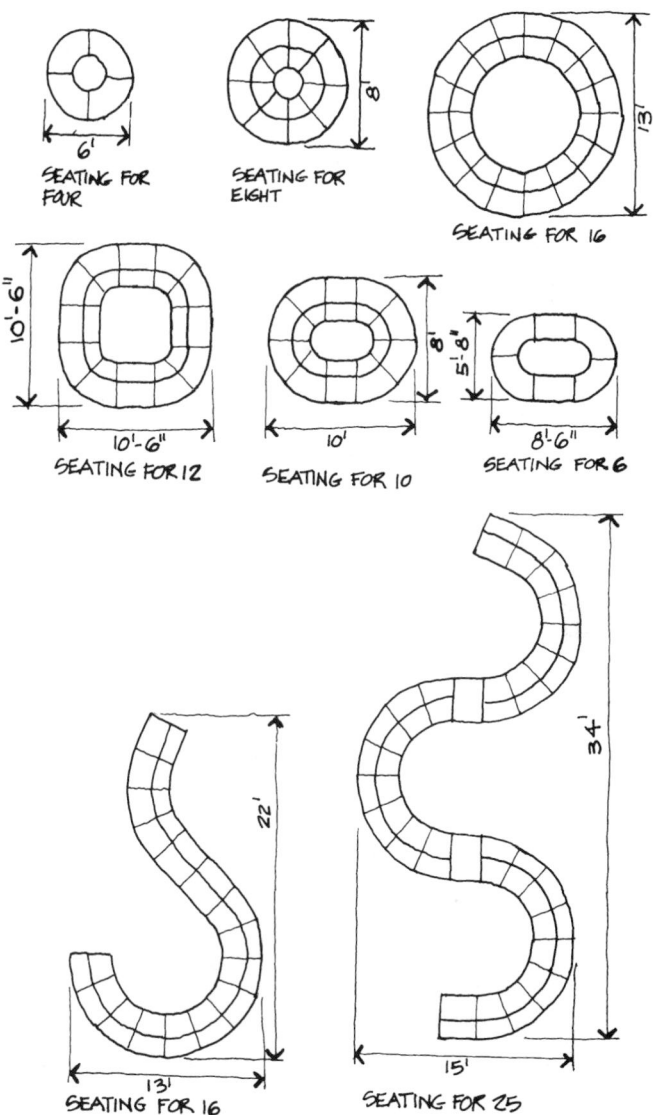

SEATING FOR FOUR

6'

SEATING FOR EIGHT

8'

SEATING FOR 16

13'

SEATING FOR 12

10'-6"

10'-6"

SEATING FOR 10

10'

8'

SEATING FOR 6

8'-6"

5'-8"

SEATING FOR 16

22'

13'

SEATING FOR 25

34'

15'

___ *c.* Living/waiting details

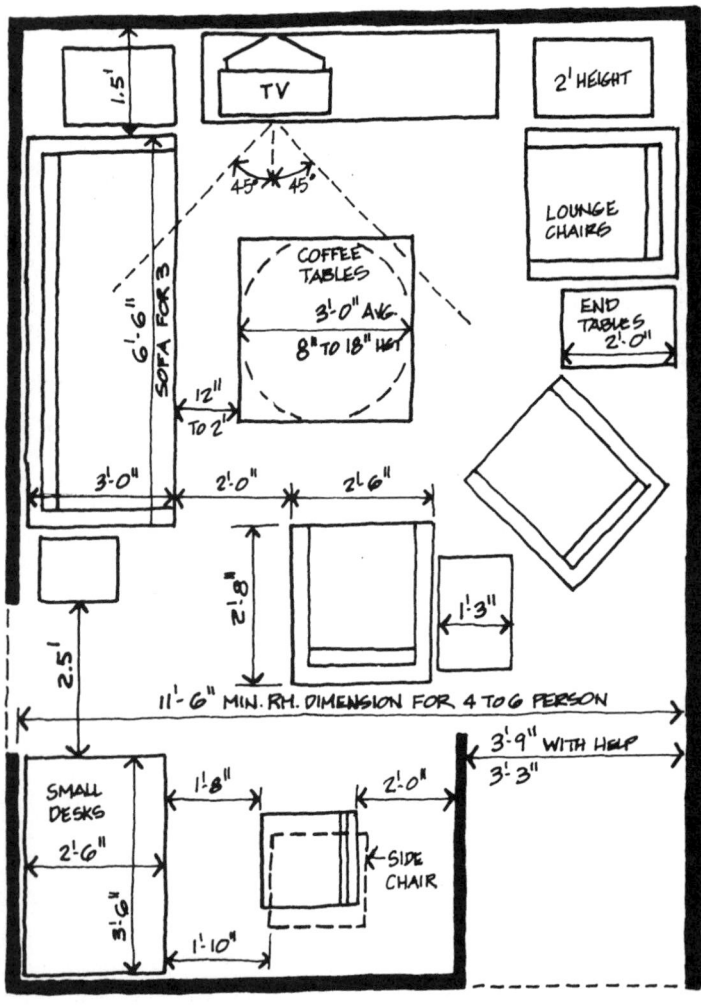

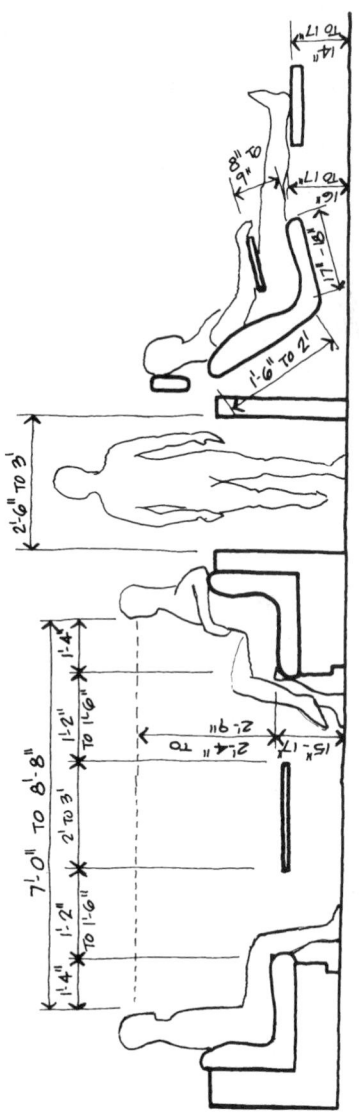

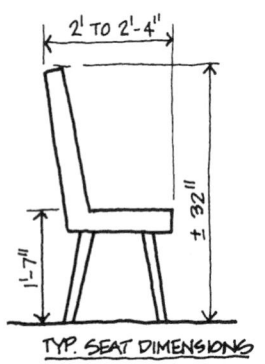

TYP. SEAT DIMENSIONS

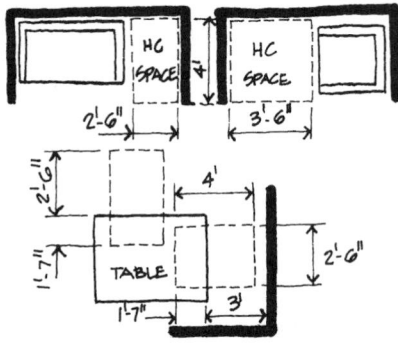

ADA ACCESSIBLE SEATING REQUIREMENTS

___ *d.* Furniture templates:
  The following are furniture templates at ¼″ = 1′0″
  scale. Use these for space planning of living/waiting
  aeras. Dimensions are in inches.

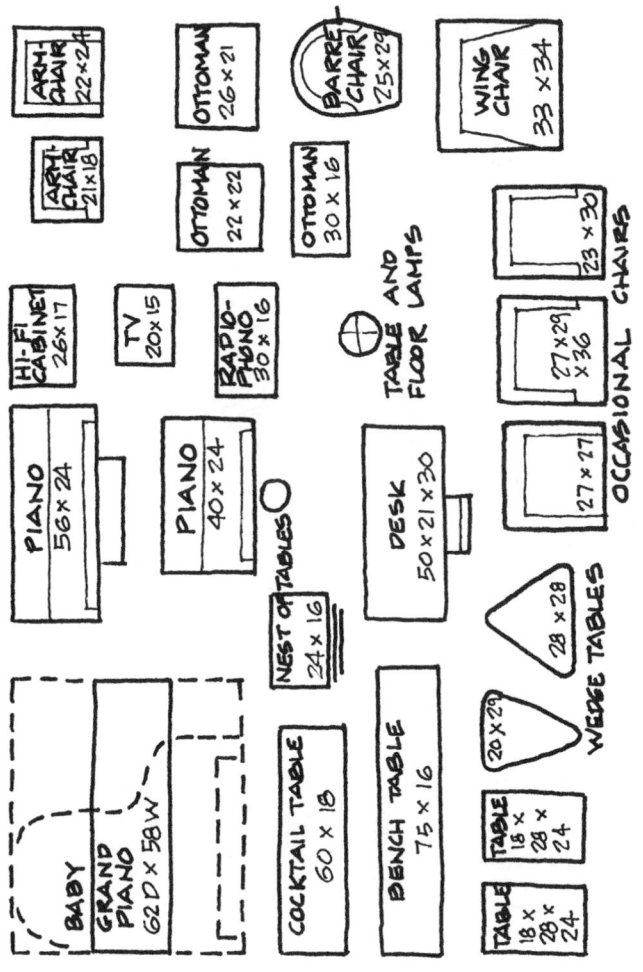

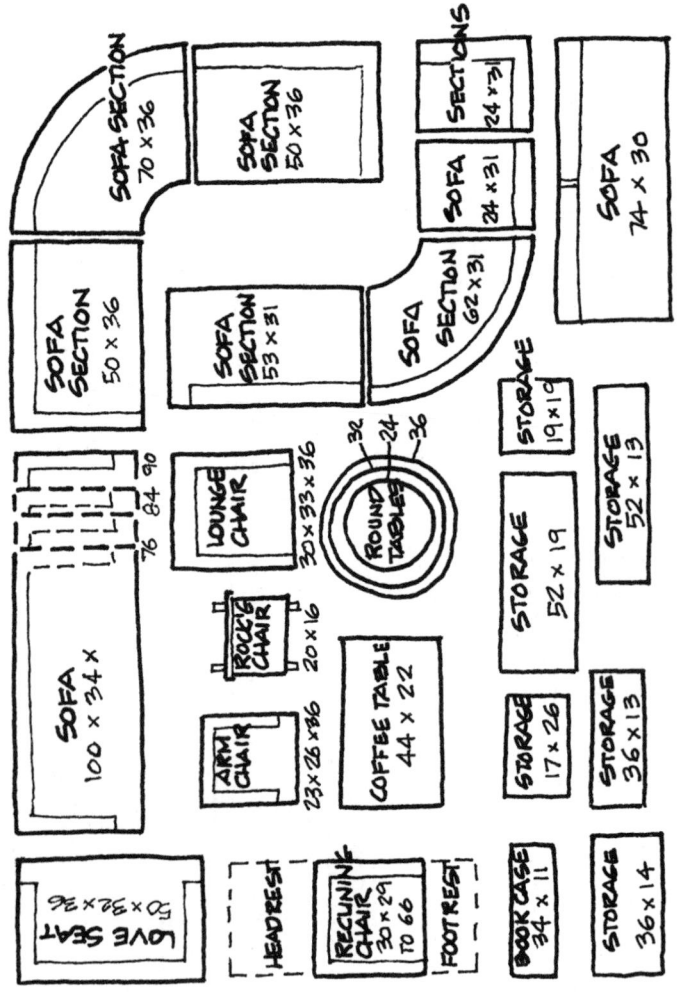

SOFA SECTION 70 x 36

SOFA SECTION 50 x 36

SOFA SECTIONS 24 x 31

SOFA 24 x 31

SOFA 74 x 30

SOFA SECTION 50 x 36

SOFA SECTION 53 x 31

SOFA SECTION 62 x 31

STORAGE 19 x 19

LOUNGE CHAIR 30 x 33 x 36

ROUND TABLES 32 - 24 - 36

STORAGE 52 x 19

STORAGE 52 x 13

76 84 90

SOFA 100 x 34 x

ROCKER CHAIR 20 x 16

ARM CHAIR 23 x 26 x 36

COFFEE TABLE 44 x 22

STORAGE 17 x 26

STORAGE 36 x 13

LOVE SEAT 50 x 32 x 36

HEADREST

RECLINING CHAIR 30 x 29 TO 66

FOOTREST

BOOK CASE 34 x 11

STORAGE 36 x 14

___ 3. <u>Bedroom/Sleeping</u>
___ *a.* Overall layouts. See p. 400 for furnishings.
Use the following layouts for general planning.
___ (1) Residential layouts

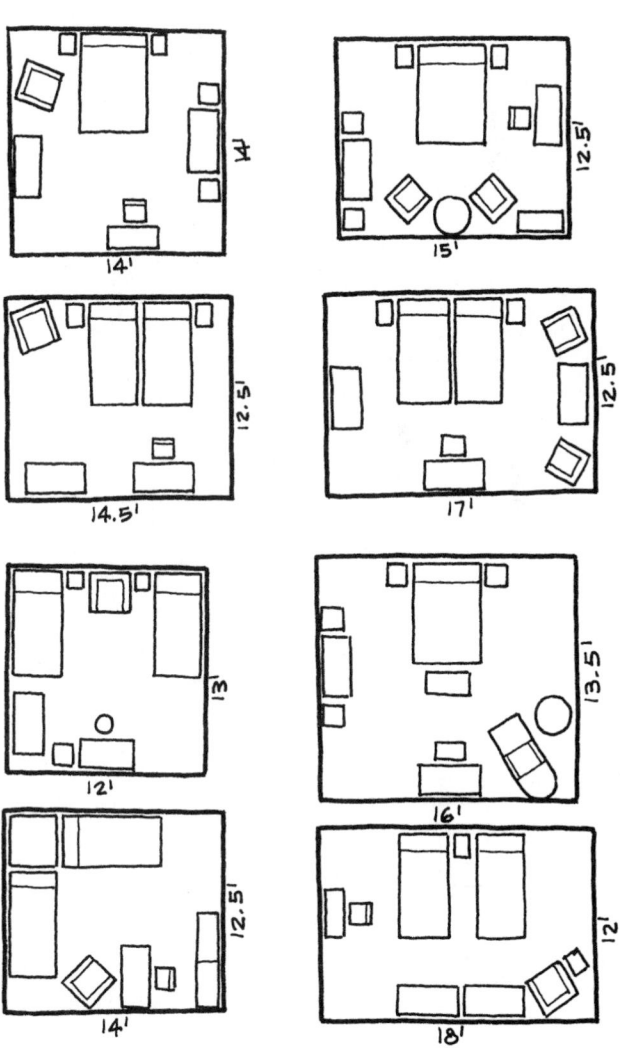

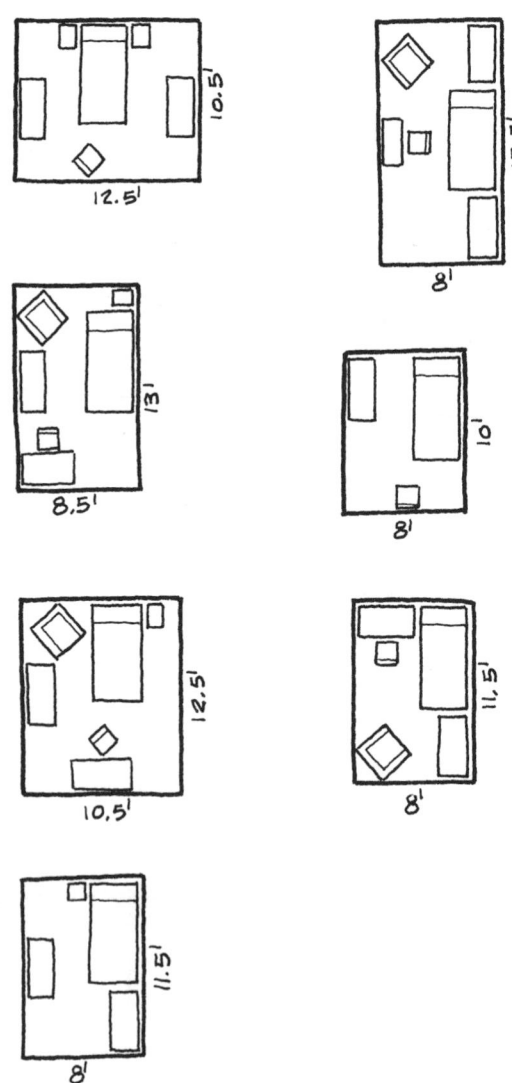

___ (2) Hotel layouts
For bathrooms (B), see p. 87.

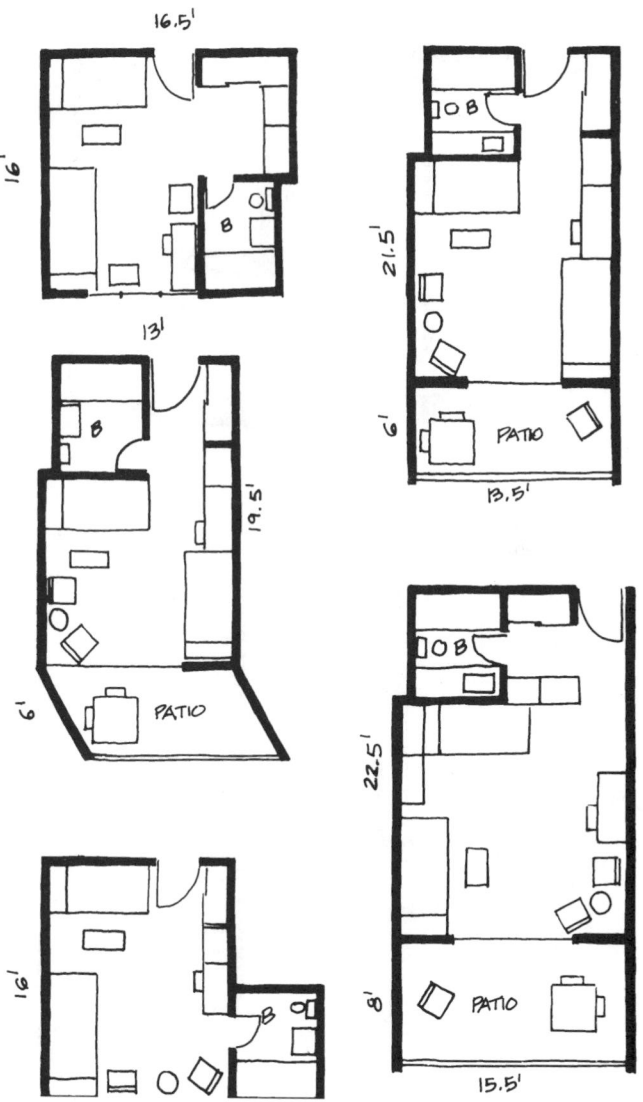

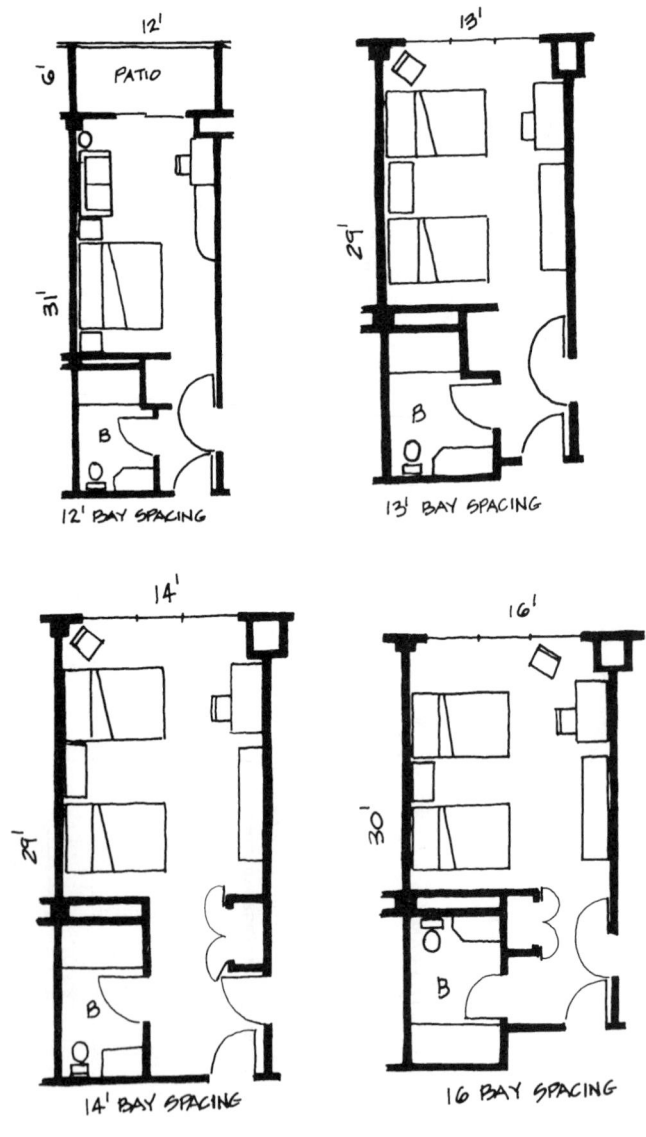

12' BAY SPACING

13' BAY SPACING

14' BAY SPACING

16 BAY SPACING

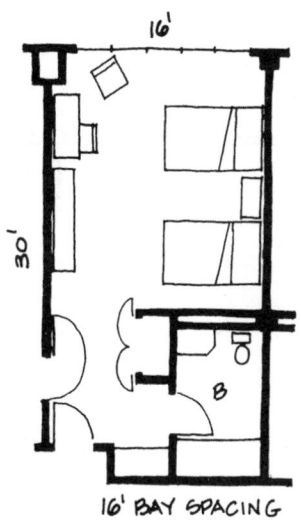

16' BAY SPACING

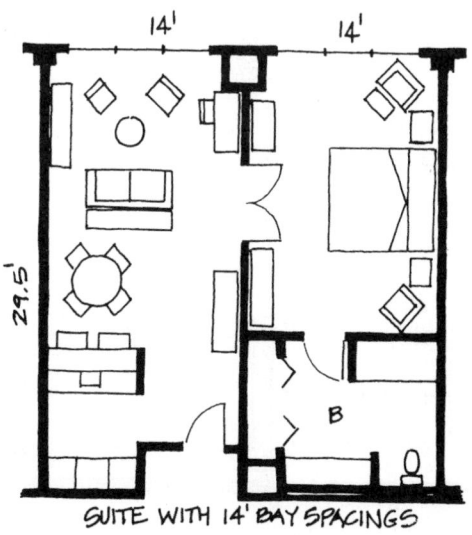

SUITE WITH 14' BAY SPACINGS

___ *b.* Details

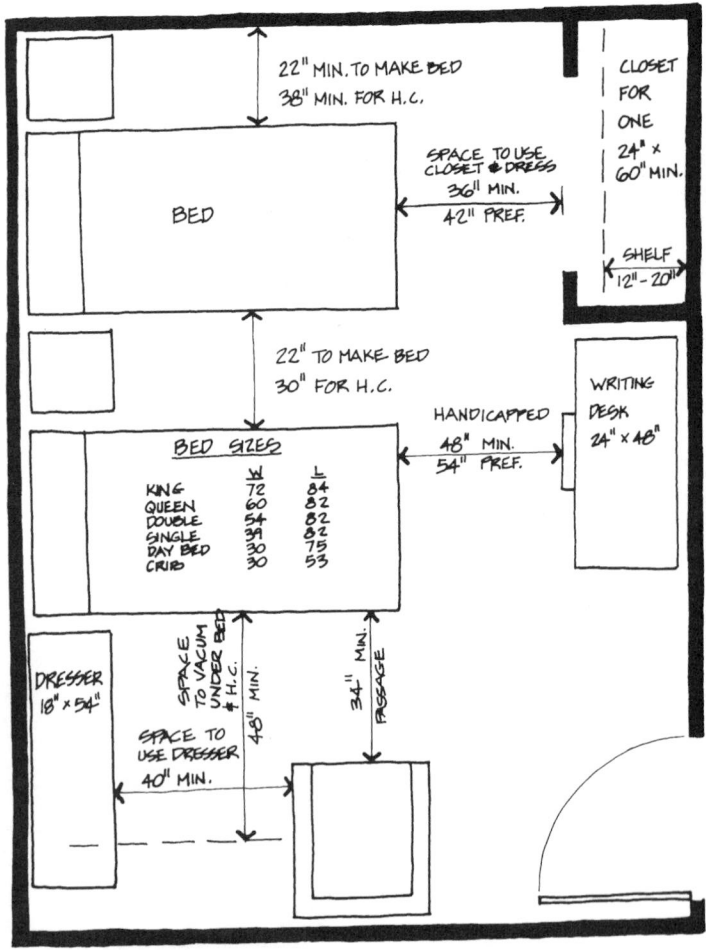

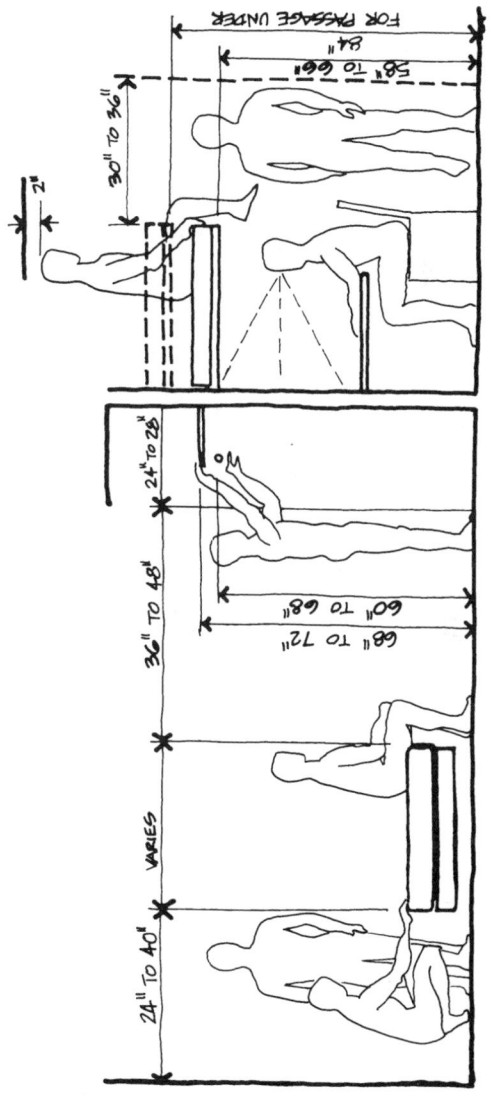

FOR PASSAGE UNDER
84"
58" TO 66"
30" TO 96"
2"

24" TO 28"
36" TO 48"
60" TO 68"
68" TO 72"

VARIES
24" TO 40"

71

___ *c.*   Bedroom templates.
The following furniture templates are at ¼″ = 1′0″ scale. Dimensions are in inches. Use these for layout of bedroom/sleeping areas.

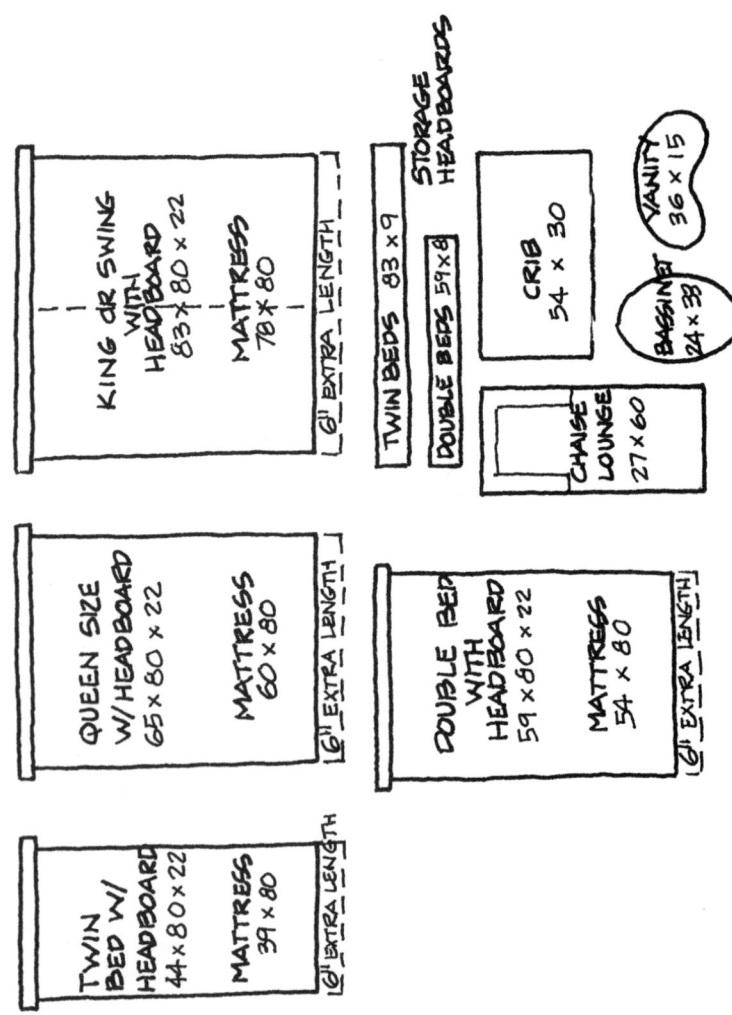

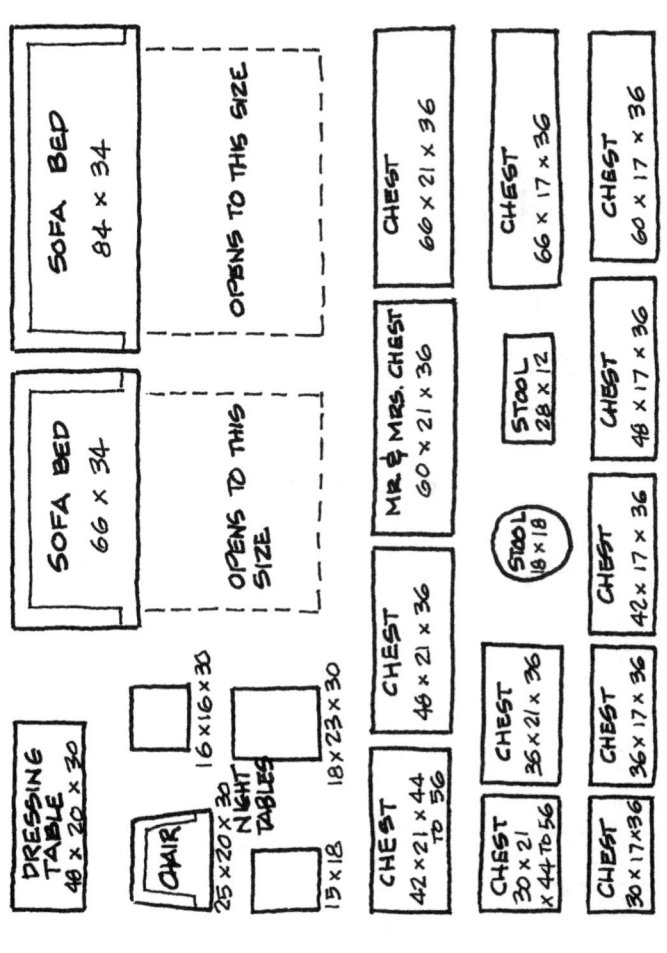

DRESSING TABLE
40 × 20 × 30

CHAIR

NIGHT TABLES
25 × 20 × 30

16 × 16 × 30

18 × 23 × 30

15 × 18

SOFA BED
66 × 34

OPENS TO THIS SIZE

SOFA BED
84 × 34

OPENS TO THIS SIZE

CHEST
42 × 21 × 44 TO 56

CHEST
46 × 21 × 36

MR. & MRS. CHEST
60 × 21 × 36

CHEST
66 × 21 × 36

CHEST
36 × 21 × 36

STOOL
18 × 18

STOOL
28 × 12

CHEST
66 × 17 × 36

CHEST
30 × 21 × 44 TO 56

CHEST
36 × 17 × 36

CHEST
42 × 17 × 36

CHEST
48 × 17 × 36

CHEST
60 × 17 × 36

CHEST
30 × 17 × 36

4. Dining/Eating

    a. Overall layouts. See p. 398 for furnishings.
Use the following layouts for general planning.

      (1) Residential layouts

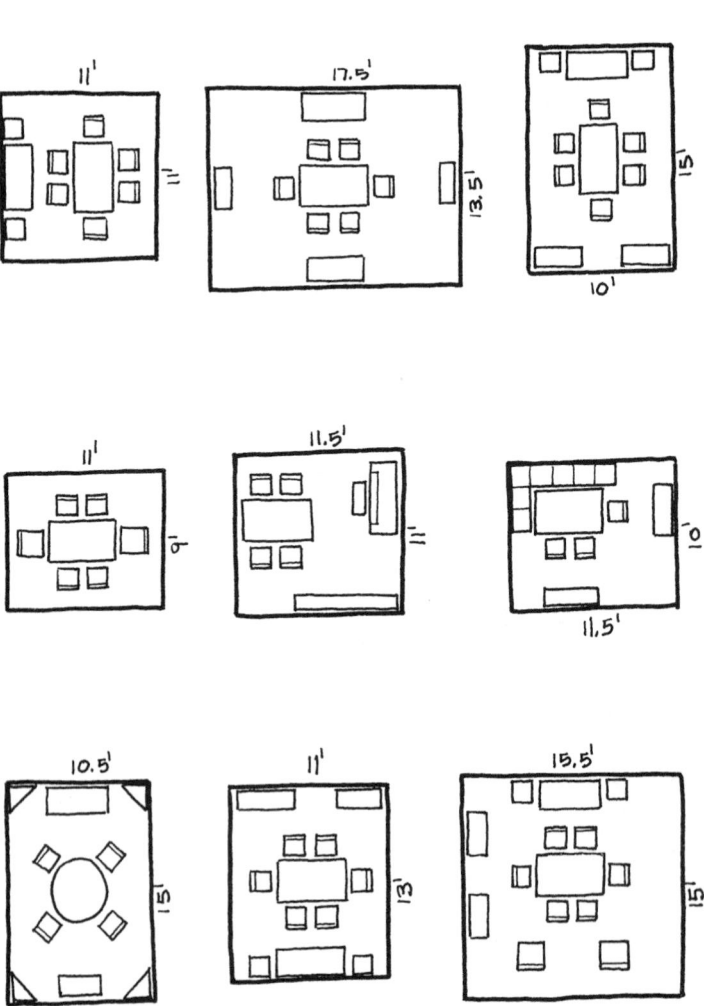

___ (2) Restaurant layouts
Restaurants usually include:
___ (*a*) General public space (waiting and toilets)
___ (*b*) Dining area
___ (*c*) Kitchen

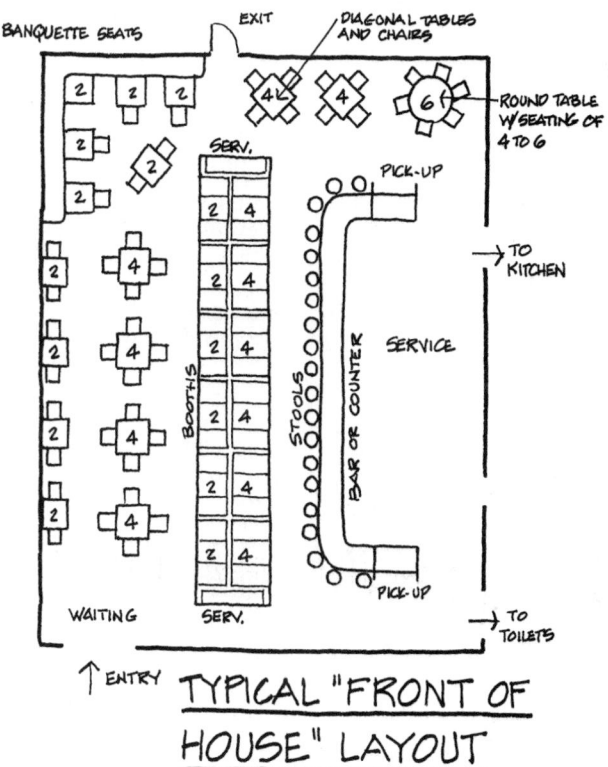

TYPICAL "FRONT OF HOUSE" LAYOUT

Estimated Areas for Total Restaurant Building

Type of Operation	Area (SF/Seat)
Table Service	25 to 35
Counter Service	20 to 25
Booth Service	20 to 30
Cafeteria Service	20 to 30

Estimated Dining Areas ("Front of House" Only)

Type	Dining Space (SF/Seat)
Table	10 to 20
Counter	15 to 20
Booth	10 to 15
Cafeteria	10 to 15
Banquet	10 to 15

Notes: 1. Toilets are taking more space in restaurants than ever before. See p. 434.
2. For occupancy count for legal exiting, see p. 169.
3. For commercial kitchens ("back of house") see p. 83.

___ *b.* Details

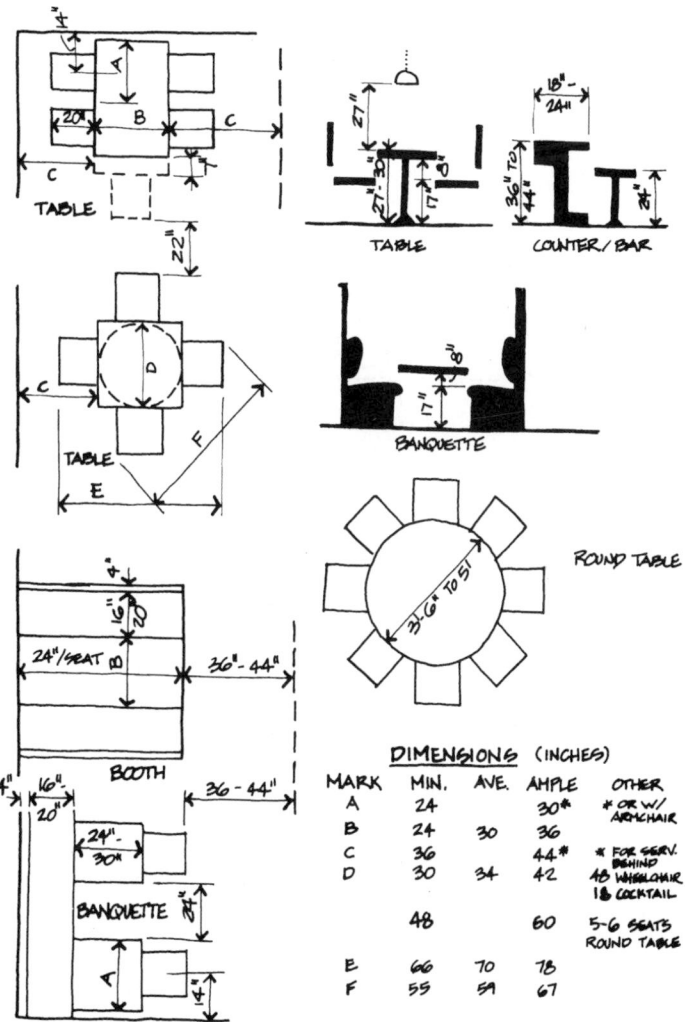

TABLE

TABLE

COUNTER / BAR

TABLE

BANQUETTE

ROUND TABLE

BOOTH

BANQUETTE

DIMENSIONS (INCHES)

MARK	MIN.	AVE.	AMPLE	OTHER
A	24		30*	* OR W/ ARMCHAIR
B	24	30	36	
C	36		44*	* FOR SERV. BEHIND
D	30	34	42	48 WHEELCHAIR 18 COCKTAIL
	48		60	5-6 SEATS ROUND TABLE
E	66	70	78	
F	55	59	67	

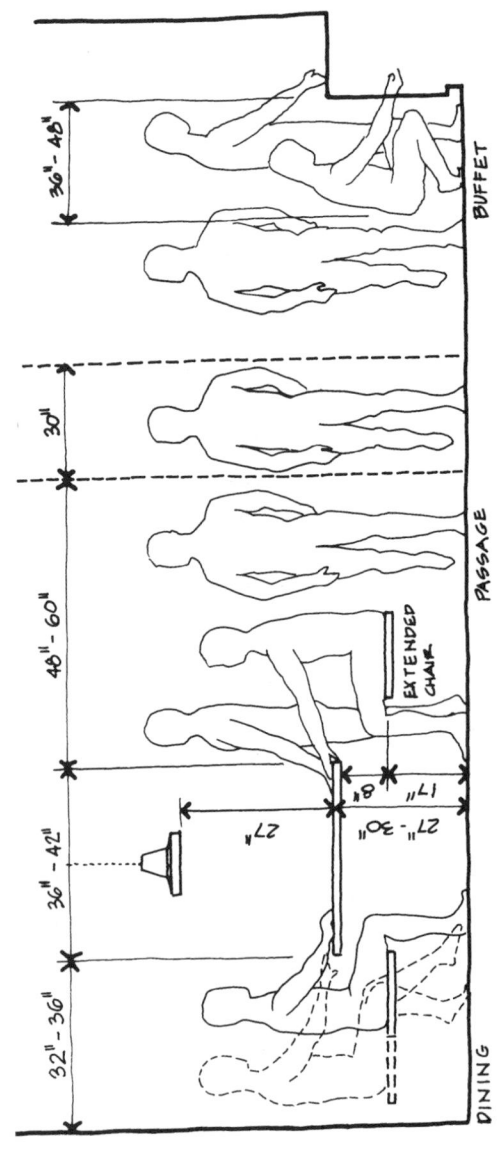

___ *c.* Dining room templates
The following furniture templates are at ¼″ = 1′0″
scale. Dimensions are in inches. Use them for layout
of residential dining areas.

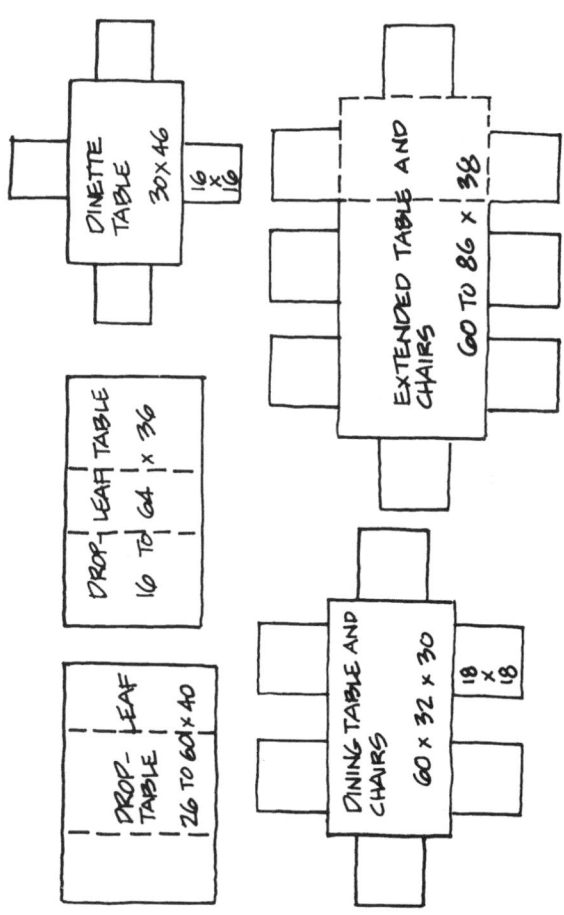

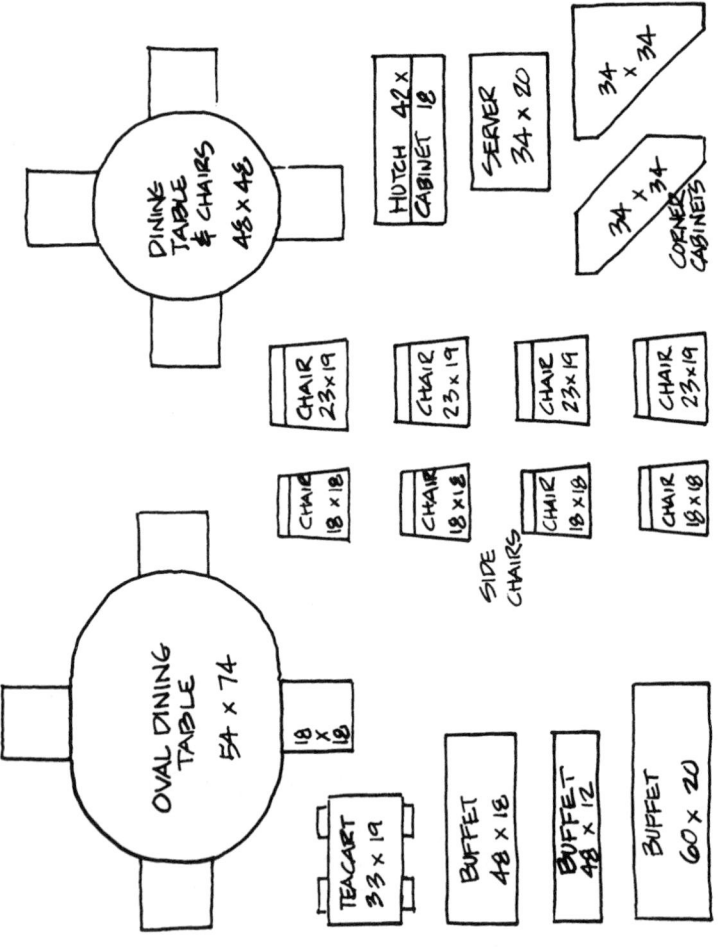

DINING TABLE & CHAIRS 48 x 48

HUTCH 42 x CABINET 18

SERVER 34 x 20

34 x 34

34 x 34

CORNER CABINETS

CHAIR 23×19

CHAIR 23×19

CHAIR 23×19

CHAIR 23×19

CHAIR 18 x 18

CHAIR 18 x 18

CHAIR 18 x 18

CHAIR 18 x 18

SIDE CHAIRS

OVAL DINING TABLE 54 x 74

18 x 18

TEACART 33 x 19

BUFFET 48 x 18

BUFFET 48 x 12

BUFFET 60 x 20

___ 5. <u>Kitchens/Cooking</u>

    ___ *a.* Overall layouts

        ___ (1) Residential

                C   = Cooktop (range/oven)

                D   = Desk

                DR = Dryer

                DW = Dishwasher

                F   = Folding

                I   = Ironing board

                P   = Pantry

R   = Refrigerator
S   = Sink
T   = Table and chairs
TR = Trash
W = Washer
WO = Wall oven

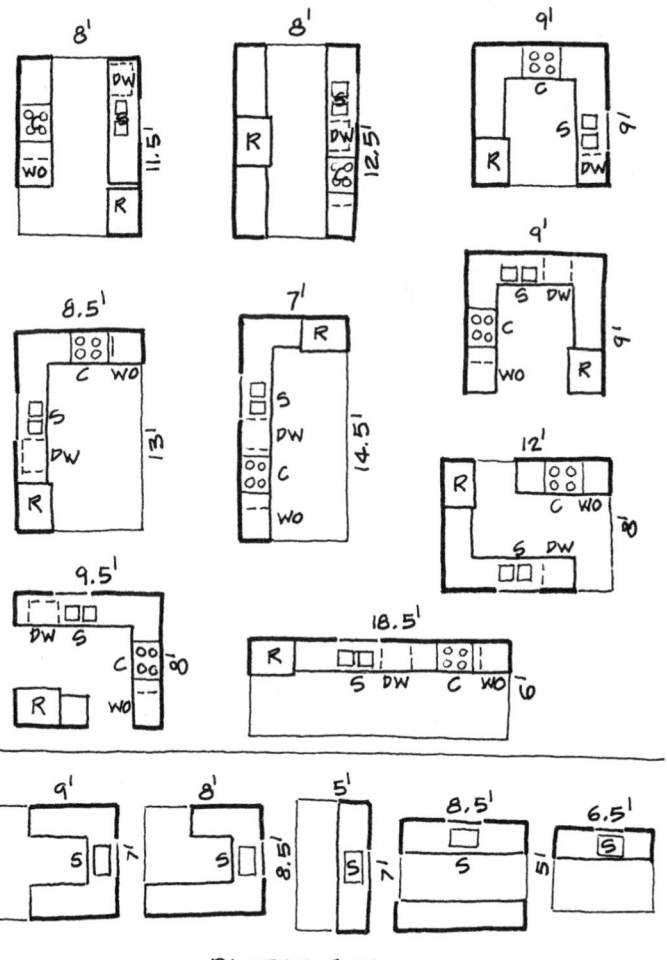

PANTRY TYPES

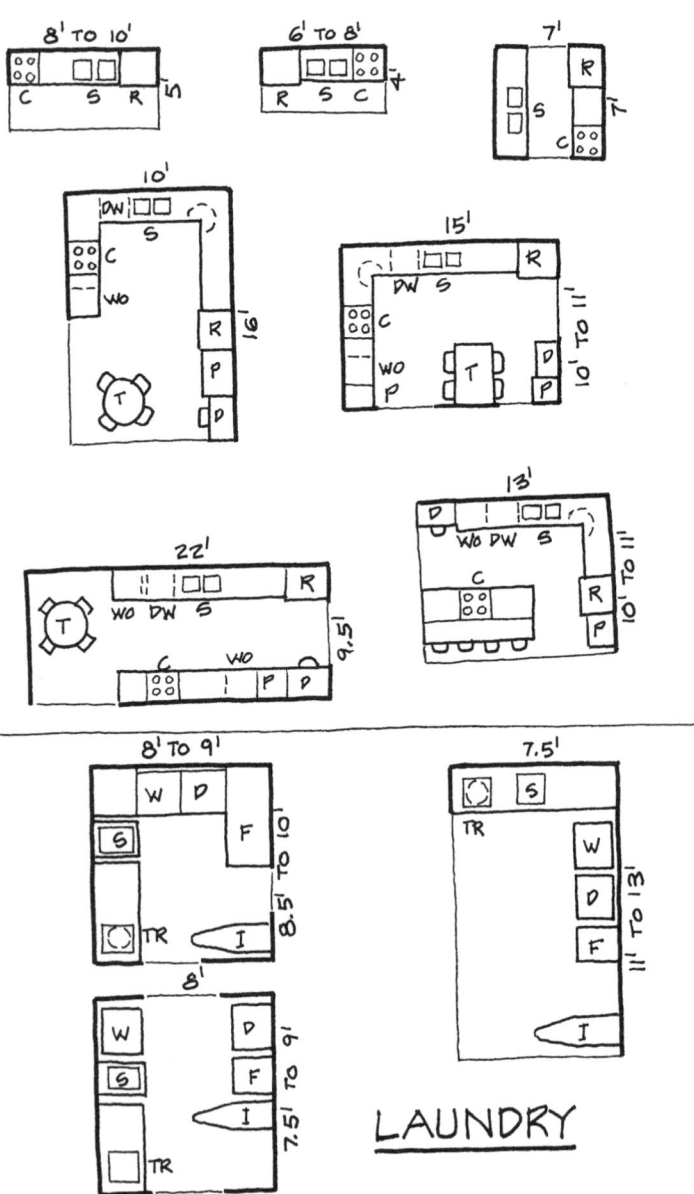

LAUNDRY

___ (2) Commercial restaurant kitchens
     ___ (*a*) Typically kitchens range roughly from one-third to one-half the total area of a restaurant.
     ___ (*b*) For general planning, the following areas are usually included in a restaurant kitchen:

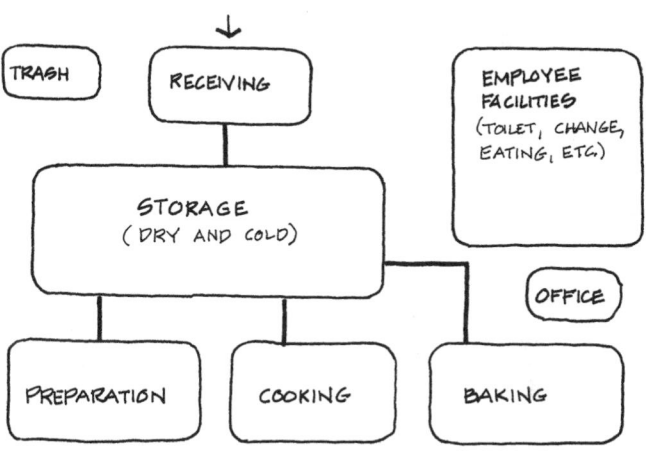

___ (*c*) Typical space allocations (%) of kitchens

Receiving	5%
Storage	20
Preparation	15
Cooking	10
Baking	10
Dishwashing	5
Aisles	15
Trash	5
Misc. (employee changing room, eating area, office etc.)	15
	100%

___ *b.* Details (residential)
Also see p. 377 for washers and dryers.

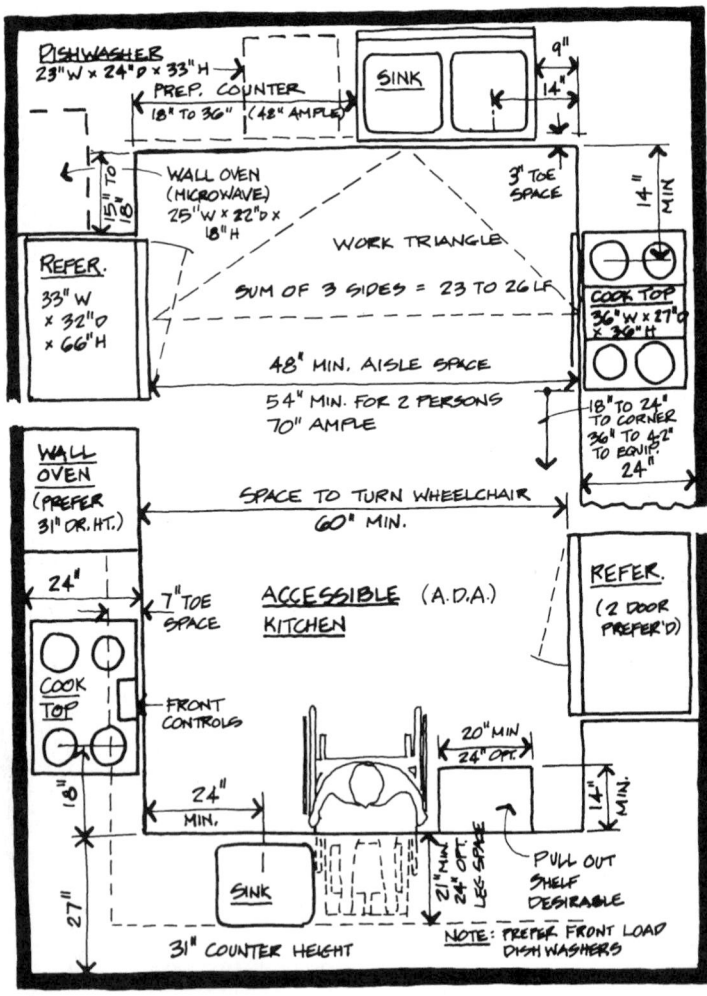

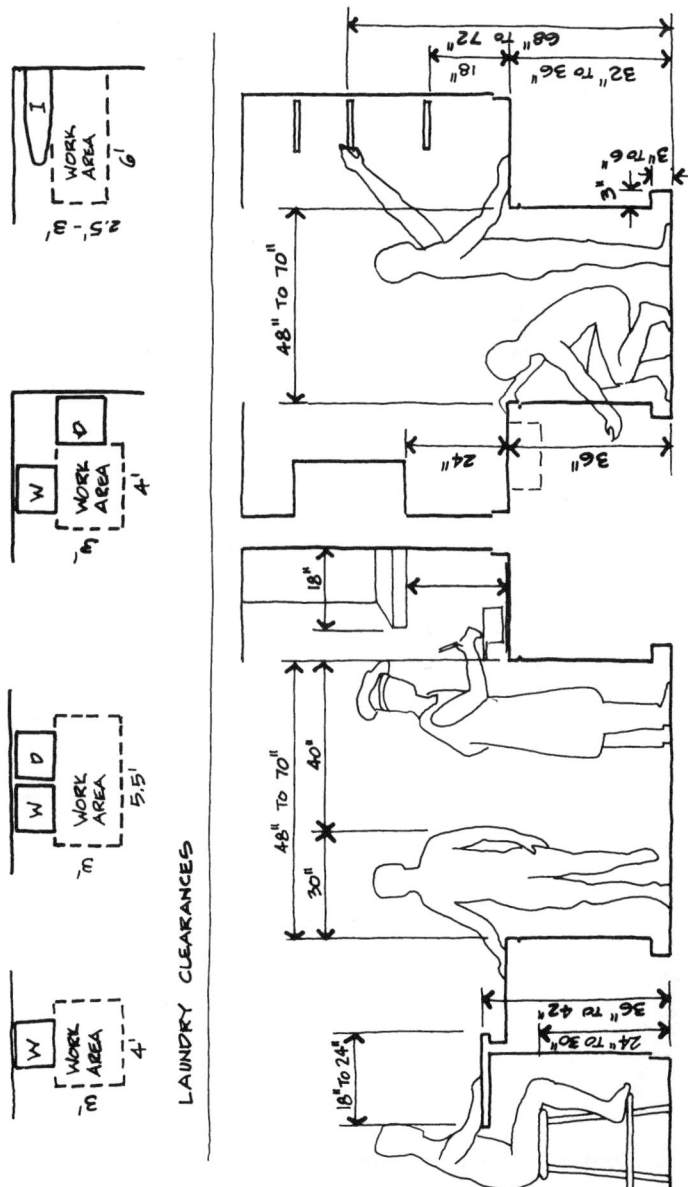

LAUNDRY CLEARANCES

___ c. Kitchen templates (residential)
The following furniture templates are at ¼″ = 1′0″
scale. Dimensions are in inches. Use them for layout
of residential kitchen areas.

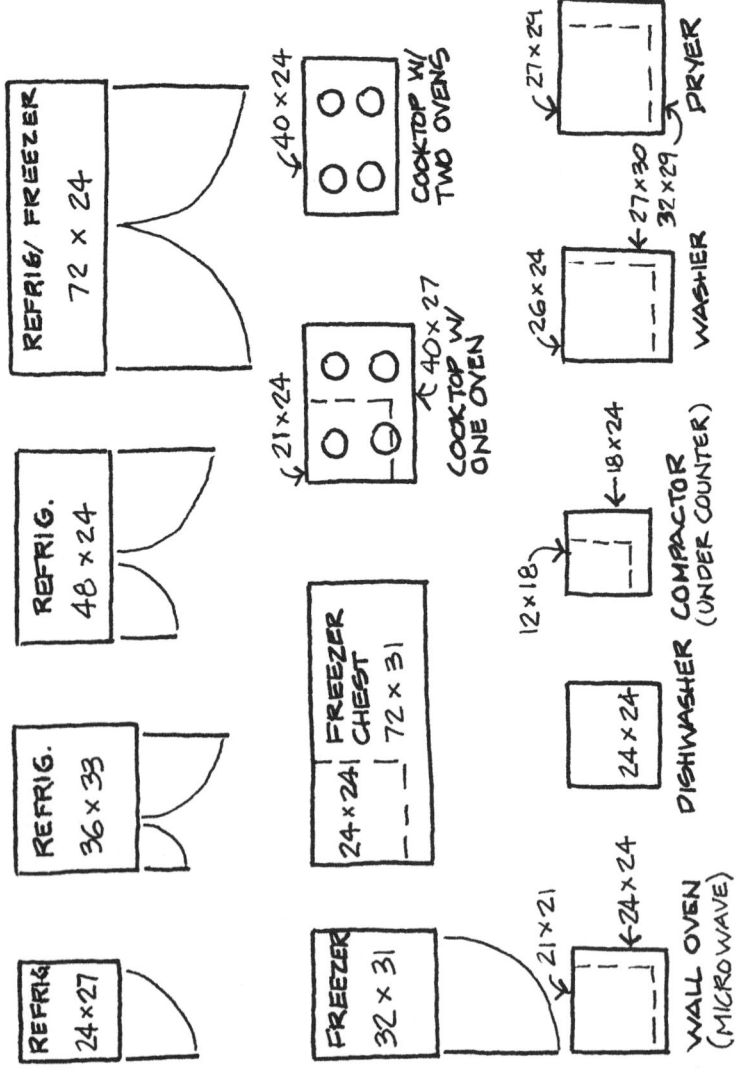

___ 6. <u>Toilets/Bathing</u> (see p. 434 for fixture costs)
  ___ *a.* Overall layouts
    ___ (1) Residential:

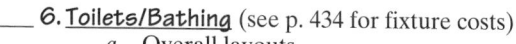

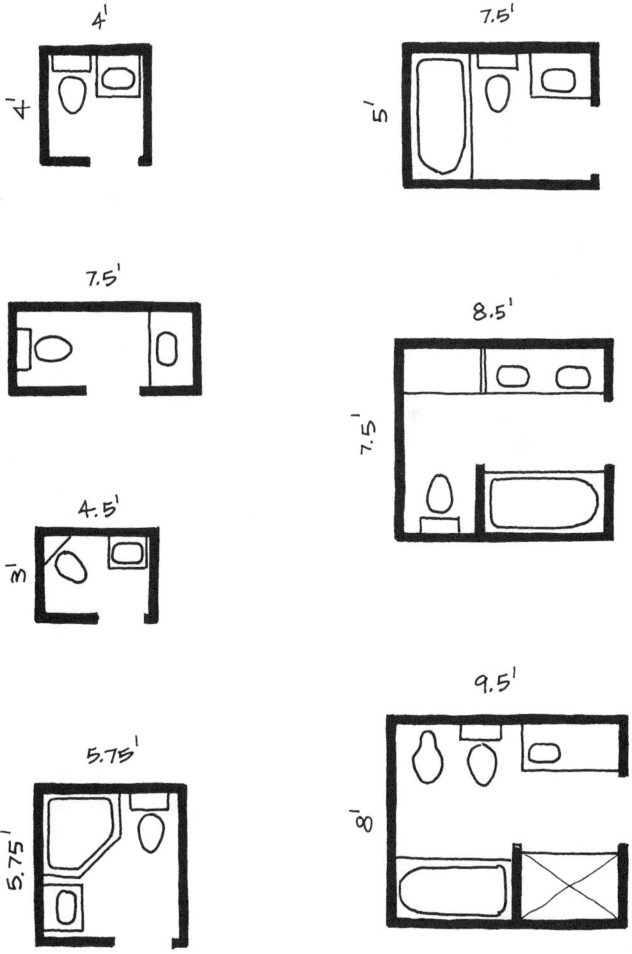

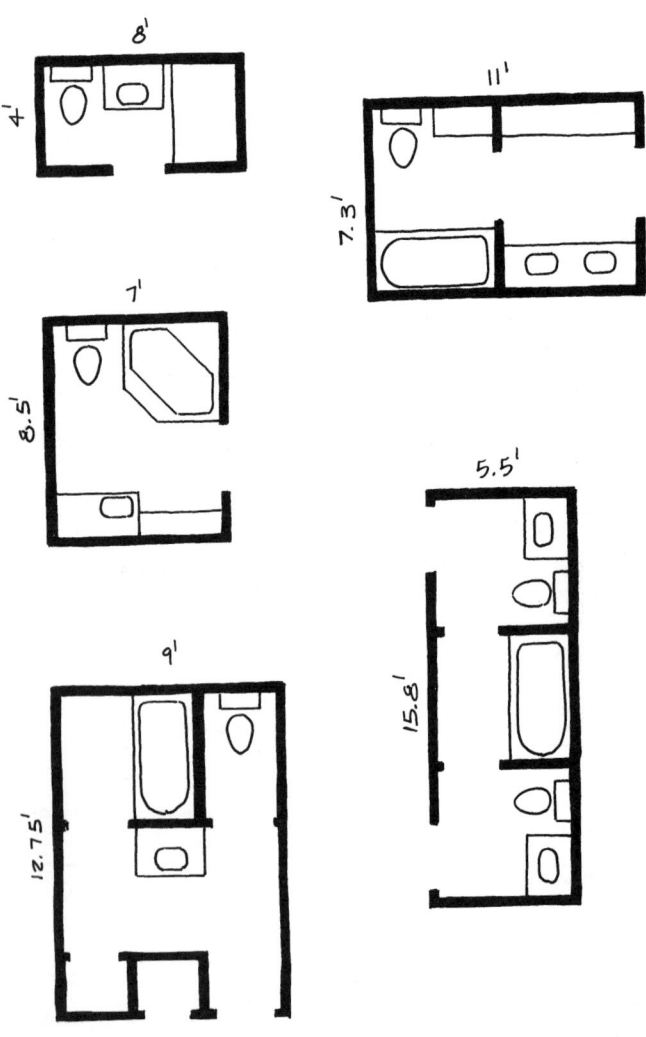

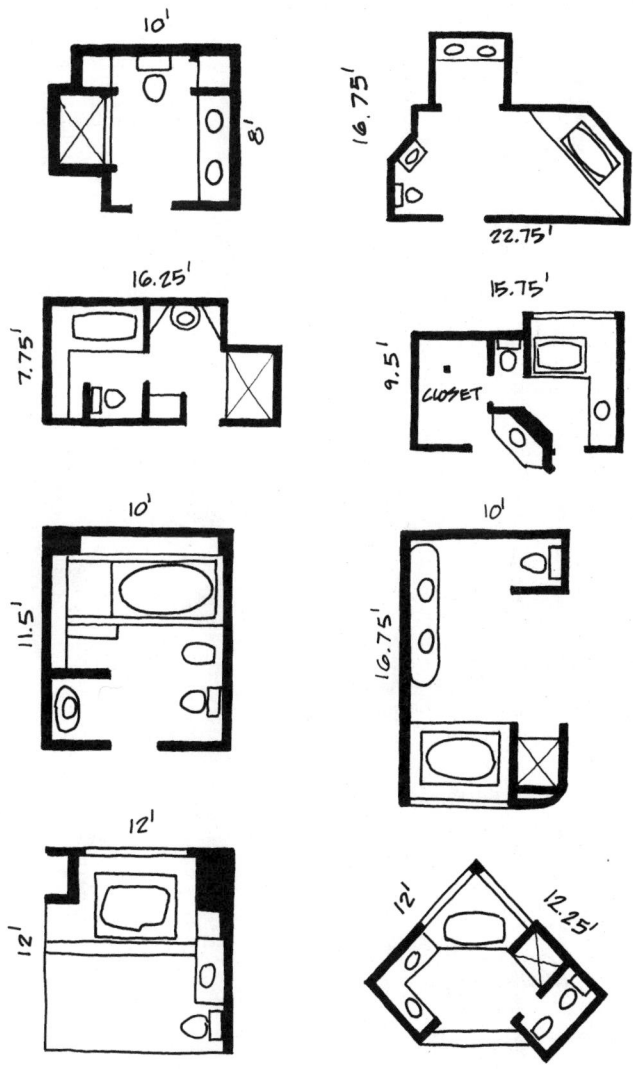

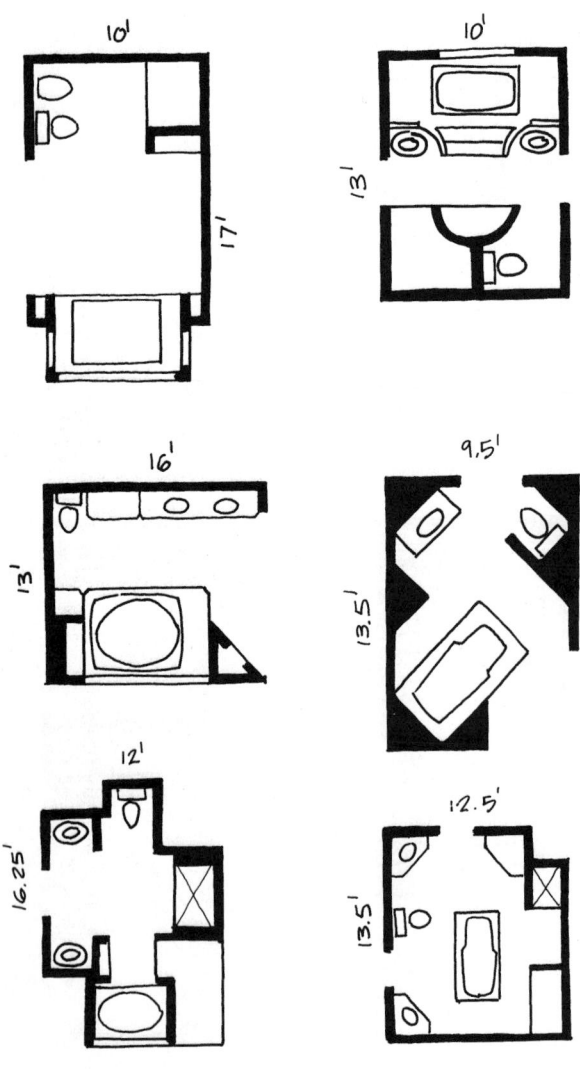

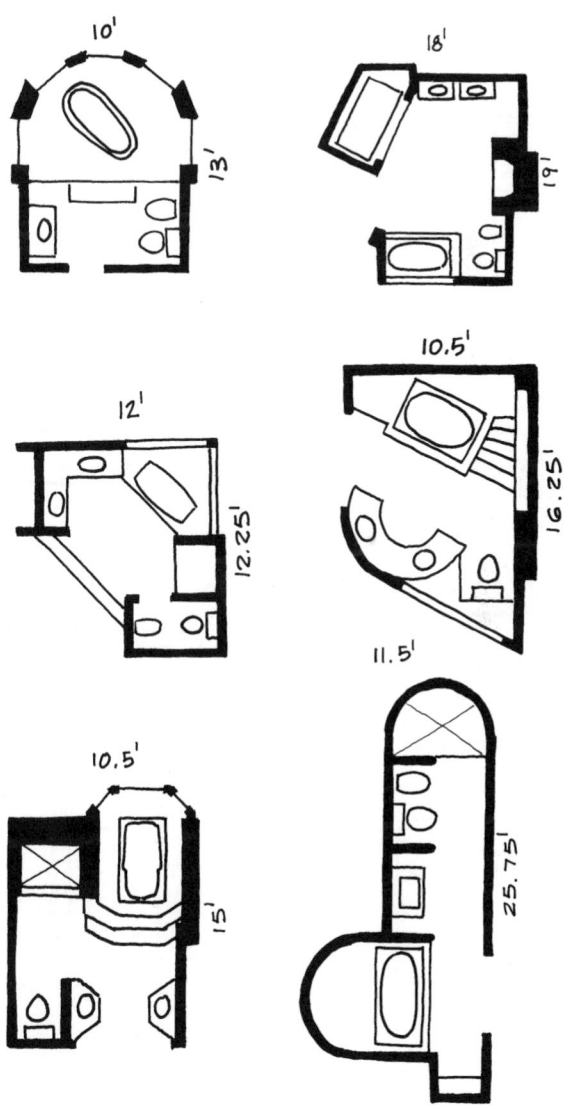

___ (2) Public toilets:

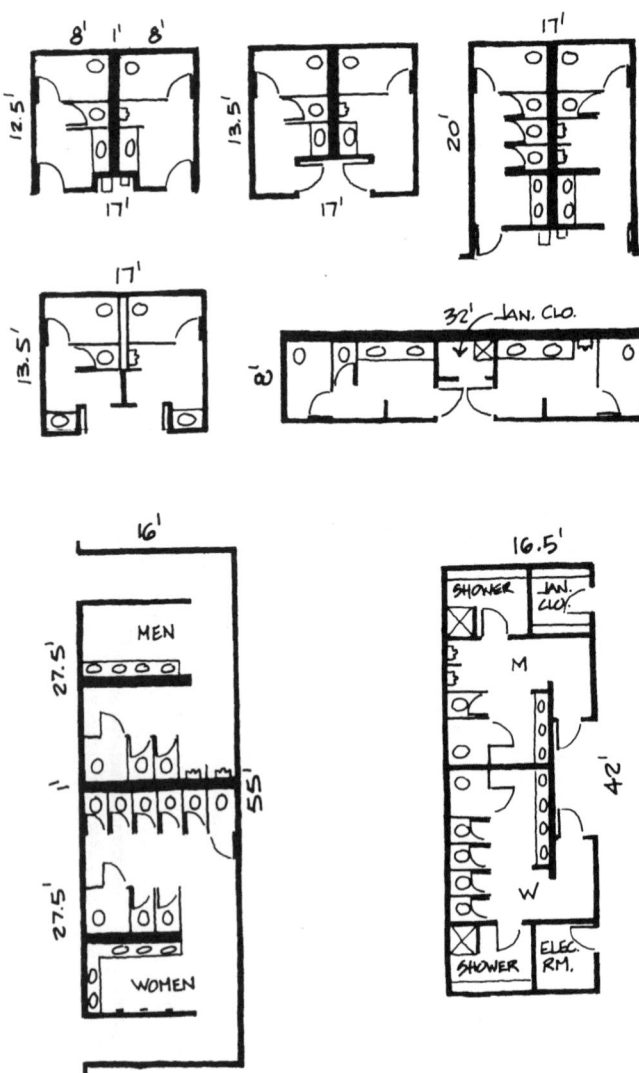

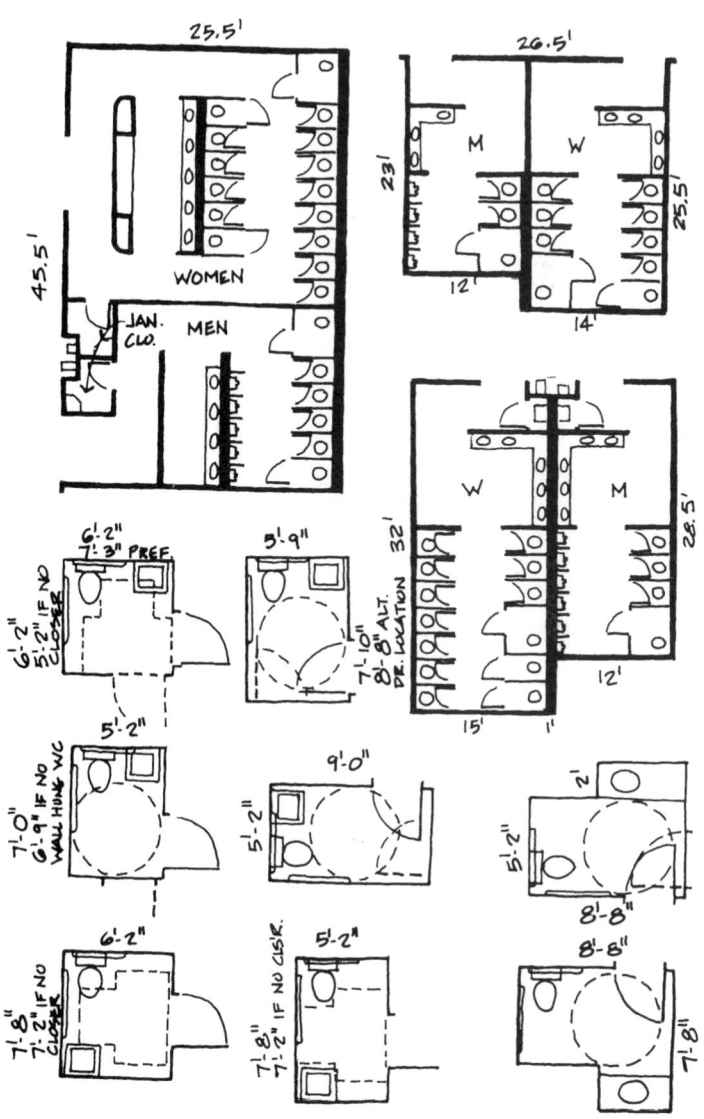

SINGLE USER TOILETS (DIM. ARE CLEAR)

___ *b.* Toilet details

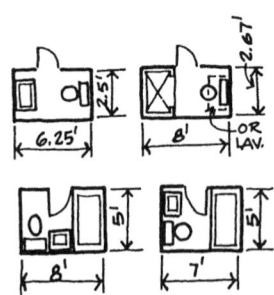

RESIDENTIAL BATHROOMS

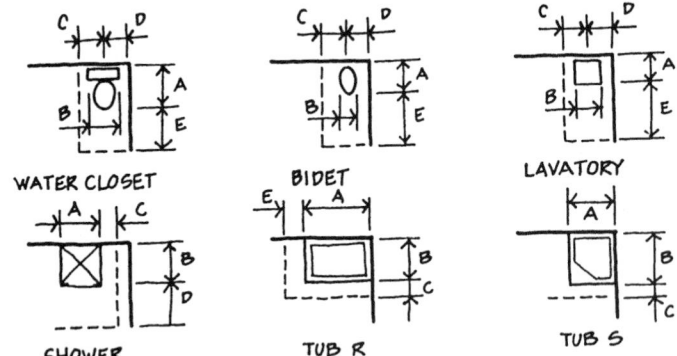

WATER CLOSET   BIDET   LAVATORY

SHOWER   TUB R   TUB S

FIXTURE SIZES AND CLEARANCES (INCHES)

FIXTURE	A		B		C		D		E	
	MIN.	LIB.	MIN.	LIB.	MIN.	LIB.	MIN.	LIB.	MIN.	LIB.
WATER C.	27	31	19	21	12	18	15	22	18	34 - 36
BIDET	25	27	14	14	12	18	15	22	18	34 - 36
LAVATORY	16	21	18	30	2	6	14	22	18	30
SHOWER	32	36	34	36	2	8	18	34		
TUB R	60 STD	72	30 STD	42	2	8	18 - 20	30 - 34	2	8
TUB S	38		39		2	4				

NOTE: FOR H.C. ACCESSIBILITY, SEE FOLLOWING PAGES.

TOILET ROOMS

SINGLE USER TOILET

PUBLIC TOILET

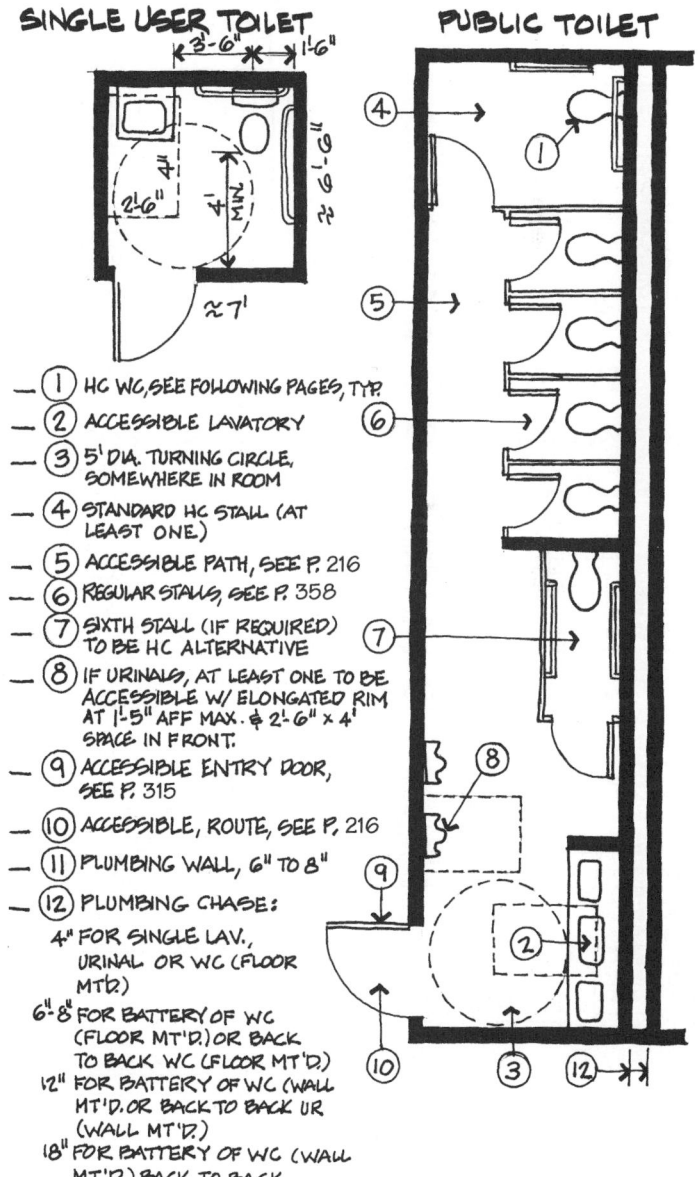

— ① HC WC, SEE FOLLOWING PAGES, TYP.

— ② ACCESSIBLE LAVATORY

— ③ 5' DIA. TURNING CIRCLE, SOMEWHERE IN ROOM

— ④ STANDARD HC STALL (AT LEAST ONE.)

— ⑤ ACCESSIBLE PATH, SEE P. 216

— ⑥ REGULAR STALLS, SEE P. 358

— ⑦ SIXTH STALL (IF REQUIRED) TO BE HC ALTERNATIVE

— ⑧ IF URINALS, AT LEAST ONE TO BE ACCESSIBLE W/ ELONGATED RIM AT 1'-5" AFF MAX. & 2'-6" x 4' SPACE IN FRONT.

— ⑨ ACCESSIBLE ENTRY DOOR, SEE P. 315

— ⑩ ACCESSIBLE, ROUTE, SEE P. 216

— ⑪ PLUMBING WALL, 6" TO 8"

— ⑫ PLUMBING CHASE:

  4" FOR SINGLE LAV., URINAL OR WC (FLOOR MT'D.)

  6"-8" FOR BATTERY OF WC (FLOOR MT'D.) OR BACK TO BACK WC (FLOOR MT'D.)

  12" FOR BATTERY OF WC (WALL MT'D. OR BACK TO BACK UR (WALL MT'D.)

  18" FOR BATTERY OF WC (WALL MT'D.) BACK TO BACK

ACCESSIBLE TOILET STALLS (ADA ANSI)

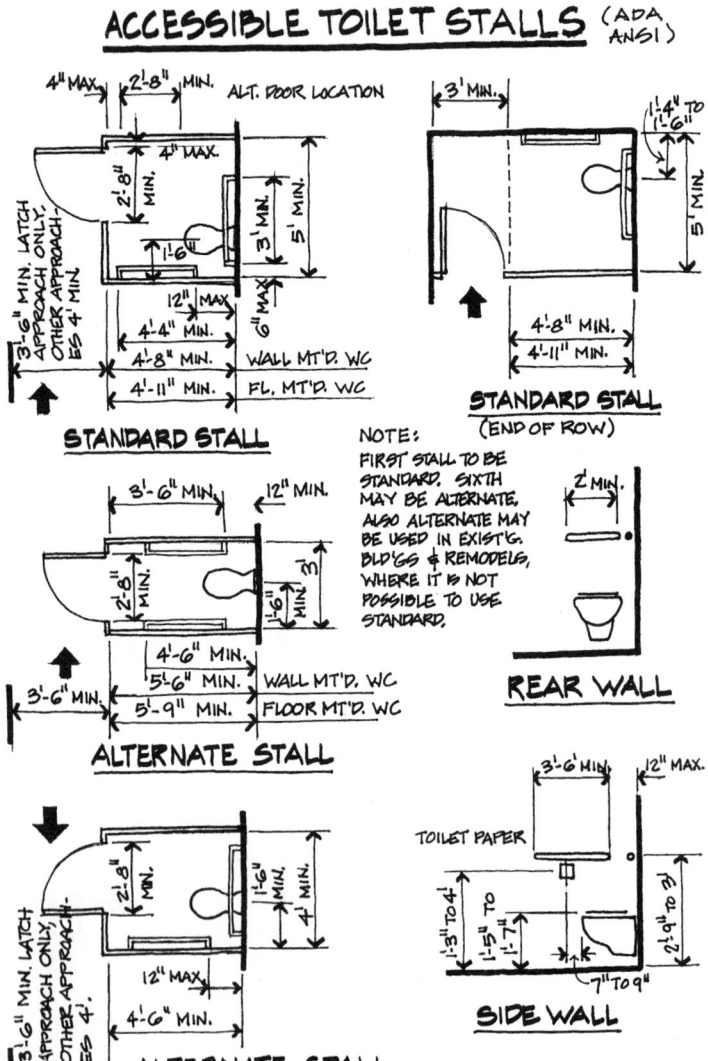

STANDARD STALL

4" MAX   2'-8" MIN   ALT. DOOR LOCATION

4" MAX
2'-8" MIN
1'-6"
3' MIN.
5' MIN.
3'-6" MIN. LATCH APPROACH ONLY. OTHER APPROACHES 4' MIN.
12" MAX
6" MAX
4'-4" MIN.
4'-8" MIN.   WALL MT'D. WC
4'-11" MIN.   FL. MT'D. WC

STANDARD STALL
(END OF ROW)

3' MIN.
1'-4" TO 1'-6"
5' MIN.
4'-8" MIN.
4'-11" MIN.

NOTE:
FIRST STALL TO BE STANDARD. SIXTH MAY BE ALTERNATE. ALSO ALTERNATE MAY BE USED IN EXIST'G. BLD'GS & REMODELS, WHERE IT IS NOT POSSIBLE TO USE STANDARD.

ALTERNATE STALL

3'-6" MIN.   12" MIN.
2'-8" MIN.
1'-6" MIN.   3'
4'-6" MIN.
3'-6" MIN.   5'-6" MIN.   WALL MT'D. WC
5'-9" MIN.   FLOOR MT'D. WC

2' MIN.

REAR WALL

ALTERNATE STALL

2'-8" MIN.
1'-6" MIN   4' MIN.
3'-6" MIN. LATCH APPROACH ONLY. OTHER APPROACHES 4'.
12" MAX.
4'-6" MIN.

TOILET PAPER

3'-6" MIN   12" MAX.
1'-3" TO 4'
1'-5" TO 1'-7"
2'-9" TO 3'
7" TO 9"

SIDE WALL

ACCESSIBLE FIXTURES (ADA)

WATER CLOSETS

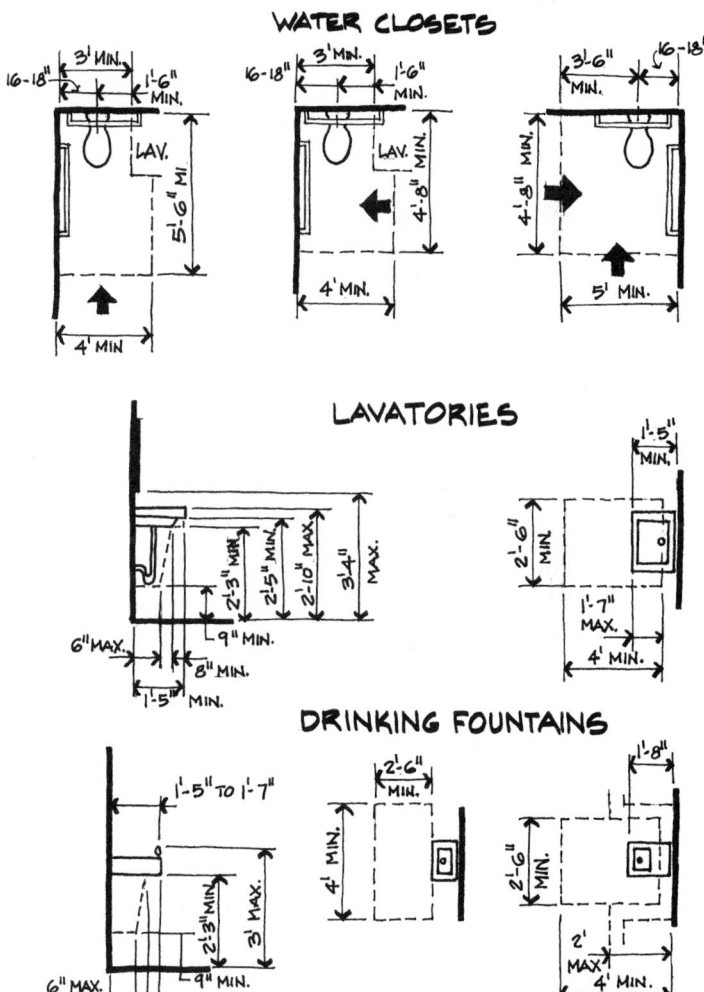

LAVATORIES

DRINKING FOUNTAINS

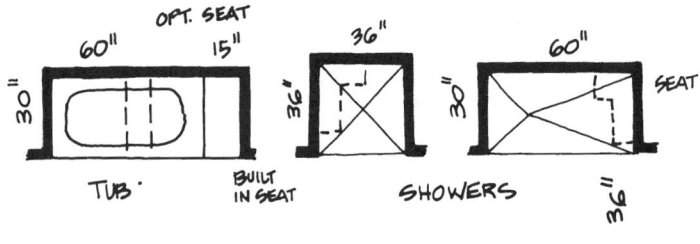

ACCESSIBLE TUBS AND SHOWERS

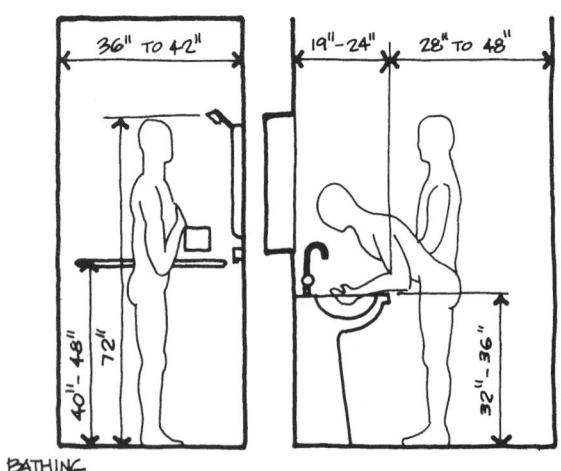

BATHING

___ *c.* Toilet and bathroom templates
The following fixture templates are at ¼″ = 1′0″
scale. Use them for layout of toilet and bathrooms.

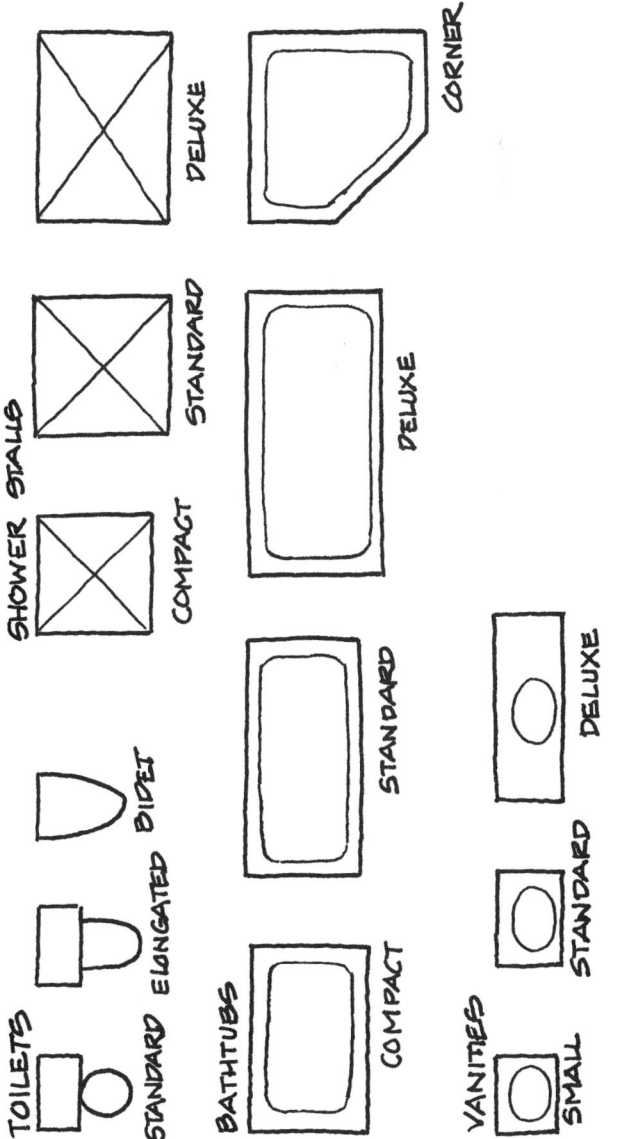

___ **7. _Office/Work_** (see p. 395 for furniture costs)

    ___ _a._ General Planning

        ___ (1) Desks should be spaced at a distance of 6 feet from the front of desk behind. Increase to 7 feet when in rows of two.

        ___ (2) Private office desk should face door.

        ___ (3) Employees performing close work should have the best lighting.

        ___ (4) Hang coats and jackets at work areas, not in special rooms.

        ___ (5) Heavy equipment should be located against walls or columns to avoid overloading floor.

        ___ (6) Circulation:

            ___ (_a_) Do not obstruct aisle or corridors for fire exiting.

            ___ (_b_) Aisles between rows of desks (secondary aisles) should be 36 inches wide.

            ___ (_c_) Aisles which carry a moderate amount of traffic should be 48 inches wide.

            ___ (_d_) Aisles leading to main exits should be 60 inches wide.

        ___ (7) Group large office functions by department:

            ___ (_a_) Management

            ___ (_b_) Finance

            ___ (_c_) Sales

            ___ (_d_) General services (central files, etc.)

            ___ (_e_) Technical services

            ___ (_f_) Production

        ___ (8) Study organization chart: The arrangement of the office functions will actually be a projection of the organization chart of the firm, located with respect for the flow of work and the physical requirements of each department.

        ___ (9) Ten guidelines for location of departments:

            ___ (_a_) Convenience to public

            ___ (_b_) Flow of work

            ___ (_c_) Equally used

            ___ (_d_) Centralized functions

            ___ (_e_) Confidential areas

            ___ (_f_) Conference rooms

            ___ (_g_) Freight elevators

___ (*h*) Shipping dock
___ (*i*) Service facilities (eating, lounge, medical, etc.)
___ (*j*) Passenger elevators
___ (10) Areas:

    ___ (*a*) Space allowance: *SF*

___	1.	Top executive	400–600
___	2.	Manager	150–200
___	3.	Assistant manager	100–125
___	4.	Supervisor	80–100
___	5.	Operator, 60-inch desk	50–60
___	6.	Operator, 55-inch desk	50–55
___	7.	Operator, 50-inch desk	45
___	8.	Standard letter file	6
___	9.	Standard legal file	7
___	10.	Letter lateral file	6.5
___	11.	Legal lateral file	7.5

___ (*b*) In general office area, an allotment of 100 SF/person for clerical work is liberal, 65 SF/person is economical, 80 SF/person is average.
___ (*c*) Private office: 100 to 300 SF is typical.
___ (*d*) Semiprivate office: 150 to 400 SF is typical.
___ (*e*) Rooms based on 15 people:
    ___ 1. Reception: 400 SF
    ___ 2. Waiting or interviewing: 200 SF
    ___ 3. Conference room: 500 SF
    ___ 4. Add approx. 10 SF for each additional person.

___ (11) Office layout can be done by *module,* ranging from 4 × 4 feet to 6 × 6 feet, adjustable in 4 to 6 inch increments. In layout of private offices, controlling factors are minimum practical layout, reconciled with exterior window and wall design. Walls need to meet at exterior window *mullions.* Also, *structural column* locations effect layout. Column spacing is usually at 25, 30, or 35 ft. centers.

___ (12) *Efficiency of layout* is measured by ratio of rentable space to total space. Average efficiency is about *70%*. Maximum possible is about 85%. The nonrentable space consists of elevators, stairs, toilets, lobbies, corridors, shafts, janitor's closets, and so forth. These are often in the central service core.

___ (13) *Floor-to-floor height* is usually about *12'.* (11' to 14'). Finished ceiling height is usually 8' to 8½'.

___ *b.* Overall layouts:

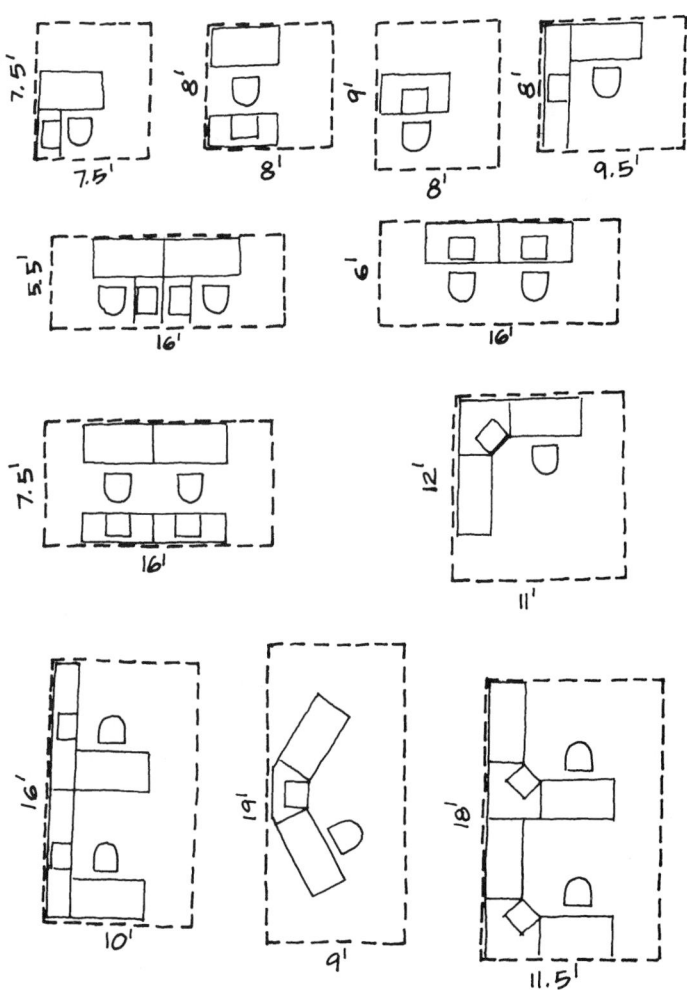

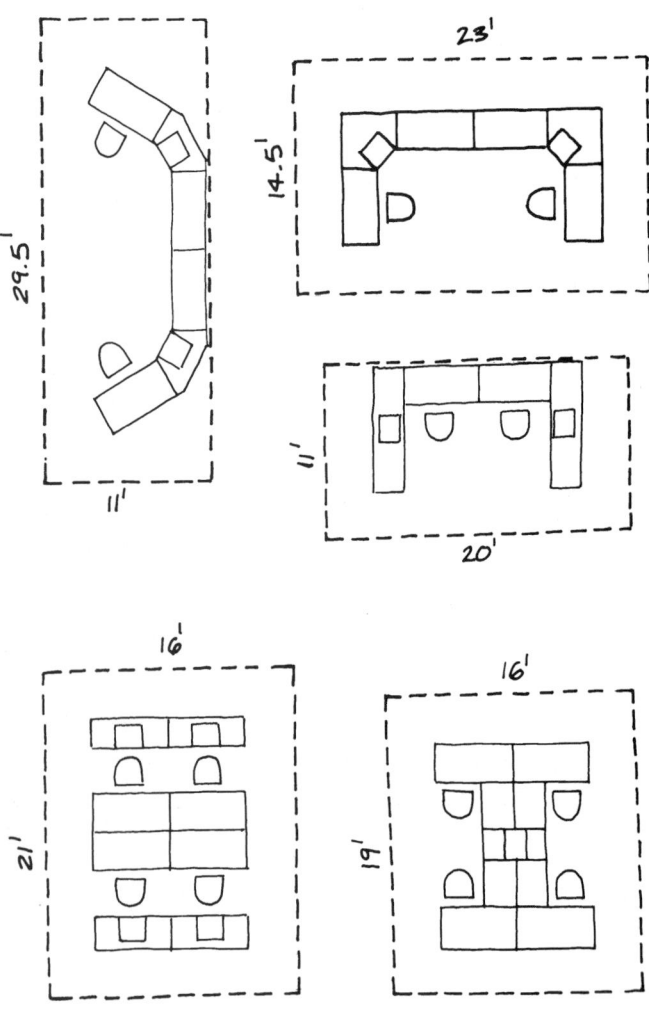

OPEN OFFICE SYSTEMS FURNITURE

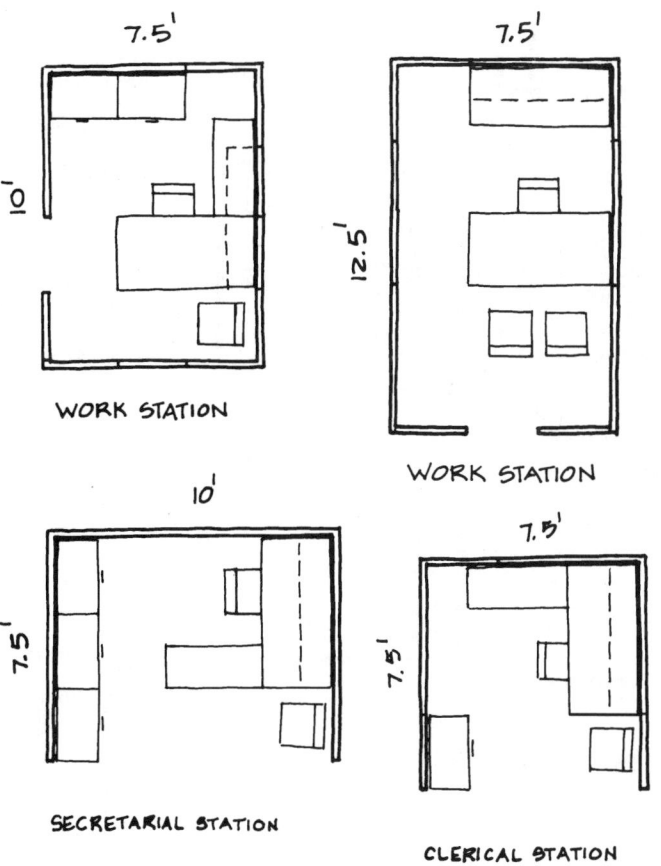

7.5'

10'

WORK STATION

7.5'

12.5'

WORK STATION

10'

7.5'

SECRETARIAL STATION

7.5'

7.5'

CLERICAL STATION

ENCLOSED OFFICE

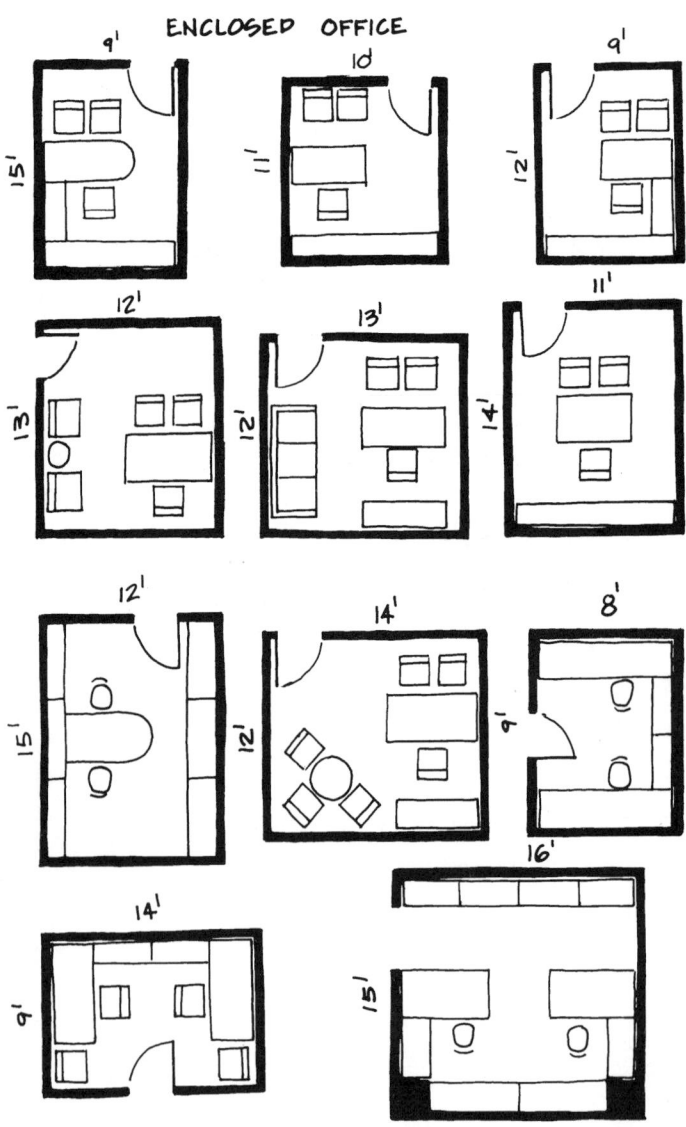

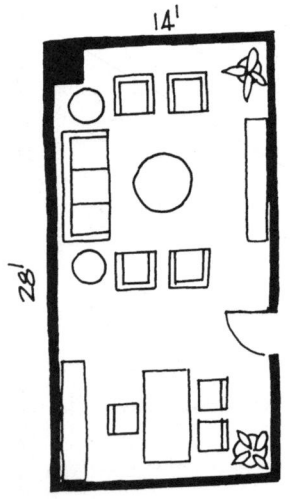

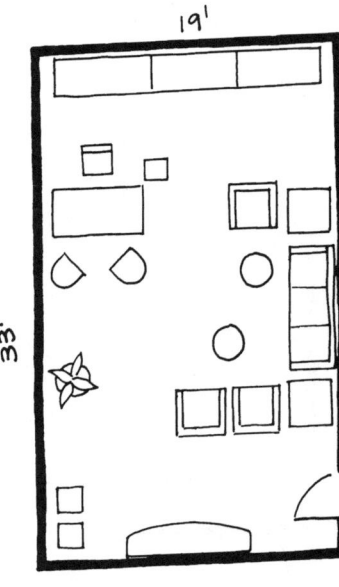

___ *c.* Details:

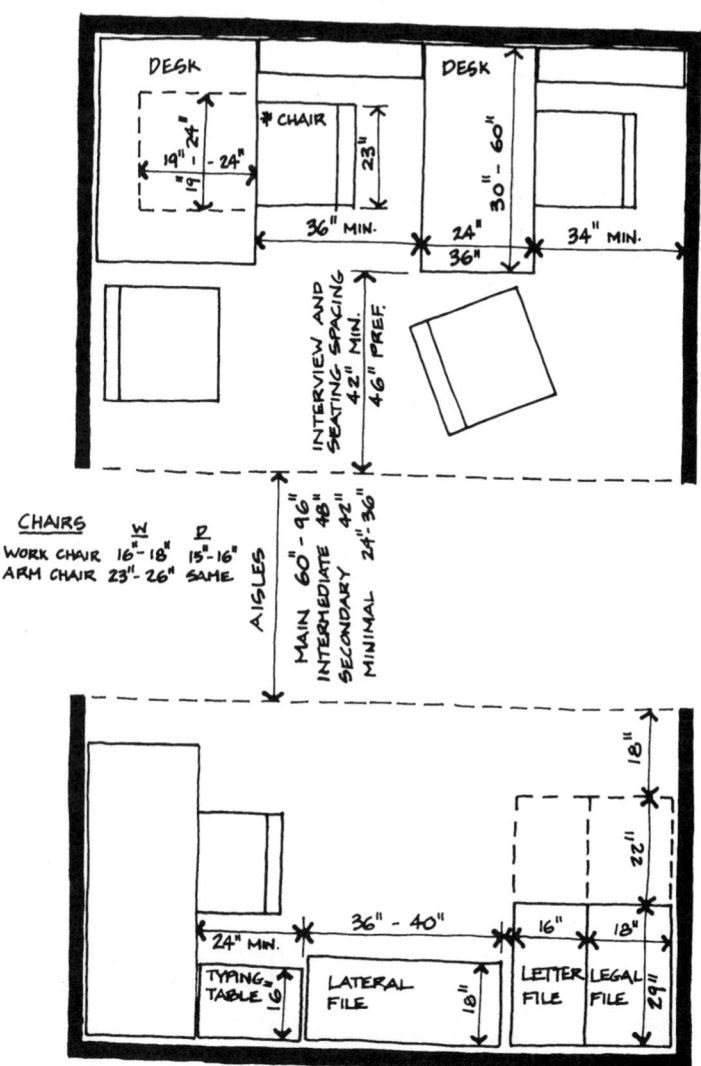

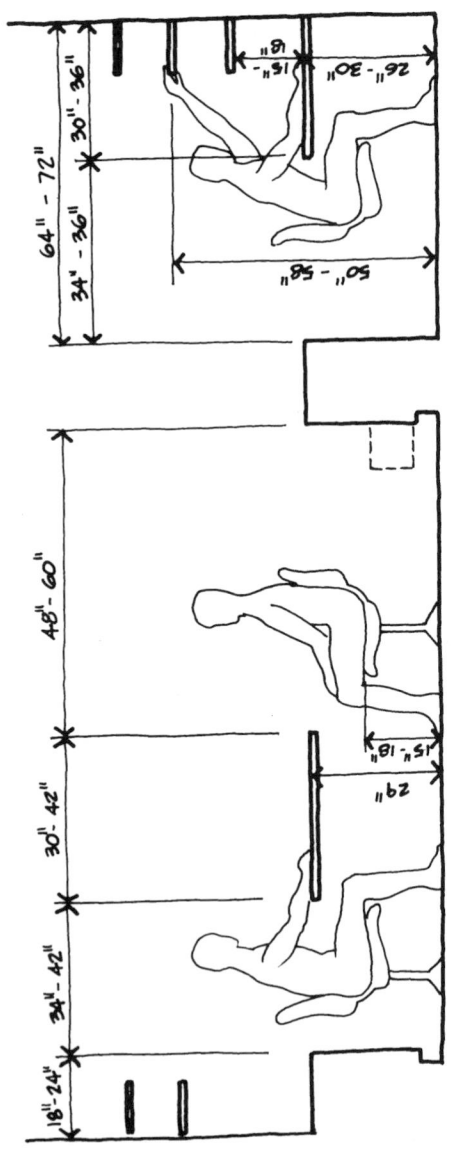

___ *d.* Office furniture templates:
   The following office furniture templates are at ¼″ =
   1′0″ scale. Use them to do office layouts.

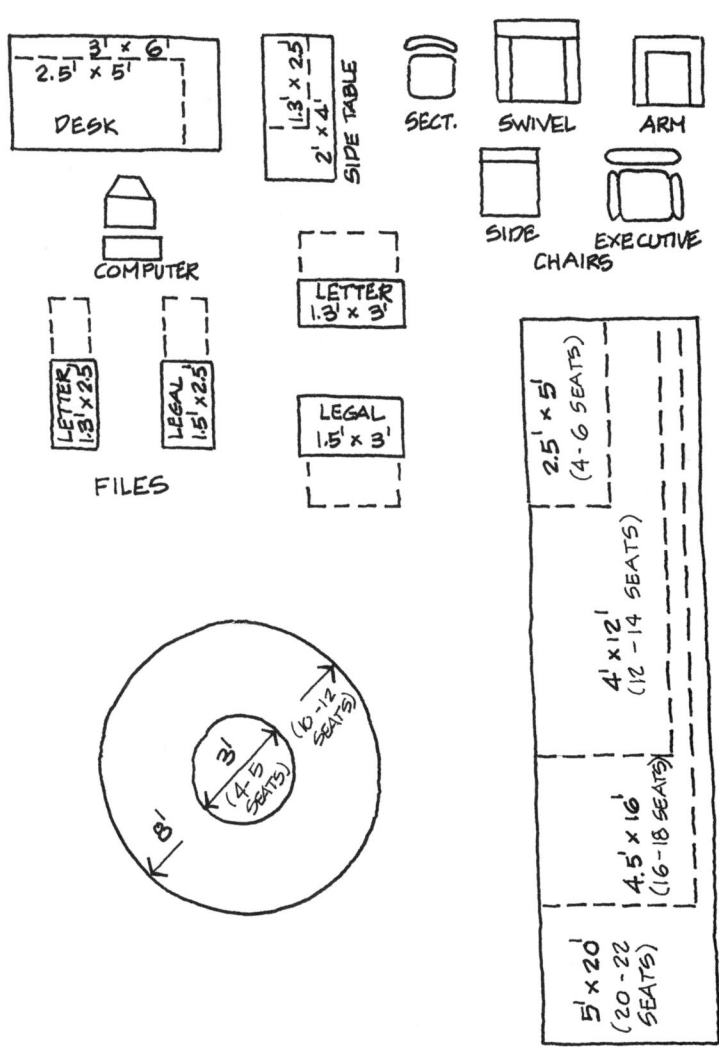

___ 8. Retail/Shopping
    ___ a. General
        ___ (1) Shops with one customer aisle only are usually 12 to 15 feet wide by 50 to 60 feet deep in large cities and 15 to 18 feet wide by 60 to 80 feet deep in smaller cities.

        ___ (2)

Aisle widths	Minimum	Desirable
Clerks	1'–8"	2' to 2'-3"
Main	4' to 6'	5'-6" to 7' up to 11'
Secondary		3' to 3'-6"

        ___ (3) Many retail stores are planned on a 4-foot module to accommodate standard retail fixtures.

        ___ (4) Divisions
            ___ (a) Entry/store front
            ___ (b) Retail
            ___ (c) Rear (stockroom, dressing rooms, office, repair, etc.) Size will vary; see sketches.

        ___ (5) Types of layouts:

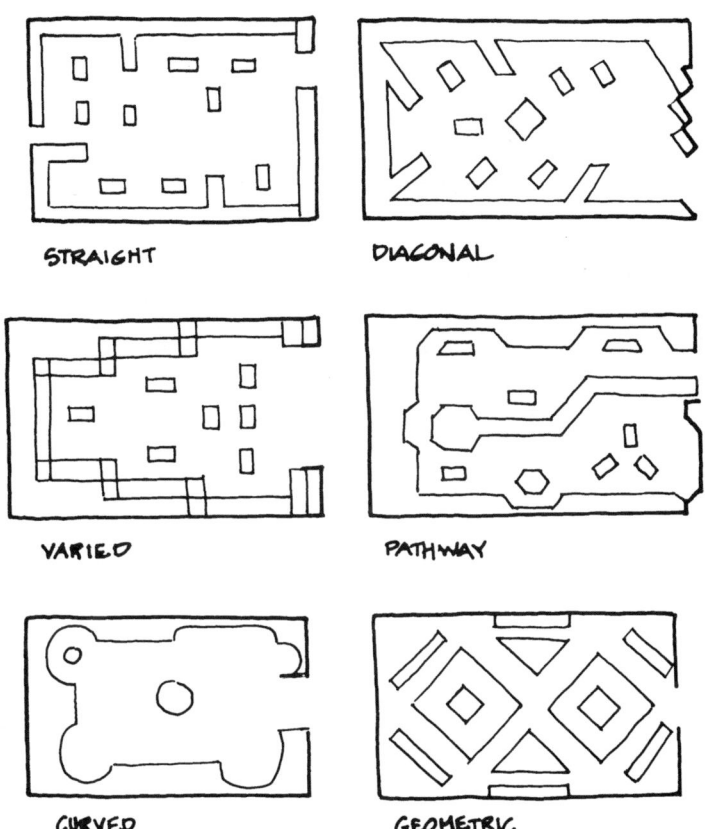

STRAIGHT

DIAGONAL

VARIED

PATHWAY

CURVED

GEOMETRIC

___ *b.* Overall layouts:

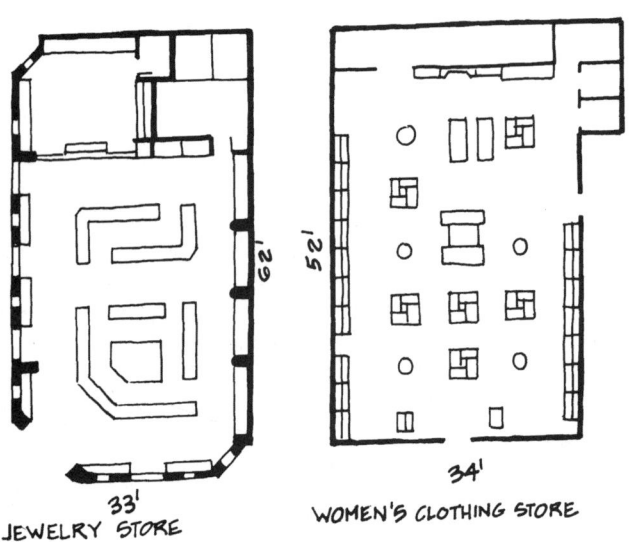

JEWELRY STORE

WOMEN'S CLOTHING STORE

MEN'S CLOTHING STORE

WOMEN'S CLOTHING STORE

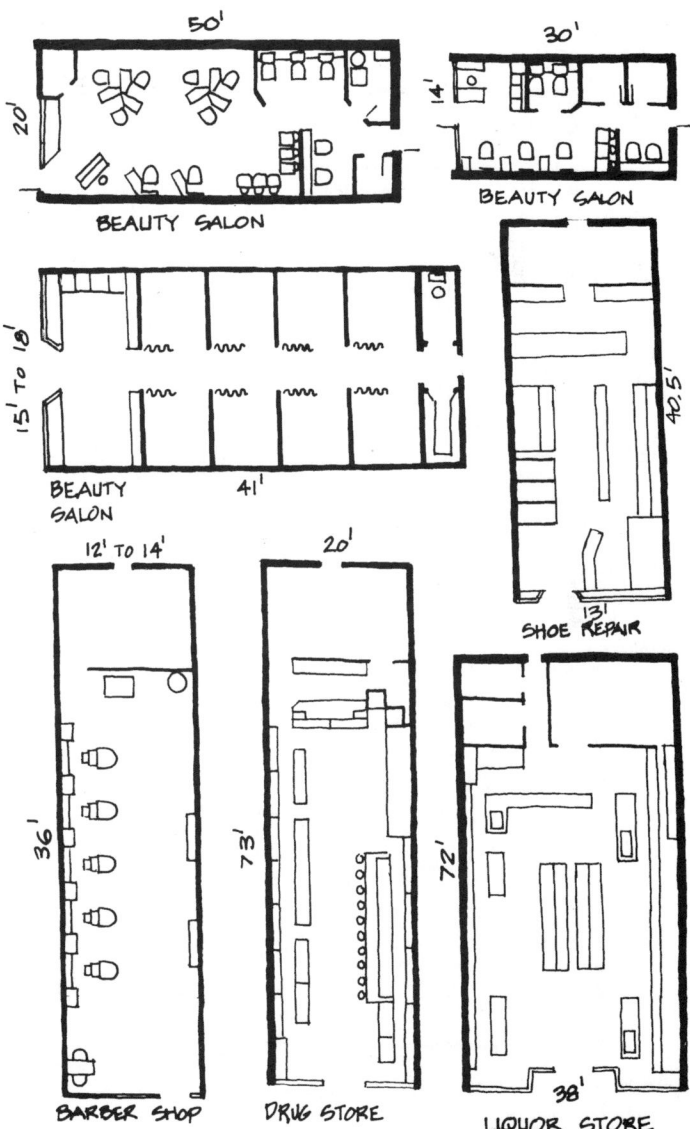

50'

20'

BEAUTY SALON

30'

14'

BEAUTY SALON

15' TO 18'

BEAUTY
SALON   41'

40.5'

13'
SHOE REPAIR

12' TO 14'

36'

BARBER SHOP

20'

73'

DRUG STORE

72'

38'
LIQUOR STORE

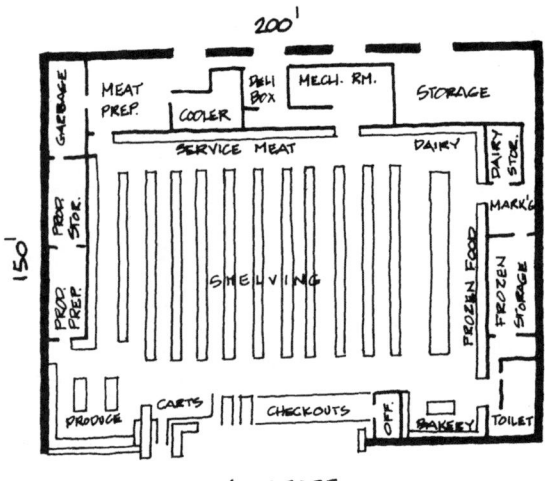

GROCERY STORE

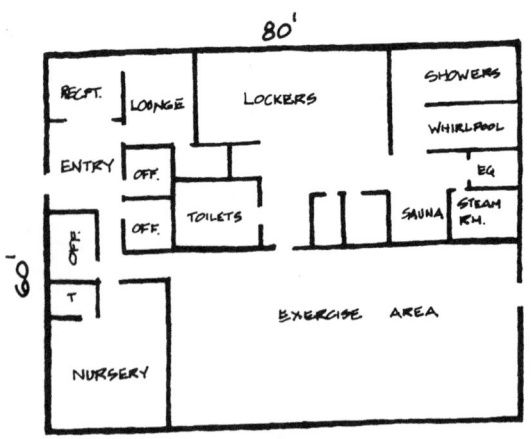

HEALTH STUDIO

___ *c.* Details:

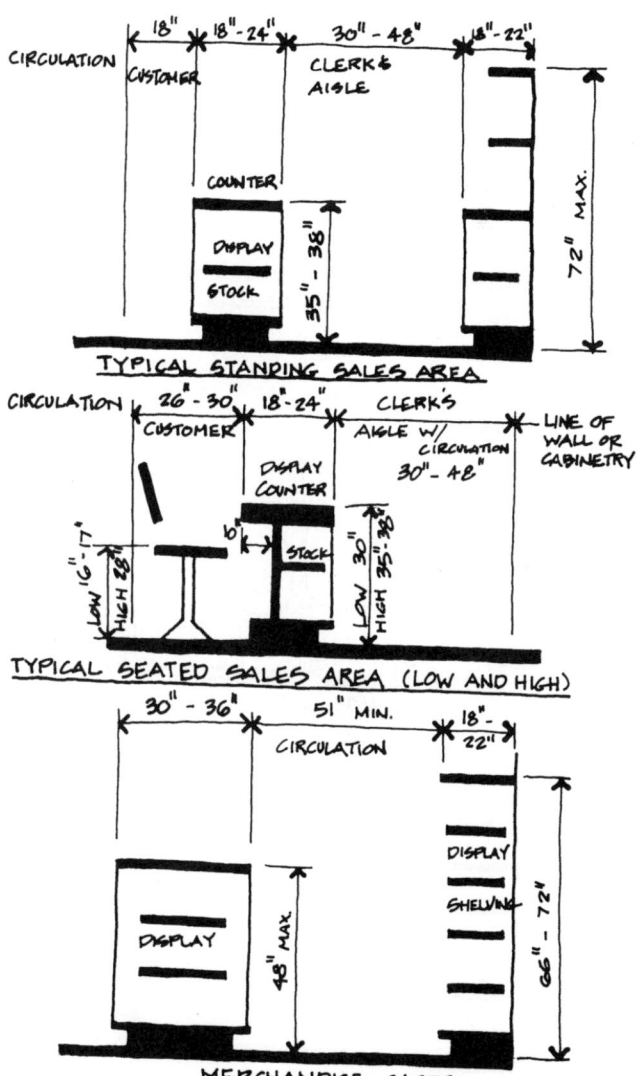

TYPICAL STANDING SALES AREA

TYPICAL SEATED SALES AREA (LOW AND HIGH)

MERCHANDISE CASES

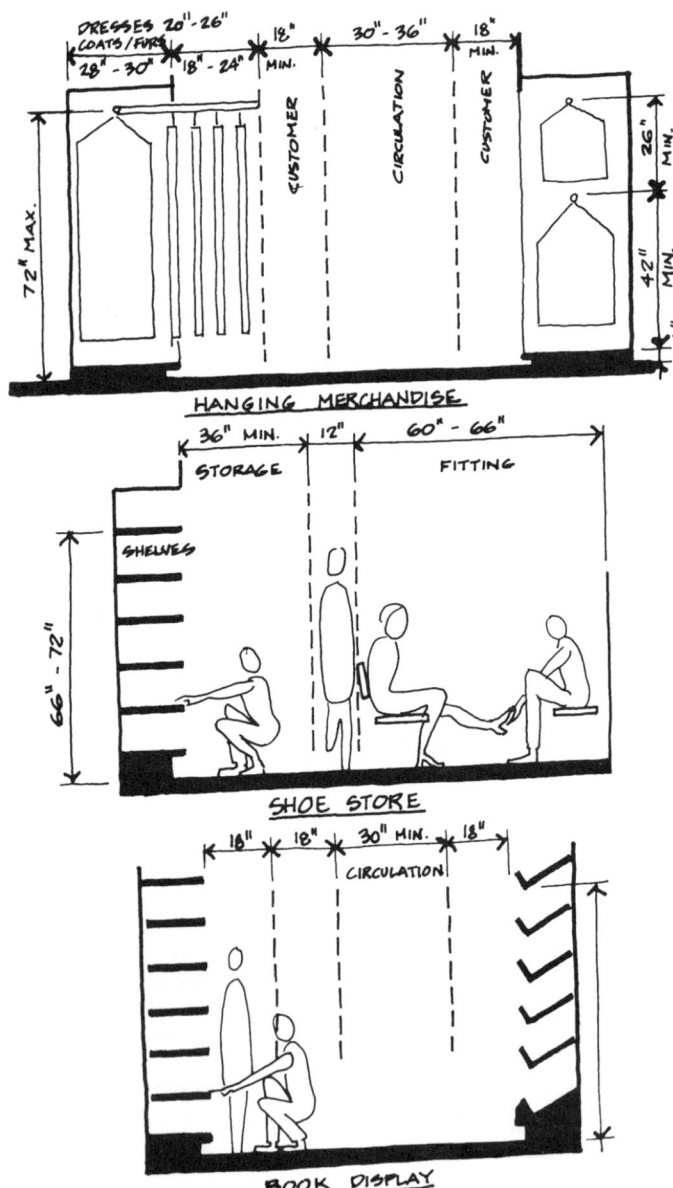

DRESSES 20"-26"
COATS/FURS
28"-30"   18"-24"   18" MIN.   30"-36"   18" MIN.

CUSTOMER   CIRCULATION   CUSTOMER

72" MAX.   26" MIN.   42" MIN.   4"

HANGING MERCHANDISE

36" MIN.   12"   60" - 66"

STORAGE   FITTING

SHELVES

66" - 72"

SHOE STORE

18"   18"   30" MIN.   18"

CIRCULATION

BOOK DISPLAY

_____ *d.* Retail templates. The following retail casework templates are at ¼″ = 1′0″ scale. Use them to do retail layouts.

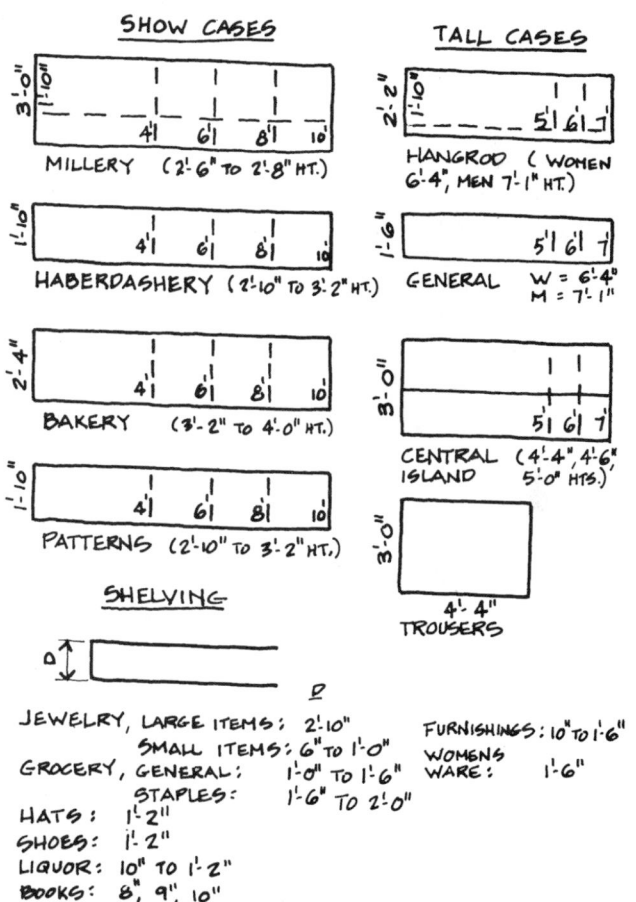

SHOW CASES

3′-0″ 1′-0″

4′ 6′ 8′ 10′

MILLERY (2′-6″ TO 2′-8″ HT.)

1′-10″

4′ 6′ 8′ 10′

HABERDASHERY (2′-10″ TO 3′-2″ HT.)

2′-4″

4′ 6′ 8′ 10′

BAKERY (3′-2″ TO 4′-0″ HT.)

1′-10″

4′ 6′ 8′ 10′

PATTERNS (2′-10″ TO 3′-2″ HT.)

SHELVING

D

TALL CASES

2′-2″ 1′-10″

5′ 6′ 7′

HANGROD (WOMEN 6′-4″, MEN 7′-1″ HT.)

1′-6″

5′ 6′ 7′

GENERAL W = 6′-4″
M = 7′-1″

3′-0″

5′ 6′ 7′

CENTRAL (4′-4″, 4′-6″
ISLAND 5′-0″ HTS.)

3′-0″

4′-4″
TROUSERS

JEWELRY, LARGE ITEMS: 2′-10″
        SMALL ITEMS: 6″ TO 1′-0″
GROCERY, GENERAL: 1′-0″ TO 1′-6″
        STAPLES: 1′-6″ TO 2′-0″
HATS: 1′-2″
SHOES: 1′-2″
LIQUOR: 10″ TO 1′-2″
BOOKS: 8″, 9″, 10″

FURNISHINGS: 10″ TO 1′-6″
WOMENS
WARE: 1′-6″

9. Reception/Counters/Circulation
___ *a.* Reception/counters

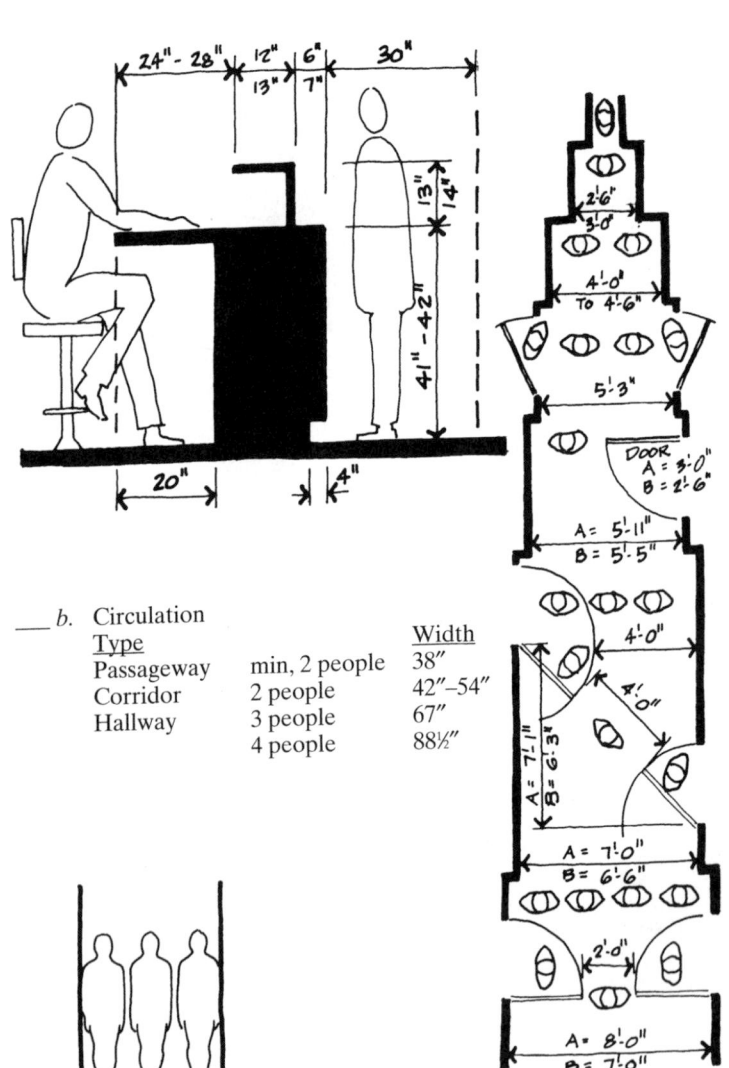

___ *b.* Circulation

Type		Width
Passageway	min, 2 people	38″
Corridor	2 people	42″–54″
Hallway	3 people	67″
	4 people	88½″

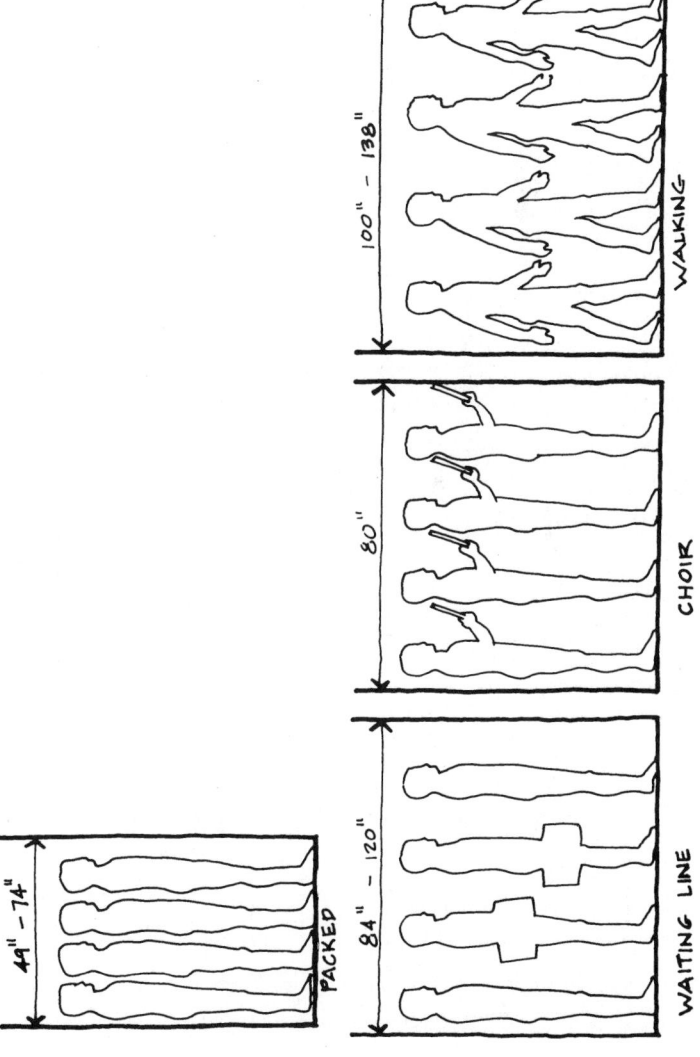

PACKED
49" – 74"

WAITING LINE
84" – 120"

CHOIR
80"

WALKING
100" – 138"

___ 10. Group Seating

___ *a.* Auditorium or theater fixed seating: Allow 7 to 12 SF (7½ SF average) per seat, including aisles and crossovers. This is sufficiently accurate for preliminary planning.

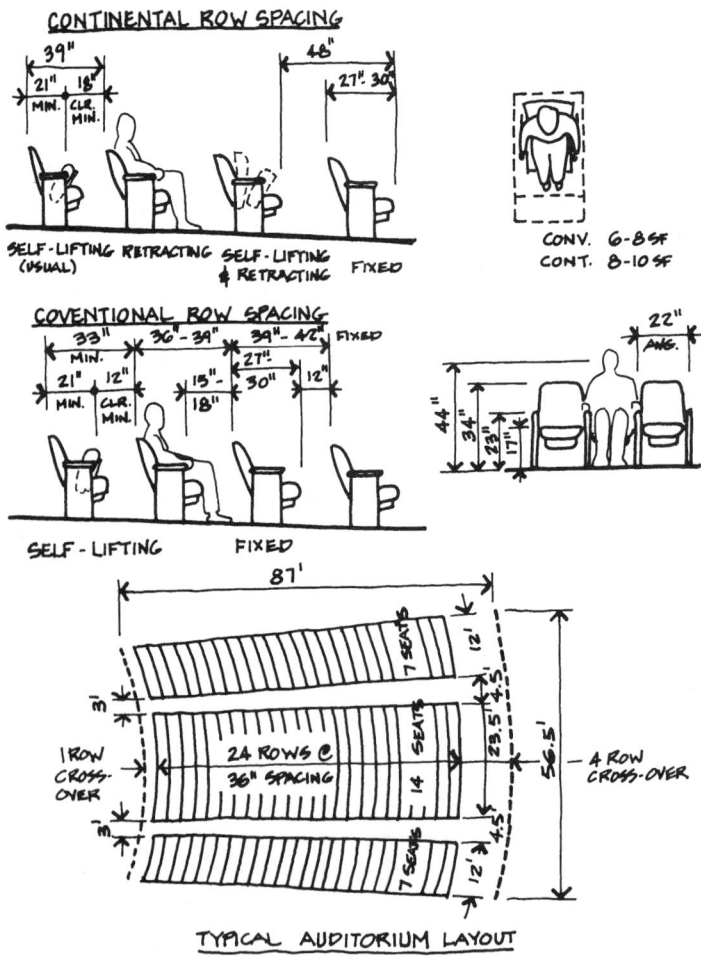

TYPICAL AUDITORIUM LAYOUT

___ *b.* Classroom with individual desks and chairs (see p. 399 for costs):

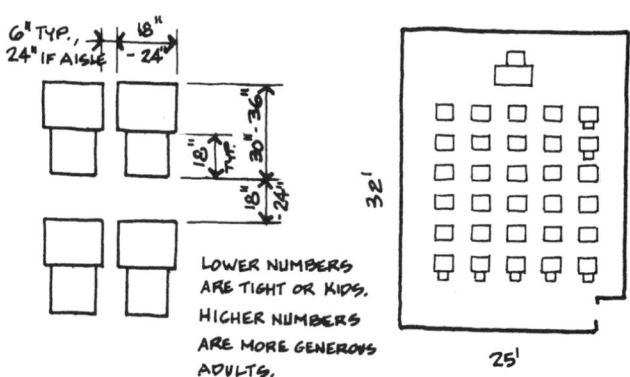

___ *c.* Group seating, tables, and chairs (see p. 399 for costs):

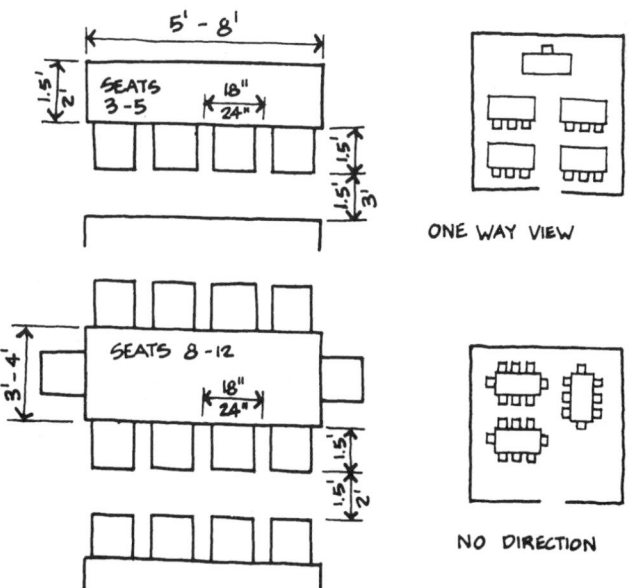

___ 11. <u>Storage</u>
    ___ *a.* Commercial: See p. 122.
    ___ *b.* Residential:
        ___ (1) <u>*Shelving:*</u> Standard shelving sizes are 6″, 8″, 10″, 12″, and 18″ deep. Shelving can be either fixed or adjustable.
        ___ (2) <u>*Drawers:*</u> 16 to 24″ deep; 12 to 36″ wide; and 2 to 8″ high. May be of wood, metal, or molded plastic.
        ___ (3) Storage requirements:
            ___ (*a*) <u>*Bedroom:*</u> 4 to 6 feet of hanging space per person. Allow 8 LF of hanging space when shared by 2 people. The average rod space per garment is about 2″ for women's clothing, 2¼″ for men's clothing, and 4″ for heavy coats. Recommended rod heights are 68″ for long rods, 63″ for adult clothing, and 32″ for children's clothing. The shelf is normally located 2″ above the rod, with another shelf 12″ higher.
            ___ (*b*) <u>*Linen:*</u> Shelves 12 to 18″ deep. Provide minimum 9 SF for 1- to 2-bedroom house and 12 SF for 3–4-bedroom house.
            ___ (*c*) <u>*Bathroom:*</u> A mirrored wall cabinet of 4 to 6″ deep is typical, plus drawers in vanity.
            ___ (*d*) Front entry closet for coats.
            ___ (*e*) Cleaning equipment: Utility closet a minimum of 24″ wide.
            ___ (*f*) Kitchen.
            ___ (*g*) Other:
                ___ 1. Books: Shelves 8″ (85%), 10″ (10%), and 12″ (5%) deep. Vertical spacing varies from 8 to 16″ (10 to 12″ most typical). Horizontally, books average 7 to 8 volumes per LF of shelf.
                ___ 2. Magazines.
                ___ 3. Card tables and chairs: 30–36″ sq. and 2 to 3″ thick

when folded. Folded chairs
average 30″ × 16″ × 3″.
___ 4. Phonograph records: 14″ ×
15″.
___ 5. Games.
___ 6. Movie and slide projectors
and screens.
___ 7. Toys.
___ 8. Sports equipment.
___ 9. Tools.
___ 10. Bulk: Usually outside, in
garage, or in attic.
___ (4) Closets: Standard closet depth is 24 to 30″
for clothing and 16 to 20″ for linens. Types:
___ (a) Reach-in (24″ minimum)
___ (b) Edge-in (additional 18″)
___ (c) Walk-in

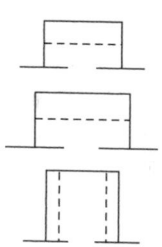

___ (5) Details:

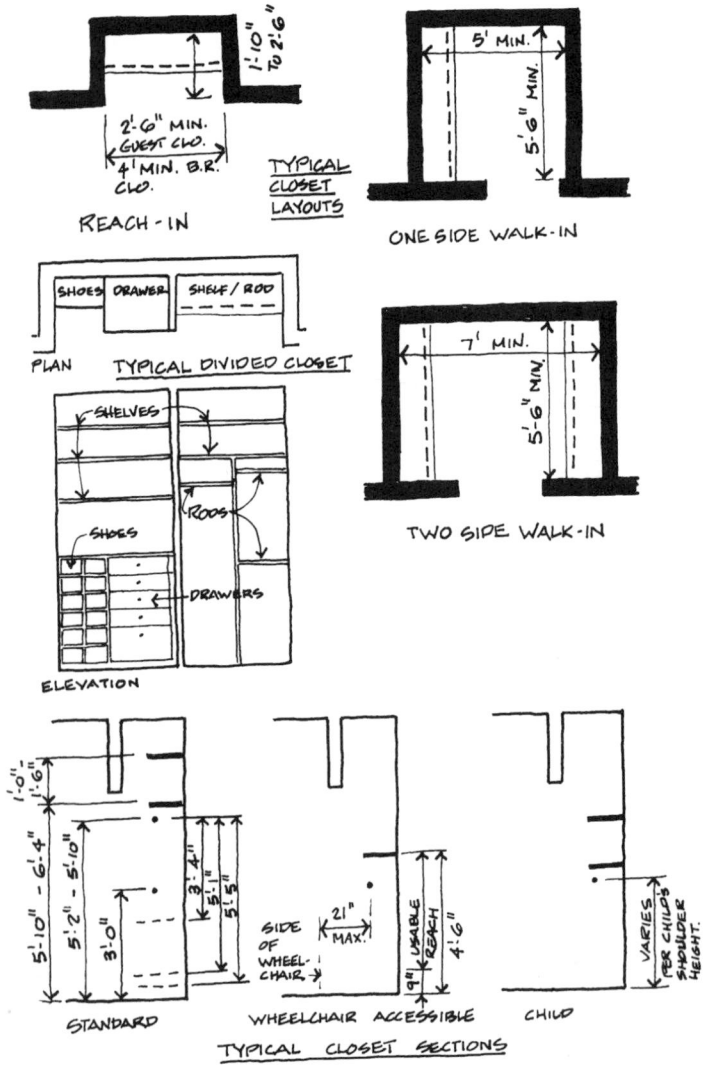

REACH-IN

TYPICAL CLOSET LAYOUTS

ONE SIDE WALK-IN

PLAN   TYPICAL DIVIDED CLOSET

TWO SIDE WALK-IN

ELEVATION

STANDARD   WHEELCHAIR ACCESSIBLE   CHILD

TYPICAL CLOSET SECTIONS

NOTES

NOTES

___ H. PRACTICAL MATH AND TABLES

(13) (18) (24) (30) (37) (38) (41A) (49)

___1. **General**: Architects or designers seldom have to be involved in higher mathematics, but they need to continually do simple math *well*.

   ___ *a.* For rough estimating (such as in this book) an accuracy of more than 90% to 95% is seldom required.

   ___ *b.* Try to have a rough idea of what the answer should be, before the calculation (i.e., does the answer make sense?).

   ___ *c.* Round numbers off and don't get bogged down in trivia.

   ___ *d.* For final exact numbers that are important (such as final building areas), go slow, and recheck calculations at least once.

___2. **Decimals of a Foot**

$1'' = .08'$	$7'' = .58'$		
$2'' = .17'$	$8'' = .67'$		
$3'' = .25'$	$9'' = .75'$		
$4'' = .33'$	$10'' = .83'$		
$5'' = .42'$	$11'' = .92'$		
$6'' = .50'$	$12'' = 1.0'$		

Decimals of an Inch

$\frac{1}{8}'' = 0.125''$	$\frac{5}{8}'' = 0.625''$
$\frac{1}{4}'' = 0.250''$	$\frac{3}{4}'' = 0.750''$
$\frac{3}{8}'' = 0.375''$	$\frac{7}{8}'' = 0.875''$
$\frac{1}{2}'' = 0.50''$	$1'' = 1.0''$

(see example on p. 153)

___3. **Simple Algebra** (see example on p. 153)

One unknown and $A = B/C$ Example: $3 = 15/5$
two knowns $B = A \times C$ $15 = 3 \times 5$
                   $C = B/A$ $5 = 15/3$

___4. **Ratios and Proportions** (see example on p. 153)

One unknown and three knowns (cross multiplication)

$$\frac{A}{B} = \frac{C}{D} \qquad \text{Example:} \quad \frac{X}{5} \bowtie \frac{10}{20} \qquad 20\,X = 5 \times 10$$

$$A \times C = B \times D \qquad X = \frac{5 \times 10}{20} = 2.5 \qquad X = 2.5$$

___5. **Exponents and Powers**

$10^6 = 10 \times 10 \times 10 \times 10 \times 10 \times 10 = 1{,}000{,}000 \; [1 + 6 \text{ zeros}]$
$10^0 = 1.0$
$10^{-6} = 0.000001 \; [6 \text{ places to left or 5 zeros in front of 1}]$

___6. **Percent Increases or Decreases** (see example on p. 154)

50% increase = $\frac{1}{2}$ increase, use $\times 1.5$
100% increase = double, use $\times 2.0$
200% increase = triple, use $\times 3.0$
Example: 20 increases to 25
To find percent increase: $25 - 20 = 5$ [amount of increase]
                            $5/20 = 0.25$ or 25% increase

___ 7. <u>Compounding</u>: A continual increase or decrease of numbers, over time, that builds on itself. Regarding construction, the % increase of cost per year compounds over the years. Thus, an item that costs $1.00 in 2000 will cost $2.10 in 2005 with 2% inflation per year. (See example on p. 155.)

___ 8. <u>Slopes, Gradients, and Angles</u>
(see p. 51)

   ___ *a.* Slope = "rise over run" or

$$\% \text{ slope} = \frac{\text{rise}}{\text{run}} \times 100$$

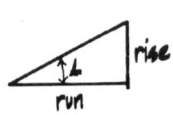

   ___ *b.* Gradient:
as ratios of rise to run
Example: expressed as 1 in 12
for a ramp or as 4″ in 12″ for
a roof

   ___ *c.* Angle
Degree angle based on rise
and run (see properties of
right angles)

___ 9. <u>Properties of Right Angles</u> (see example on p. 154)

45° angle: $a^2 + b^2 = c^2$

or $c = \sqrt{a^2 + b^2}$

For other right angles use simple
trigonometry:

   Sin angle = opp/hyp
   Cos angle = adj/hyp
   Tan angle = opp/adj

Use calculators with trig. functions or
table on p. 132.

___ 10. <u>Properties of Non-Right Angles</u>
Use law of sines:

   a/sin A = b/sin B = c/sin C
   a/b = sin A/sin B, etc.

___ 11. <u>Properties of Circles</u>
A circle is divided into 360 equal parts, called *degrees* (°). One degree is an angle at the center of a circle which cuts off an arc that is ¹⁄₃₆₀ of the circumference. Degrees are subdivided into 60 minutes ('). Minutes are subdivided into 60 seconds (″). See p. 143.

___ 12. <u>Geometric Figures</u>
Use the formulas on pp. 142 to calculate areas and volumes.

Table of Slopes, Grades, Angles

% Slope	Inch/ft	Ratio	Deg. from horiz.
1	⅛	1 in 100	
2	¼	1 in 50	
3	⅜		
4	½	1 in 25	
5	⅝	1 in 20	3
6	¾		
7	⅞		
8	approx. 1	approx. 1 in 12	
9	1⅛		
10	1¼	1 in 10	6
11	1⅜	approx. 1 in 9	
12	1½		
13	1⅝		
14	1¾		
15			8.5
16	1⅞		
17	2	approx. 2 in 12	
18	2⅛		
19	2¼		
20	2⅜	1 in 5	11.5
25	3	3 in 12	14
30	3.6	1 in 3.3	17
35	4.2	approx. 4 in 12	19.25
40	4.8	approx. 5 in 12	21.5
45	5.4	1 in 2.2	24
50	6	6 in 12	26.5
55	6⅝	1 in 1.8	28.5
60	7¼	approx. 7 in 12	31
65	7¾	1 in 1½	33
70	8⅜	1 in 1.4	35
75	9	1 in 1.3	36.75
100	12	1 in 1	45

Trigonometry Tables

Deg	Sin	Cos	Tan	Deg	Sin	Cos	Tan	Deg	Sin	Cos	Tan
1	.0175	.9998	.0175	31	.5150	.8572	.6009	61	.8746	.4848	1.8040
2	.0349	.9994	.0349	32	.5299	.8480	.6249	62	.8829	.4695	1.8807
3	.0523	.9986	.0524	33	.5446	.8387	.6494	63	.8910	.4540	1.9626
4	.0698	.9976	.0699	34	.5592	.8290	.6745	64	.8988	.4384	2.0503
5	.0872	.9962	.0875	35	.5736	.8192	.7002	65	.9063	.4226	2.1445
6	.1045	.9945	.1051	36	.5878	.8090	.7265	66	.9135	.4067	2.2460
7	.1219	.9925	.1228	37	.6018	.7986	.7536	67	.9205	.3907	2.3559
8	.1392	.9903	.1405	38	.6157	.7880	.7813	68	.9272	.3746	2.4751
9	.1564	.9877	.1584	39	.6293	.7771	.8098	69	.9336	.3584	2.6051
10	.1736	.9848	.1763	40	.6428	.7660	.8391	70	.9397	.3420	2.7475
11	.1908	.9816	.1944	41	.6561	.7547	.8693	71	.9455	.3256	2.9042
12	.2079	.9781	.2126	42	.6691	.7431	.9004	72	.9511	.3090	3.0777
13	.2250	.9744	.2309	43	.6820	.7314	.9325	73	.9563	.2924	3.2709
14	.2419	.9703	.2493	44	.6947	.7193	.9657	74	.9613	.2756	3.4874
15	.2588	.9659	.2679	45	.7071	.7071	1.0000	75	.9659	.2588	3.7321
16	.2756	.9613	.2867	46	.7193	.6947	1.0355	76	.9703	.2419	4.0108
17	.2924	.9563	.3057	47	.7314	.6820	1.0724	77	.9744	.2250	4.3315
18	.3090	.9511	.3249	48	.7431	.6691	1.1106	78	.9781	.2079	4.7046
19	.3256	.9455	.3443	49	.7547	.6561	1.1504	79	.9816	.1908	5.1446
20	.3420	.9397	.3640	50	.7660	.6428	1.1918	80	.9848	.1736	5.6713
21	.3584	.9336	.3839	51	.7771	.6293	1.2349	81	.9877	.1564	6.3138
22	.3746	.9272	.4040	52	.7880	.6157	1.2799	82	.9903	.1392	7.1154
23	.3907	.9205	.4245	53	.7986	.6018	1.3270	83	.9925	.1219	8.1443
24	.4067	.9135	.4452	54	.8090	.5878	1.3764	84	.9945	.1045	9.5144
25	.4226	.9063	.4663	55	.8192	.5736	1.4281	85	.9962	.0872	11.4301
26	.4384	.8988	.4877	56	.8290	.5592	1.4826	86	.9976	.0698	14.3007
27	.4540	.8910	.5095	57	.8387	.5446	1.5399	87	.9986	.0523	19.0811
28	.4695	.8829	.5317	58	.8480	.5299	1.6003	88	.9994	.0349	28.6363
29	.4848	.8746	.5543	59	.8572	.5150	1.6643	89	.9998	.0175	57.2900
30	.5000	.8660	.5774	60	.8660	.5000	1.7321	90	1.000	.0000	∞

Note: Deg = degrees of angle; Sin = sine; Cos = cosine; Tan = tangent.

13. Perspective Sketching

Use the following simple techniques of using 10′ cubes and lines
at 5′ with diagonals for quick perspective sketching:

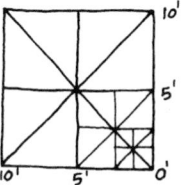

____ *a.* The sketches shown on p. 134
show two techniques:
The *first* establishes diagonal
Vanishing Points (VP) on the
Horizon Line (HL) at certain distances from the
VPs, also on the HL. 10′ cubes are established by pro-
jecting diagonals to the VPs. The *second* technique
has 10′ cubes and lines at the 5′ half-points. Diagonals through the half-points continue the 5′ and 10′
module to the VPs. The vertical 5′ roughly equals eye
level, and establishes the HL. Half of 5′ or 2.5′ is a
module for furniture height and width.

____ *b.* The sketch shown on p. 135 illustrates the most common way people view buildings. That is, close up, at
almost a one-point perspective. To produce small
sketches, set right vertical measure at ½″ apart. Then,
about 10½″ to left, set vertical measure at ⅜″ apart.
This will produce a small sketch to fit on 8½ × 11
paper. Larger sketches can be done using these proportions.

PERSPECTIVE SKETCHING

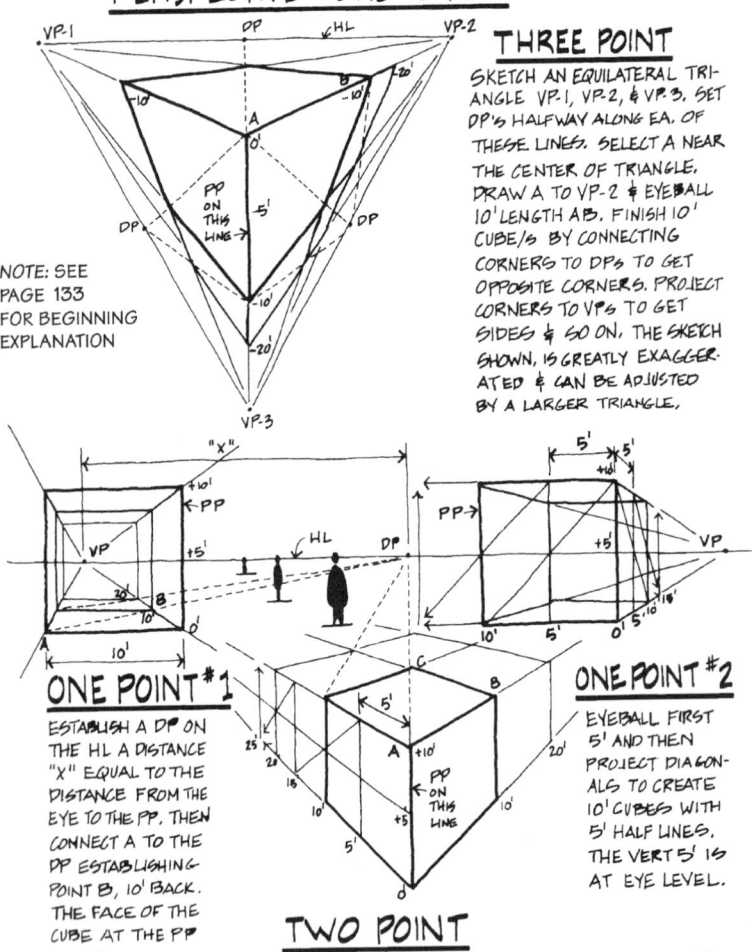

THREE POINT

SKETCH AN EQUILATERAL TRI-ANGLE VP-1, VP-2, & VP-3. SET DP's HALFWAY ALONG EA. OF THESE LINES. SELECT A NEAR THE CENTER OF TRIANGLE. DRAW A TO VP-2 & EYEBALL 10' LENGTH AB. FINISH 10' CUBE/s BY CONNECTING CORNERS TO DPs TO GET OPPOSITE CORNERS. PROJECT CORNERS TO VPs TO GET SIDES & SO ON, THE SKETCH SHOWN, IS GREATLY EXAGGER-ATED & CAN BE ADJUSTED BY A LARGER TRIANGLE,

NOTE: SEE
PAGE 133
FOR BEGINNING
EXPLANATION

ONE POINT #1

ESTABLISH A DP ON THE HL A DISTANCE "X" EQUAL TO THE DISTANCE FROM THE EYE TO THE PP, THEN CONNECT A TO THE DP ESTABLISHING POINT B, 10' BACK. THE FACE OF THE CUBE AT THE PP IS EYEBALLED.

ONE POINT #2

EYEBALL FIRST 5' AND THEN PROJECT DIAGON-ALS TO CREATE 10' CUBES WITH 5' HALF LINES. THE VERT 5' IS AT EYE LEVEL.

TWO POINT

#1 EYEBALL LVP & RVP AND THEN DP, ALL ON THE HL. CONNECT A TO DP. EYEBALL FIRST 10' ON LINE AB. CONNECT B TO LVP. THE INTERSECTION ESTABLISH-ES POINT C, 10' BACK, & SO ON.

#2 5' DIAGONALS CAN ALSO BE USED BY EYEBALLING THE FIRST 5'.

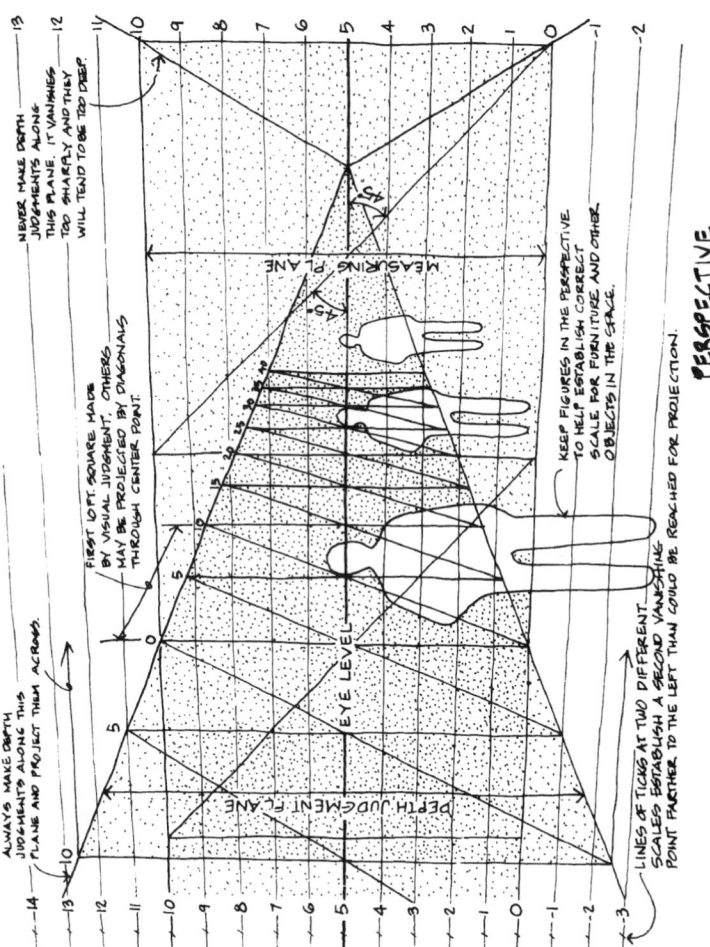

PERSPECTIVE

ANGLES (41A)

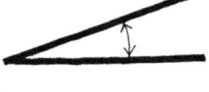

ACUTE ANGLE THIS IS AN ANGLE THAT MEASURES MORE THAN 0° BUT LESS THAN 90°

RIGHT ANGLE AN ANGLE THAT MEASURES EXACTLY 90°. THE LINES AT RIGHT ANGLES ARE PERPENDICULAR TO ONE ANOTHER.

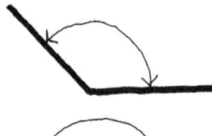

OBTUSE ANGLE THIS IS AN ANGLE OF MORE THAN 90° BUT LESS THAN 180°.

STRAIGHT ANGLE AN ANGLE THAT MEASURES 180° AND FORMS A STRAIGHT LINE.

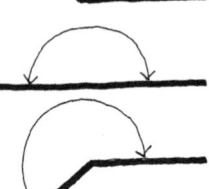

REFLEX ANGLE THIS ANGLE MEASURES MORE THAN 180° BUT LESS THAN 360°.

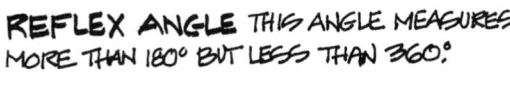

COMPLEMENTARY ANGLES THESE ARE TWO ANGLES THAT ADD UP TO 90°.

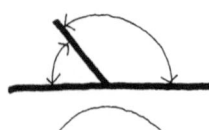

SUPPLEMENTARY ANGLES TWO ANGLES WHOSE SUM IS 180°.

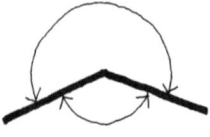

CONJUGATE ANGLES TWO ANGLES WHOSE SUM IS 360°.

GEOMETRIC SHAPES (41A)

	NAME OF POLYGON	NUMBER OF SIDES	EACH INTERNAL ANGLE	SUM OF INTERNAL ANGLES
	TRIANGLE	3	60°	180°
	SQUARE	4	90°	360°
	PENTAGON	5	108°	540°
	HEXAGON	6	120°	720°
	HEPTAGON	7	128.6°	900°
	OCTAGON	8	135°	1080°
	NONAGON	9	140°	1260°
	DECAGON	10	144°	1440°
	UNDECAGON	11	147.3°	1620°
	DODECAGON	12	150°	1800°

TRIANGLES

SHOWN BELOW ARE SIX TYPES OF TRIANGLE. THE SUM OF THE INTERNAL ANGLES OF ANY FLAT TRIANGLE IS ALWAYS 180°.

AN EQUILATERAL TRIANGLE HAS ALL THE SIDES OF THE SAME LENGTH. ALL THE INTERNAL ANGLES ARE EQUAL.

AN ISOSCELES TRIANGLE HAS TWO SIDES WHICH ARE OF THE SAME LENGTH AND TWO ANGLES WHICH ARE OF EQUAL SIZE.

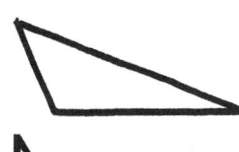

A SCALENE TRIANGLE HAS ALL THE SIDES OF DIFFERENT LENGTHS AND HAS ALL THE ANGLES OF DIFFERENT ANGLES.

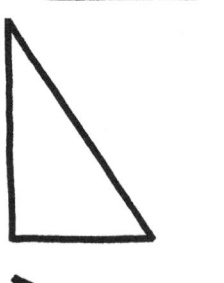

A RIGHT-ANGLE TRIANGLE IS A TRIANGLE WHICH CONTAINS ONE RIGHT ANGLE OF 90°.

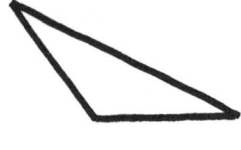

AN OBTUSE-ANGLE TRIANGLE IS A TRIANGLE WHICH CONTAINS ONE OBTUSE ANGLE — AN ANGLE OVER 90°.

AN ACUTE-ANGLE TRIANGLE IS A TRIANGLE WHICH CONTAINS THREE ACUTE ANGLES. THESE ARE EACH LESS THAN 90°.

QUADRILATERALS (41A)

A QUADRILATERAL IS A FOUR-SIDED POLYGON OR A FOUR-SIDED PLANE FIGURE.

A SQUARE HAS ALL THE SIDES THE SAME LENGTH AND ALL THE ANGLES ARE RIGHT ANGLES.

A RECTANGLE HAS OPPOSITE SIDES WHICH ARE THE SAME LENGTH. ALL THE ANGLES ARE RIGHT ANGLES.

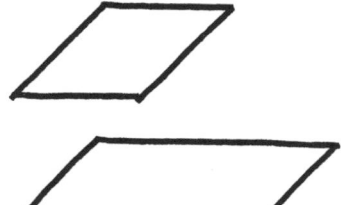

A RHOMBUS HAS ALL SIDES OF THE SAME LENGTH BUT NONE OF THE ANGLES ARE RIGHT ANGLES.

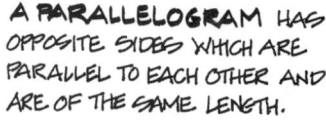

A PARALLELOGRAM HAS OPPOSITE SIDES WHICH ARE PARALLEL TO EACH OTHER AND ARE OF THE SAME LENGTH.

A TRAPEZOID HAS ONE PAIR OF OPPOSITE SIDES WHICH ARE PARALLEL.

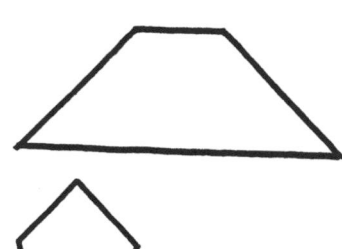

A KITE HAS ADJACENT SIDES OF THE SAME LENGTH. THE DIAGONALS INTERSECT AT RIGHT ANGLES.

SOLIDS (41A)

SOLID SHAPES ARE THREE - DIMENSIONAL. THIS MEANS THEY HAVE LENGTH, WIDTH, AND DEPTH. A POLYHEDRON IS A SOLID SHAPE WITH POLYGONS FOR FACES, OR SIDES. OPENING OUT A POLYHEDRON GIVES A SHAPE CALLED A NET.

ALL THE FACES OF A REGULAR SOLID ARE IDENTICAL RE-GULAR POLYGONS OF EQUAL SIZE. A REGULAR POLYHEDRON WILL FIT INTO A SPHERE WITH ALL THE VERTICES, OR EDGES, TOUCHING THE SPHERE. THERE ARE FIVE REGULAR SOLIDS WHICH ARE SHOWN BELOW WITH THEIR NETS.

A TETRAHEDRON HAS FOUR EQUILATERAL TRIANGLES FOR FACES.

 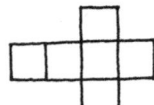

A CUBE HAS SIX FACES, EACH OF WHICH IS A SQUARE.

 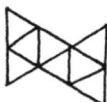

AN OCTAHEDRON HAS EIGHT SIDES AND IS MADE FROM EIGHT EQUILATERAL TRIANGLES.

 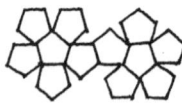

A DODECAHEDRON IS A SOLID WITH 12 SIDES, EACH OF WHICH IS A PENTAGON.

 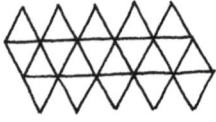

AN ICOSAHEDRON HAS 20 SIDES CONSISTING OF EQUILATERAL TRIANGLES.

GEOMETRY OF AREA   (41A)

ABBREVIATIONS
a = LENGTH OF TOP
b = LENGTH OF BASE
h = PERPENDICULAR HEIGHT
r = LENGTH OF RADIUS

$$\pi = 3.1416$$

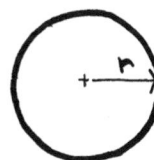

CIRCLE

$$AREA = \pi \times r^2$$

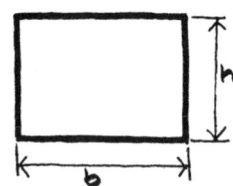

RECTANGLE

$$AREA = b \times h$$

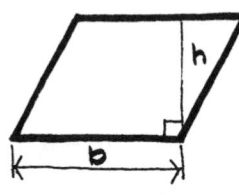

PARALLELOGRAM

$$AREA = b \times h$$

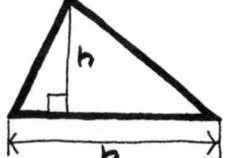

TRIANGLE

$$AREA = \tfrac{1}{2} \times b \times h$$

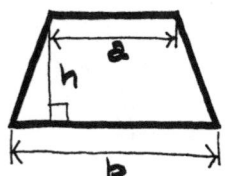

TRAPEZOID

$$AREA = \frac{(a+b) \times h}{2}$$

GEOMETRY OF VOLUME

ABBREVIATIONS
b = BREADTH OF BASE
h = PERPENDICULAR HEIGHT
l = LENGTH OF BASE
r = LENGTH OF RADIUS

$$\pi = 3.1416$$

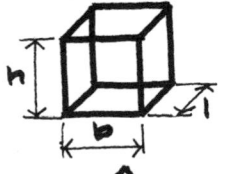

CUBE

$$AREA = b \times h \times l$$

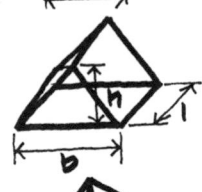

PRISM

$$AREA = \frac{b \times h \times l}{2}$$

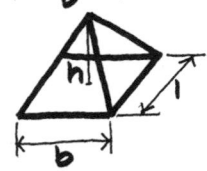

PYRAMID

$$AREA = \frac{b \times h \times l}{3}$$

CYLINDER

$$AREA = \pi \times r^2 \times l$$

SPHERE

$$AREA = \frac{4 \times \pi \times r^3}{3}$$

CONE

$$AREA = \frac{\pi \times r^2 \times h}{3}$$

PROPERTIES OF THE CIRCLE

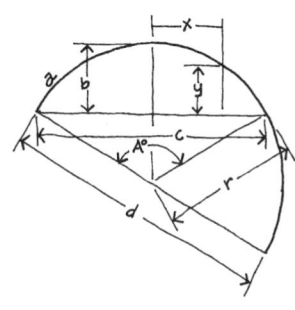

Circumference $= 6.28318\,r = 3.1416\,d$
Diameter $= 0.31831$ circumference
Area $= 3.1416\,r^2$

Arc $a = \dfrac{\pi r A^\circ}{180^\circ} = 0.017453\,r A^\circ$

Angle $A^\circ = \dfrac{180^\circ a}{\pi r} = 57.2957\overline{8}\,\dfrac{a}{r}$

Radius $r = \dfrac{4b^2 + c^2}{8b}$

Chord $c = 2\sqrt{2br - b^2} = 2\,r\sin\dfrac{A}{2}$

Rise $b = r - \tfrac{1}{2}\sqrt{4r^2 - c^2} = \dfrac{c}{2}\tan\dfrac{A}{4}$

$= 2\,r\sin^2\dfrac{A}{4} = r + y - \sqrt{r^2 - x^2}$

$y = b - r + \sqrt{r^2 - x^2}$

$x = \sqrt{r^2 - (r + y - b)^2}$

Diameter of circle of equal periphery as square = 1.27324 side of square
Side of square of equal periphery as circle = 0.78540 diameter of circle
Diameter of circle circumscribed about square = 1.41421 side of square
Side of square inscribed in circle = 0.70711 diameter of circle

CIRCULAR SECTOR

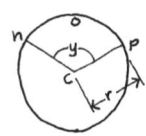

$r =$ radius of circle $y =$ angle ncp in degrees

Area of Sector ncpo $= \tfrac{1}{2}$ (length of arc nop $\times r$)

$= $ Area of Circle $\times \dfrac{y}{360}$

$= 0.0087266 \times r^2 \times y$

CIRCULAR SEGMENT

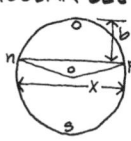

$r =$ radius of circle $x =$ chord $b =$ rise

Area of Segment nop = Area of Sector ncpo – Area of triangle ncp

$= \dfrac{(\text{length of arc nop} \times r) - x(r - b)}{2}$

Area of Segment nsp = Area of Circle – Area of Segment nop

TEMPERATURE

MULTIPLY	BY	TO GET
DEG. C	1.8	DEG. F
DEG. F	0.5556	DEG. C

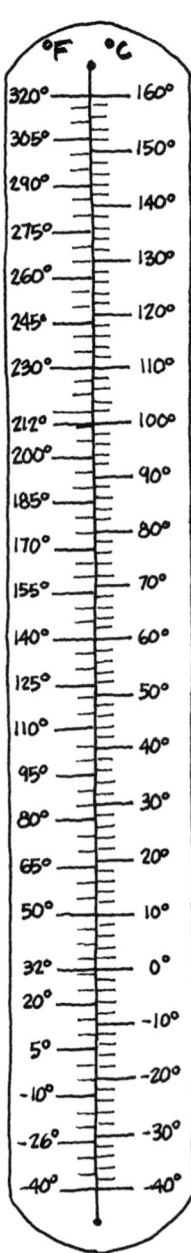

LENGTHS

METERS* m	INCHES in.	FEET ft.	YARD yd.	RODS r.	CHAINS ch.	MILES		KILO-METERS km
						STATUTE	NAUTICAL	
1	39.37	3.28	1.09	0.199	0.05	$0.{}^{3}6214$	$0.{}^{3}05396$	0.001
0.025	1	0.083	0.028	$0.{}^{2}051$	$0.{}^{2}13$	$0.{}^{4}158$	$0.{}^{4}137$	$0.{}^{4}0254$
0.305	12	1	0.333	0.06	0.015	$0.{}^{3}189$	$0.{}^{3}165$	$0.{}^{3}305$
0.914	36	3	1	0.18	0.045	$0.{}^{3}568$	$0.{}^{3}443$	$0.{}^{3}914$
5.029	198	16.5	5.5	1	0.25	$0.{}^{2}313$	$0.{}^{2}271$	$0.{}^{2}503$
20.117	792	66	22	4	1	0.013	0.0109	0.020
1609.35	63360	5280	1760	320	80	1	0.868	1.609
1853.25	72962.5	6080.2	2026.7	368.5	92.12	1.15	1	1.853
1000	39370	3280.8	1093.6	198.8	49.71	0.621	0.540	1

* 1 METER (m) = 10 DECIMETERS (dm.) = 100 CENTIMETERS (cm) = 1000 MILLIMETERS (mm)

NOTE : NOTATIONS $0.{}^{2}0$, $0.{}^{3}0$, $0.{}^{4}$, ETC., INDICATE THE NUMBER OF ZEROS.

EXAMPLE ; 1 METER = $0.{}^{3}6214$ = 0.0006214 STATUTE MILES.

AREAS

SQUARE METERS SM	SQUARE INCHES SI	SQUARE FEET SF	SQUARE YARDS SY	SQUARE RODS SR	ACRES AC	HECTARES HA	SQUARE MILES STATUTE	SQUARE KILOMETER SQ KM
1	1550.0	10.76	1.196	0.039	0.0^3247	0.0001	0.0^6386	0.0^51
0.0^365	1	0.0^269	0.0^377	0.0^226	0.0^616	0.0^765	0.0^925	0.0^965
0.093	144	1	0.111	0.0^237	0.0^423	0.0^593	0.0^736	0.0^793
0.836	1296	9	1	0.033	0.0^321	0.0^484	0.0^632	0.0^684
25.293	39204	272.25	30.25	1	0.006	0.0^325	0.0^598	0.0^426
4046.87	6272640	43560	4840	160	1	0.405	0.0^216	0.0^241
10000	15499969	107639	11959.9	395.37	2.47104	1	0.0^239	0.01
2589999		27878400	3097600	1024000	640	259	1	2.59
1000000		10763867	1195985	39536.6	247.104	100	0.386	1

VOLUMES

Cubic Deci-meter or Liters	Cubic Inches	Cubic Feet	Cubic Yards	U.S. Quarts		U.S. Gallons		U.S. Bushels
				Liquid	Dry	Liquid	Dry	
1	61.02	0.035	$0.^{2}013$	1.057	0.908	0.264	0.227	0.028
0.016	1	$0.^{3}058$	$0.^{4}21$	0.017	0.015	$0.^{2}43$	$0.^{2}72$	$0.^{2}47$
28.32	1728	1	0.037	29.92	25.714	7.481	6.429	0.804
764.56	46656	27	1	807.90	694.28	201.97	173.57	21.70
0.946	57.75	0.033	$0.^{2}124$	1	0.859	0.25	0.215	0.027
1.1012	67.20	0.039	$0.^{2}144$	1.1637	1	0.291	0.25	0.031
3.786	231	0.134	$0.^{2}495$	4	3.437	1	0.859	0.107
4.405	268.8	0.196	$0.^{2}576$	4.655	4	1.164	1	0.125
35.24	2150.4	1.244	0.0461	37.24	32	9.309	8	1

U.S. DRY MEASURE: 1 BUSHEL = 4 PECKS = 8 GALLONS = 32 QUARTS = 64 PINTS
U.S. LIQUID MEASURE: 1 GALLON = 4 QUARTS = 8 PINTS = 32 GILLS = 128 FLUID OUNCES
1 U.S. GALLON = 0.83268 IMPERIAL GALLON

WEIGHTS

KILO-GRAMS KG	GRAINS	OUNCES TROY	OUNCES AVOIR	POUNDS TROY	POUNDS AVOIR	TONS NET (SHORT) 2000 LBS	TONS GROSS (LONG) 2240 LBS	TONS METRIC 1000 KG
1	15432.4	32.15	35.27	2.679	2.205	$0.^{2}1102$	$0.^{3}984$	0.001
$0.^{4}648$	1	$0.^{2}0208$	$0.^{3}023$	$0.^{3}174$	$0.^{3}143$	$0.^{7}714$	$0.^{7}638$	$0.^{7}648$
0.031	480	1	1.097	0.083	0.069	$0.^{4}343$	$0.^{4}306$	$0.^{4}311$
0.028	437.5	0.911	1	0.076	0.063	$0.^{4}313$	$0.^{4}279$	$0.^{4}284$
0.373	5760	12	13.166	1	0.823	$0.^{3}411$	$0.^{3}367$	$0.^{3}373$
0.454	7000	14.58	16	1.215	1	0.0005	$0.^{3}446$	$0.^{3}454$
907.185	14000000	291666.7	32000	2430.56	2000	1	0.893	0.907
1016.05	15680000	326666.7	35840	2722.22	2240	1.12	1	1.016
1000	15432356	32150.7	35274	2679.23	2204.62	1.102	0.984	1

1 LONG HUNDREDWEIGHT (CWT.) = 1/20 TON = 4 QUARTERS = 8 STONE = 112 LBS = 50.8 Kg

DENSITIES

GRAMS PER CU. CENTIMETER g/cm³	POUNDS PER CU. INCH lb./in.³	POUNDS PER CU. FOOT lb./ft³	POUNDS PER CU. YARD lb./yd³	KILOGRAMS PER CU. METER kg/m³	POUNDS PER BUSHEL, U.S.	POUNDS PER GALLON, DRY, U.S.	POUNDS PER GALLON, LIQUID, U.S.	KILOGRAMS PER HECTOLITER kg/hl
1	0.036	62.43	1685.56	1000	77.689	9.711	8.345	100
27.68	1	1728	46656	27679.7	2150.4	268.8	231	2767.97
0.016	$0.0^{3}579$	1	27	16.02	1.24	0.156	0.134	1.602
$0.0^{3}59$	$0.0^{4}21$	0.037	1	0.593	0.046	$0.0^{2}576$	$0.0^{2}495$	0.059
0.001	$0.0^{4}36$	0.062	1.686	1	0.078	$0.0^{2}97$	$0.0^{2}83$	0.10
0.013	$0.0^{3}47$	0.804	21.696	12.87	1	0.125	0.107	1.287
0.103	$0.0^{2}37$	6.429	173.57	102.97	8	1	0.859	10.297
0.119	$0.0^{2}43$	7.481	201.97	119.83	9.31	1.164	1	11.98
0.01	$0.0^{3}36$	0.624	16.86	10	0.777	0.097	0.083	1

PRESSURE

PASCALS N/m²	BARS 10⁵ N/m²	POUNDS PER IN²	ATMOS-PHERES	COLUMNS OF MERCURY (0°C, g=9.807 m/s²)		COLUMNS OF WATER (15°C, g=9.807 m/s²)	
				cm	in	cm	in
1	0.0^41	0.0^3145	0.0^41	0.0^375	0.0^3295	0.0102	0.004
100000	1	14.5	0.99	75	29.53	1020.7	401.8
6894.8	0.0689	1	0.068	5.17	2.04	70.37	27.7
101326	1.01	14.696	1	76	29.92	1034	407.1
1333	0.013	0.193	0.013	1	0.39	13.61	5.357
3386	0.034	0.49	0.033	2.54	1	34.56	13.61
97.98	0.0^398	0.014	0.0^397	0.073	0.029	1	0.39
248.9	0.0^225	0.036	0.0^225	0.187	0.073	2.54	1

(19A)

150

POWER

HORSE-POWER	KILO-WATTS	METRIC HORSE-POWER	kgf·m PER SEC.	FT-LBf PER SEC.	KILO-CALORIES PER SEC.	B.T.U. PER SEC.
1	0.746	1.014	76.04	550	0.178	0.707
1.341	1	1.36	102.0	737.6	0.239	0.948
0.986	0.736	1	75	542.5	0.176	0.697
0.013	0.0^298	0.013	1	7.23	0.0^223	0.0^293
0.0^218	0.0^214	0.0^218	0.138	1	0.0^332	0.0^213
5.615	4.187	5.692	426.9	3088	1	3.968
1.415	1.055	1.424	107.6	778.2	0.252	1

ENERGY OR WORK

JOULES (NEWTON-METER)	KILOGRAM-METERS	FOOT-POUNDS	KILOWATT-HOURS	METRIC HORSE POWER-HOURS	HORSE-POWER-HOURS	LITER-ATMOS-PHERES	KILO-CALORIES	BRITISH THERMAL UNITS
1	0.102	0.738	$0.0^{6}278$	$0.0^{6}378$	$0.0^{6}37$	$0.0^{2}987$	$0.0^{3}24$	$0.0^{3}948$
9.807	1	7.233	$0.0^{6}272$	$0.0^{6}310$	$0.0^{6}37$	0.0968	$0.0^{2}234$	$0.0^{2}93$
1.356	0.138	1	$0.0^{6}377$	$0.0^{6}512$	$0.0^{6}505$	0.0134	$0.0^{3}324$	0.0013
3600000	367100	2655000	1	1.36	1.34	35528	859.9	3412
2648000	270000	1952900	0.736	1	0.986	26131	632.4	2510
2684500	2737500	1980000	0.746	1.014	1	26493	641.2	2544
101.33	10.33	74.74	$0.0^{4}28$	$0.0^{4}38$	$0.0^{4}38$	1	0.024	0.096
4186.8	426.9	3088	$0.0^{2}116$	$0.0^{2}158$	$0.0^{2}156$	41.32	1	3.968
1055	107.6	778.2	$0.0^{3}29$	$0.0^{3}399$	$0.0^{3}393$	10.41	0.252	1

EXAMPLE (DECIMALS)

PROBLEM: CONVERT 5'-3" TO A DECIMAL NUMBER.
SOLUTION: SEE PAGE 129 (ITEM 2).
3" = .25', SO: 5' + .25' = <u>5.25'</u>

EXAMPLE (DECIMALS)

PROBLEM: CONVERT ¼" TO DECIMALS OF AN INCH.
SOLUTION: SEE PAGE 129 (ITEM 2). ¼" = <u>0.25"</u>

EXAMPLE: (SIMPLE ALGEBRA)

PROBLEM: A ROOM OF 15' x 15' MUST BE
LAYED OUT IN A QUARTER GRID.
WHAT IS THE GRID DIMENSION?

SOLUTION: SEE PAGE 129 (ITEM 3).
15' ÷ 4 = <u>3.75'</u> OR 3'-9" GRID = X

EXAMPLE: (PROPORTIONS)

PROBLEM: A 20' LONG ROOM HAS A SLOPING
CEILING WITH A RATIO OF 1 HIGH
TO 12 LONG. IF THE LOW WALL IS 8' HIGH,
WHAT IS THE HEIGHT OF THE FAR HIGH WALL?

SOLUTION:

SEE PAGE 129 (ITEM 4).

$\frac{A}{B} = \frac{C}{D}$ OR $\frac{1}{12} = \frac{X}{20}$

CROSS MULTIPLY
$12X = 1 \times 20$
$X = 20 \div 12$ OR 1.67'

THE HIGH WALL IS: 8' + 1.67' = <u>9.67'</u>
OR APPROX. 9'-8"

EXAMPLE: (PERCENTAGE)

PROBLEM: THE ORIGINAL PROGRAM FOR A TENANT IMPROVEMENT CALLED FOR A 5000 SF SPACE. IT WAS LATER INCREASED TO 8500 SF. WHAT IS THE % INCREASE IN SIZE?

SOLUTION:

SEE PAGE 129 (ITEM 6).

8500 SF − 5000 SF = 3500 SF INCREASE

3500 SF ÷ 5000 SF = 0.7 OR <u>70% INCREASE</u>

EXAMPLE: (SLOPE)

PROBLEM: A RAMP IS 3' HIGH AND 28' LONG. WHAT IS THE SLOPE?

SOLUTION:

SEE PAGE 130 (ITEM 8).

$\% \text{ SLOPE} = \dfrac{\text{RISE}}{\text{RUN}} = \dfrac{3'}{28'} = 0.107'$ OR <u>10.7% SLOPE</u>

EXAMPLE: (RIGHT ANGLES)

PROBLEM: A CEILING SLOPES AN ADDITIONAL 15' ABOVE THE NORMAL 8' HEIGHT. WHAT IS THE LENGTH OF CEILING, IF THE ROOM IS 50' LONG?

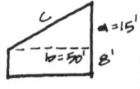

SOLUTION:

SEE PAGE 130 (ITEM 9).

$c = \sqrt{a^2 + b^2} = \sqrt{15^2 + 50^2} = \underline{52.2'}$

EXAMPLE: (CONVERSION)

PROBLEM: CONVERT A 835 SF APARTMENT INTO SQUARE METERS.

SOLUTION:

REFERRING TO PAGE 146 , 1SF = 0.093 SM.
THEREFORE : 835 SF × 0.093 SM/SF = <u>77.7 SM</u>

EXAMPLE: (DECIMALS, CONVERSION)

PROBLEM: CONVERT 5'· 3 1/4" TO MILLIMETERS.

SOLUTION:

REFER TO PAGE 129 TO CONVERT TO FEET
IN DECIMALS : 1/4" = 0.25". INCHES = 3 + .25.
CONVERT DIMENSIONS TO FEET IN DECIMELS:
1/4 OF DIFFERENCE BETWEEN 3" (.25') AND
4" (.33') = 0.08' ÷ 4 = 0.02'. THEREFORE,
5' + .25' + 0.02' (1/4") = 5.27'

REFER TO PAGE 145 TO CONVERT TO METERS :
1 FT = 0.305 M. THEREFORE, 5.27' × 0.305
= 1.607 M . SINCE 1M = 1000 MM (SEE
BOTTOM OF PAGE 145), 1.607 × 1000
= <u>1607 MILLIMETERS</u>

EXAMPLE: (COMPOUNDING)   SEE PAGE 130.

PROBLEM: IF AN ITEM COSTS $5⁰⁰/SF IN 2000,
WHAT WILL IT COST IN 2003 W/2% INCREASE
PER YEAR?

SOLUTION:

2000		$5⁰⁰/SF
2001	(× 1.02 =)	5¹⁰
2002	(× 1.02 =)	5²⁰
2003	(× 1.02 =)	$5³⁰/SF

NOTES

NOTES

NOTES

___ I. BUILDING LAWS

___1. <u>Zoning</u> (45)

When doing a tenant improvement in or a remodel of an existing building, where the use is changing, the interior designer will need to check with the local zoning or planning department. An example of change of use is changing an office to a restaurant. Issues that may need to be resolved are:

___ *a.* Zone
___ *b.* Allowable use
___ *c.* Prohibited uses or special-use permit
___ *d.* Restrictions on operation of facility
___ *e.* Maximum building coverage
___ *f.* Floor area ratio
___ *g.* Restrictions due to adjacent zone(s)
___ *h.* Required parking
___ *i.* Required loading zone
___ *j.* Parking layout restrictions
___ *k.* Signage (exterior)
___ *l.* Special submittals required for approval and/or hearings
___ *m.* Although not part of the zoning ordinance, private covenants, conditions, and restrictions (CC&Rs) that "run" with the land should be checked.
___ *n.* The interior designer may need to retain an architect or civil engineer for this type of work.

NOTES

NOTES

NOTES

___ 2. **Code Requirements for Residential Construction** (27B)
(2000 International Residential Code [IRC])

Use the following checklist for residences. The IRC applies to one- and two-family dwellings and multiple single-family dwellings (townhouses) not more than three stories in height with a separate means of egress, and their accessory structures.

___ *a.* Location on lot

     ___ (1) Openings must be 3′ from property line.

     ___ (2) Walls less than 3′ must be 1-hour construction.

     ___ (3) Windows not allowed in exterior walls with a fire separation distance of less than 3′ to the closest interior lot line (usually with parapet).

___ *b.* Separation between abutting dwelling units must be a minimum of 1-hour construction (½ hour, if sprinklered).

___ *c.* Windows and ventilation

     ___ (1) Habitable rooms must have natural light and ventilation by exterior windows with area of at least 8% of floor area and 4% must be openable.

     ___ (2) Bath and laundry-type rooms must have ventilation by operable exterior windows with area of not less than 3 SF (½ to be openable).

     ___ (3) In lieu of natural ventilation and light, mechanical ventilation and artificial lighting may be used.

     ___ (4) Any room may be considered as a portion of an adjoining room when at least ½ of area of the common wall is open and provides an opening of at least $\frac{1}{10}$ of floor area of interior room, but not less than 25 SF.

___ *d.* Room dimensions

     ___ (1) At least one room shall have at least 120 SF.

     ___ (2) Other habitable rooms, except kitchens, shall have at least 70 SF.

     ___ (3) Kitchens shall be 50 SF, min.

     ___ (4) Habitable rooms shall be 7′-0″ min. in any direction.

___ *e.* Ceiling heights

     ___ (1) 7′-0″ min.

     ___ (2) Where exposed beams not less than 4′ apart, bottoms may be at 6′-6″.

___ (3) Basements, 6'-8" min. (6'-4" to obstructions).

___ (4) At sloped ceilings, the min. ceiling height is required at only ½ the area, but never less than 5' height.

___ *f.* Sanitation

    ___ (1) Every dwelling unit (DU) shall have a kitchen with a sink.

    ___ (2) Every DU shall have a bath with a WC, lavatory, bathtub, or shower.

    ___ (3) Every sink, lavatory, bathtub, or shower shall have hot and cold running water.

___ *g.* Fire warning system (smoke alarms)

    ___ (1) Each dwelling must have smoke detectors in each sleeping room and the corridor to sleeping rooms, at each story (close proximity to stairways), and basement.

    ___ (2) In new construction, smoke detectors are to be powered by building wiring but equipped with backup battery.

    ___ (3) If there are additions or alterations (particularly sleeping rooms being added), the entire building shall have smoke detectors.

    ___ (4) In existing buildings, smoke detectors may be solely battery-operated.

___ *h.* Exits

    ___ (1) Doors

        ___ (*a*) At least one entry door shall be 3' wide by 6'-8" high.

        ___ (*b*) There must be a floor or landing at each side of each door, not more than 1.5" below door.

        ___ (*c*) At exterior, doors may open at the top step; if door swings away from step and step or landing is not lower than 8", the landing must be the width of stair or door and 36" deep.

    ___ (2) Emergency exits

        ___ (*a*) Sleeping rooms and basements with habitable space shall have at least one door or operable window.

        ___ (*b*) The window shall be operable from the inside and have a minimum clear opening of 5 SF at grade or 5.7 SF (24" high min., 20" wide min.) and sill shall not be higher than 44" above floor.

    ___ (*c*) Bars, grilles, or grates may be installed provided they are operable from inside.

    ___ (*d*) Windows, below grade, shall have a window well. The window shall be 9 SF clear opening, min., and 36″ min. dimension. When the well is deeper than 44″, must provide ladder or steps.

___ *i.* Stairs

    ___ (1) Min. width = 36″

    ___ (2) Max. rise: 7¾″

    ___ (3) Min. run (tread): 10″

    ___ (4) Variation in treads and risers = ⅜″ max.

    ___ (5) Winders: require 10″-wide tread at 12″ out from narrow side, but never less than 6″ width at any point.

    ___ (6) Spiral stairs to have 26″ min. clear width. Tread at 12″ from center to be 7½″. Max. riser = 9½″. Min. headroom = 6′-6″.

    ___ (7) Handrails

        ___ (*a*) At least one, at open side, continuous, and terminations to posts or walls

        ___ (*b*) Height: 34″ to 38″ above tread nosing

        ___ (*c*) Clearance from walls: 1½″

        ___ (*d*) Width of grip: 1¼″ to 2⅜″

    ___ (8) Headroom: 6′-8″ min.

    ___ (9) Guardrails at floor or roof openings, more than 30″ above grade. Height = 36″ min. If open, submembers must be spaced so a 4″ dia. sphere cannot pass through.

___ *j.* Garages and carports

    ___ (1) Must separate from DU (dwelling unit) with ½″ gypboard on garage side and 1⅜″ SC wood or 20 min. doors.

    ___ (2) No openings to sleeping areas allowed.

    ___ (3) Carports (open on at least two sides) do not apply (for above).

    ___ (4) Floors must slope to garage door opening.

___ *k.* Fireplaces: See p. 361.

___ *l.* Glazing: See p. 333.

___ *m.* Electrical: See p. 495.

___ *n.* Residential Accessibility (per IBC and ADA [ANSI]):

___ (1) Facilities *not* required to be accessible:
     ___ (*a*) Detached 1- and 2-story DU (this section).
     ___ (*b*) R-1 (boarding houses and hotels, occupancies with not more than 5 rooms for rent.
     ___ (*c*) R-2 and -3 (apartments and residential care homes) with 3 or fewer DU in a building.
     ___ (*d*) Existing residential buildings.
     ___ (*e*) Where unfeasible due to steep grade (see IBC).

___ (2) Facilities required to be accessible:
     ___ (*a*) Types of accessible units (see ANSI):
         ___ Type A are to be fully accessible.
         ___ Type B are to be minimally accessible.
     ___ (*b*) Scoping:
         ___ Occupancies R-2 and -3 with more than 5 DUs, every DU to be Type B, except:
         ___ R-2 with more than 20 DU: 20% (but at least one) to be Type A. Where no elevator, need only be on ground floor. Must have 20% of ground floor DUs as Type B.
         ___ Sleeping accommodations (for all R, not exempted):

Accessible units	Total units
1	1–25
2	26–50
4	51–75
5	76–100
7	101–150
8	151–200
10	201–300
12	301–400
13	401–500
3%	501–1000
30 + 2 for ea 100 over 1000	over 1000

NOTES

NOTES

___ 3. Building Code   ③⑦A

Configuring a tenant improvement or interior remodel in an existing building that meets fire safety code requirements is one of the interior designer's first responsibilities. The two biggest building law problems with remodel projects is new construction messing up fire-rated assemblies and achieving accessibility. The interior designer should acquire plans of the existing building. They will probably identify the code basics (occupancy type, construction type, occupancy load for exiting, whether the building is sprinklered, etc.). Because the interior designer may not be familiar with all of the intricacies of the code, he or she should work closely with the building official or an architect. Also, obtaining previously issued permits from the building department for the building or previous tenant improvement can be a good source of information.

This handbook uses the 2009 International Building Code (IBC) as a guide.

NOTE: This section and the following tables are from the *Architects' Portable Handbook,* 4th Ed.

Steps in preliminary code check are:

___ *a.* *Occupant Load:* Determining the occupant load from IBC Table A (p. 191), in some cases, will help determine the occupancy classification. When starting a project, a listing of architectural program areas by name, along with their floor area, occupant classification, and occupant load, should be compiled. Total the occupant load to help determine the final overall occupancy classification and for design of the exiting. The occupancy load can always be increased, provided the design of the building follows suit.

___ *b.* *Occupancy Classification:* The building code classifies buildings by occupancy in order to group similar life-safety problems together. Table B (p. 193) provides a concise definition of all occupancy classifications. In some cases, refer to the code for more detail. If the remodeling work results in a change of use or occupancy, then the existing building might have to be brought up to the current code. However, provisions are adjustable if the new use is less hazardous than the existing use.

___ c. *Construction Type:* The interior designer should be aware of the building's construction type as this might affect interior work. Construction types are based on whether or not the building construction materials are combustible or noncombustible, and on the number of hours that a wall, column, beam, floor, or other structural element can resist fire. Wood is an example of a material that is combustible. Steel and concrete are examples of noncombustible materials. Steel, however, will lose structural strength as it begins to soften in the heat of a fire.

There are two ways that construction can be resistant to fire. First it can be *fire-resistive*—built of a monolithic, noncombustible material like concrete or masonry. Second, it can be *protected*—encased in a noncombustible material such as steel columns or wood studs covered with gypsum plaster, wallboard, etc.

Note that the term "hourly rating" refers to the tested time a fire can "eat through" and collapse an element of construction.

The following are the construction types per the IBC:

___ *Type I* and *Type II* (*fire-resistive*) construction is noncombustible, built from concrete, masonry, and/or encased steel, and is used when substantial hourly ratings (2 to 3 hours) are required.

___ *Type II-B* construction is of the same materials as already mentioned, but the hourly ratings are lower. Light steel framing would fit into this category.

___ *Type III-A, IV, V-A* construction has noncombustible exterior walls of masonry or concrete, and interior construction of any allowable material including wood, but may require 1-hour fire protection.

___ *Type IV* construction is combustible *heavy timber* framing. It achieves its rating from the large size of the timber (2″ to 3″ thickness, min., actual). The outer surface chars, creating a fire-resistant layer that protects the remaining wood. Exterior walls must be of noncombustible materials.

___ *Type V-B* construction is of light wood framing and is generally nonrated.

Note: As the type number goes up, the cost of construction goes down, so generally use the lowest construction type (highest number) the code allows.

___ d. *Fireproofing:* See Table D for specific requirements of each construction type. See below on how to achieve these ratings.

___ (1) **On interior projects** you must make sure you do not effect the fire rating of existing structural elements, or fire barriers. For example, removing the gypsum wall board from a steel column can remove its fire protection. Penetrating walls and floors with piping and ductwork is a big problem with remodel projects. Specific fire-rated assemblies must be used. Thicknesses (in inches) of fire resistance structural materials will give approximate hourly ratings, as follows:

ITEM	NON- COMBUSTIBLE						HEAVY TIMBER	LIGHT WOOD FRAME
	4 HOUR	3 HOUR	2 HOUR	1½ HOUR	1 HOUR	0 HOUR		
STEEL, STRUCTURAL✳ LT. GA. JOISTS STUDS	←	SEE	NOTE	3	BELOW SEE NOTE 3C	→		↑
CONCRETE, COLUMNS WALLS SLABS POST TENSION FLOOR PRE-CAST CONC. COL. BEAMS WALLS SLABS PLANKS TEE BMS	6-8"	14" 6½" 6¼" 6¼" 12" 9½" 6½" 6¼" 8"+2" TOP'G.	12" 6" 5" 5" 10" 7" 6" 5" 8" 3¾"	10" 5" 4½" 4½" 3" 7" 5" 4½" 8" 2¾"	8" 3½" 3½" 3½" 6" 4" 3½" 3½" 8" 1¾"←TOPPING			SEE NOTE 3, BELOW
BRICK MASONRY WALLS, VAULTS, & DOMES (RISE NOT LESS THAN 1/12 SPAN	6-8"	8" 8"	6" 8"	6" 6"	4" 4"			
C.M.U. MASONRY WALLS	8"SOLID	8"	8"	6"	4"			
WOOD: COLUMNS FLOOR ROOF BEAMS, FLOOR ROOF TRUSSES,FLOOR ROOF							8×8 6×8 6×10 4×6 8×8 4×6	
WOOD DECK, FLOOR ROOF							3"+1" 1⅛"-2"	↓

✳ AT 25' ABOVE FLOOR, OPEN STEEL STRUCTURE DOES NOT NEED FIRE PROTECTION.

___ (2) **Fire-resistive materials** may be applied to structural members to protect from fire. Use the above table, as well as the following:

   ___ *a.* Concrete: $1'' \approx 2$ hr. $2''$ to $3'' \approx 4$ hr.

   ___ *b.* Solid masonry: $2'' \approx 1$ hr., add $1''$/hr to $4'' \approx 4$ hr.

   ___ *c.* Plaster: $1'' \approx 1$ hr., add $1''$/hr.

   ___ *d.* Vermiculite (spray-on): $1'' \approx 4$ hr.

   ___ *e.* Gypsum wallboard: 2 layers $\frac{1}{2}''$ type "X" or 1 layer of $\frac{5}{8}''$ type "X" $\approx \frac{3}{4}$ to 1 hr.

Costs: Spray-on vermiculite: \$3.50/SF surface/inch thickness

___ (3) **Flame Spread:** Surface finishes can ignite and lead to flame and smoke. The IBC requires finish materials to resist the spread of flame as follows:

   ___ (*a*) Maximum flame-spread class, below, gives required finish classifications:

INTERIOR FINISHES   TABLE 803.4

TABLE 803.4
INTERIOR WALL AND CEILING FINISH REQUIREMENTS BY OCCUPANCY[k]

GROUP	SPRINKLERED[l]			UNSPRINKLERED		
	Vertical exits and exit passageways[a,b]	Exit access corridors and other exitways	Rooms and enclosed spaces[c]	Vertical exits and exit passageways[a,b]	Exit access corridors and other exitways	Rooms and enclosed spaces[c]
A-1 & A-2	B	B	C	A	A[d]	B[e]
A-3[f], A-4, A-5	B	B	C	A	A[d]	C
B, E, M, R-1, R-4	B	C	C	A	B	C
F	C	C	C	B	C	C
H	B	B	C[g]	A	A	B
I-1	B	C	C	A	B	B
I-2	B	B	B[h,i]	A	A	B
I-3	A	A[j]	C	A	A	B
I-4	B	B	B[h,i]	A	A	B
R-2	C	C	C	B	B	C
R-3	C	C	C	C	C	C
S	C	C	C	B	B	C
U	No restrictions			No restrictions		

For SI: 1 inch = 25.4 mm, 1 square foot = 0.0929 m².

a. Class C interior finish materials shall be permitted for wainscotting or paneling of not more than 1,000 square feet of applied surface area in the grade lobby where applied directly to a noncombustible base or over furring strips applied to a noncombustible base and fireblocked as required by Section 803.3.1.
b. In vertical exits of buildings less than three stories in height of other than Group I-3, Class B interior finish for unsprinklered buildings and Class C interior finish for sprinklered buildings shall be permitted.
c. Requirements for rooms and enclosed spaces shall be based upon spaces enclosed by partitions. Where a fire-resistance rating is required for structural elements, the enclosing partitions shall extend from the floor to the ceiling. Partitions that do not comply with this shall be considered enclosing spaces and the rooms or spaces on both sides shall be considered one. In determining the applicable requirements for rooms and enclosed spaces, the specific occupancy thereof shall be the governing factor regardless of the group classification of the building or structure.
d. Lobby areas in A-1, A-2 and A-3 occupancies shall not be less than Class B materials.
e. Class C interior finish materials shall be permitted in places of assembly with an occupant load of 300 persons or less.
f. For churches and places of worship, wood used for ornamental purposes, trusses, paneling or chancel furnishing shall be permitted.
g. Class B material required where building exceeds two stories.
h. Class C interior finish materials shall be permitted in administrative spaces.
i. Class C interior finish materials shall be permitted in rooms with a capacity of four persons or less.
j. Class B materials shall be permitted as wainscotting extending not more than 48 inches above the finished floor in exit access corridors.
k. Finish materials as provided for in other sections of this code.
l. Applies when the vertical exits, exit passageways, exit access corridors or exitways, or rooms and spaces are protected by a sprinkler system installed in accordance with Section 903.3.1.1 or Section 903.3.1.2.

_____ (b) Flame-spread classification from above gives index below:

Class	Flame-spread index
A	0 to 25
B	26 to 75
C	76 to 200

___ (*c*) Use finishes to meet above requirements. Check finishes product literature for classifications or index.
  ___ (1) For woods, see p. 300.
  ___ (2) Aluminum: 5 to 10
  ___ (3) Masonry or Concrete: 0
  ___ (4) Gypsum wallboard: 10 to 25
  ___ (5) Carpet: 10 to 600
  ___ (6) Mineral-fiber sound-absorbing panels: 10 to 25
  ___ (7) Vinyl tile: 10 to 50
  ___ (8) Chemically treated wood fiberboard: 20 to 25
  ___ (9) Certain intumescent paints can reduce the flame spread of combustible finishes to as low as class A.

___ (**d**) **Floor Finishes:** Most floor finishes present little if any hazard due to flame spread. Carpet is the exception.
  ___ Types:
      Class I (radiant flux of 0.45 W/cm^2 or more) more resistant to flame spread. This is usually of a low pile and/or natural fiber.
      Class II (radiant flux of 0.22 W/cm^2) less resistant. This is usually of a high pile and/or synthetic fiber.
      Sprinklers can allow Class II where I is required.

___ (**e**) **Trim and Decorations on Walls and Ceilings:**
  ___ Must be at least Class C.
  ___ Limited to 10% of area (except sprinklered auditoriums may be up to 50%).
  ___ At Group I-3 occupancies, only noncombustible materials allowed.

_____ At Groups A, E, I, R-1, and R-2 dormitories, only flame-resistant or noncombustible materials allowed.

_____ (f) **Fire Loads:** Interior building contents that will start or contribute to a fire. These typically range from 10 (residential) to 50 PSF (office), and can be reduced 80% to 90% by use of metal storage containers for paper.

_____ (g) **Fabrics and Furnishings:** See page 383.

_____ e. *Occupancy Separations:* Different occupancies have different vulnerability to fire and therefore sometimes, must be separated by fire barriers. For hourly ratings, these are determined from Table E. Most buildings will have some mix of occupancies. If one of the occupancies is a minor area and accessory to the major one, the whole building can often be classified by the major occupancy. It will then have to meet the requirements of the more restrictive occupancy, but no separating walls will be necessary.

_____ f. *Sprinkler Requirements:* The interior designer should note if the existing building does or does not have sprinklers. If they exist, they will almost always be legally required for the tenant improvement or remodel. If they do not exist and the occupancy type is being changed, they may be required. Also, added local ordinances may require sprinklers over and above what the model code says. A sprinkler system can be used to *substitute for 1-hour construction* if it is not otherwise required by the code. However, in most cases, exit access corridors, exit stairs, shafts, area separation walls and similar structures must maintain their required fire protection.

A sprinkler system is the most effective way to provide life and fire safety in a building. The IBC requires fire sprinklers in the following situations:

_____ *Assembly occupancies:* Sprinklers to be provided throughout Group A areas as well as all floors between Group A and level of exit discharge, where:

_____ A-1 uses exceeding 12,000 SF, or occupant load > 300.

___ A-2 uses where fire area exceeds 5000 SF, or occupant load > 300.

___ A-3 uses where fire area exceeds 12,000 SF or occupant load > 300.

___ A-4 same as A-3, with exemption for participant sports areas where main floor located at level exit discharge of main entrance and exit.

___ A-5 concession stands, retail areas, press boxes, and other accessory use areas > 1000 SF.

___ *Educational occupancies,* where floor area exceeds 20,000 SF or schools below exit discharge level, except where each classroom has one exit door at ground level.

___ *Commercial and industrial uses:*

     ___ Groups F-1 and S-1 where their fire areas > 12,000 SF, or where > 3 stories, or where > 24,000 SF total of all floors.

     ___ Repair garages under the following conditions:

         ___ Buildings ≥ 2 stories, including basements where fire area > 10,000 SF.

         ___ Buildings of one story where fire area > 12,000 SF.

         ___ Buildings with repair garage in basement.

     ___ S-2 enclosed parking garage, unless located under Group R-3.

     ___ Buildings used for storage of commercial trucks or buses where fire area exceeds 5000 SF.

___ *Hazardous (H) occupancies.*

___ *Institutional (I) occupancies.*

___ *Mercantile (M) occupancies where:*

     ___ Fire area > 12,000 SF.

     ___ More than 3 stories high.

___ Total fire area on all floors (including mezzanines) > 24,000 SF.

___ *Residential occupancies:*

___ R-1 hotel uses, regardless of number of stories or guest rooms (except motels where guest rooms not more than 3 stories above lowest level of exit discharge and each guest room has at least one door direct to exterior egress).

___ R-2 apartment house, congregate residence uses more than 2 stories in height, including basements, or having more than 16 DUs per fire area.

___ R-4 residential care or assisted living facilities having more than 8 occupants per fire area.

___ *All occupancies without sufficient fire department access through outside wall at basement or floor in excess of 1500 SF* (Group R-3 and Group U excluded). Sufficient access is 20 SF of openings with a minimum dimension of 30″ per 50 LF of wall. If these openings are on only one side, the floor dimension cannot exceed 75′ from the opening. Furthermore, sprinklers are required in buildings with a floor level that is located 55′ above the lowest level of fire truck access (excluding airport control towers, open parking garages, and F-2 occupancies).

___ *Miscellaneous:* Rubbish and linen chutes; nitrate film storerooms; and combustible-fiber storage vaults, atriums, and stages.

See p. 438 for sprinkler installations.

___ g. *Fire Areas, Walls, Barriers, and Partitions:*

___ *Structural Walls:* The interior designer will need to determine which, if any, interior walls may be "bearing" (usually exterior walls are). The best way to do this is obtaining the plans of the building. An architect or structural engineer may need to be consulted.

___ *Fire-Resistance-Rated-Elements:*

___ *Fire Walls:* A rated wall (to enclose fire areas) with protected openings that extends continuously from foundation to or through roof (with some exceptions). A collapse from fire on one side will not allow collapse on the other. Used for fire compartments to allow larger buildings. See Table F for required ratings of fire walls.

___ *Fire Barriers:* A rated wall used for required occupancy separations (see Table E). Note that the rating can be reduced 1 hour (except H and I-2 occupancies) when a sprinkler system is used (even if sprinkler is required). Also used to enclose vertical exit enclosures, exit passageways, horizontal exits, and incidental use areas. Floors that support barriers must be of the same rating (except for some incidental use separations). If building is sprinklered, there is no limit on openings or fire doors; otherwise, the openings are limited to 25% of the length of wall (a single opening is limited to 120 SF). This also applies to fire walls. Openings must be rated per Table G on p. 202.

___ *Fire Partitions:*

    ___ One-hour rated walls for:

        ___ Protected *corridors*, where required by Table H (see p. 203).

        ___ Separate *dwelling units* (DUs).

        ___ Separating *guest rooms* in R-1.

        ___ Separating *tenant spaces* in covered malls.

    ___ Half-hour-rated walls in *sprinklered* buildings for:

        ___ Construction Types II-B, III-B, or V-B.

        ___ Separations of dwelling units or guest rooms.

    ___ Floors supporting partition must be of same rating (except for ½-hour-rated walls).

    ___ Openings must be 20 minutes for corridor walls and 45 minutes for others.

    ___ Where required to be noncombustible, they are usually metal studs or masonry. Where allowed to be combustible, the framing may be of wood but the clading is rated (usually ⅝" Type X gypsum board). The typical 1-hour partition is of wood (or metal) studs w/rated gypsum board each side.

___ *Nonrated Partitions:*

> ___ Sometimes fire-retardant treated wood is allowed in noncombustible construction types.

> ___ Wood panels or similar light construction up to 6 ft. (when higher must enclose top with glass) are usually allowed. Light transmitting plastic may be used as or in interior partitions.

> ___ Folding portable or movable partitions do not need to be rated but their surfaces need to conform to flame-spread requirements (see p. 172). These should not block any exiting.

___ *h.* *Exiting and Stairs:* At the conceptual stage of design, the most important aspects of the building code requirements are the number and distribution of exits.

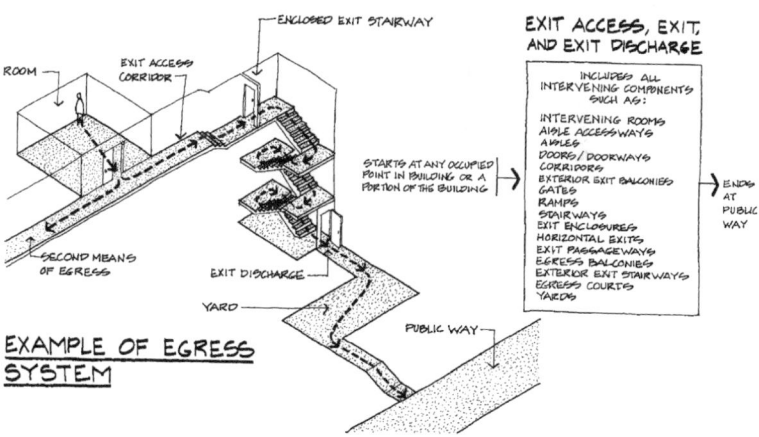

EXAMPLE OF EGRESS SYSTEM

A *means of egress* is a continuous path of travel from any point in a building or structure to the open air outside at ground level (public way). It consists of three separate and distinct parts:

___ 1. Exit access

___ 2. The exit

___ 3. The exit discharge

The *exit access* leads to an exit. See Tables M and N, where only one is required; otherwise a minimum of two exits are almost always required. Other general requirements:

___ 1. Exit width determined by Table J, p. 204, but corridor width is usually no less than 44″. It can be 36″ for fewer than 51 people. School corridors must be 6′ wide. Hospitals 8′ wide. Large residential care homes, 5′ wide.

___ 2. Dead-end corridors are usually limited to 20′ long (in some cases 50′).

___ 3. When more than one exit is required, the occupant should be able to go toward either exit from any point in the corridor system.

___ 4. Corridors used for exit access usually require 1-hour construction, unless sprinklered.

___ 5. Maximum travel distance from any point to an exit is per Table L on p. 205.

___ 6. Handrails or fully open doors cannot extend more than 7″ into the corridor.

___ 7. Doors at their worst extension into the corridor cannot obstruct the required width by more than half.

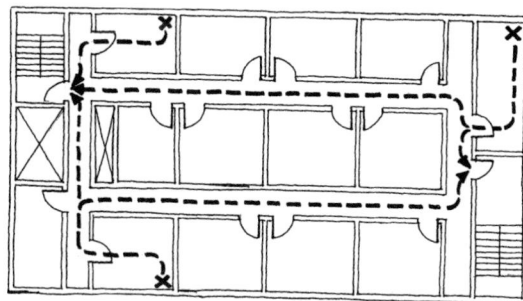

EXIT ACCESS ON UPPER OFFICE FLOOR ▬ ▬ ▶

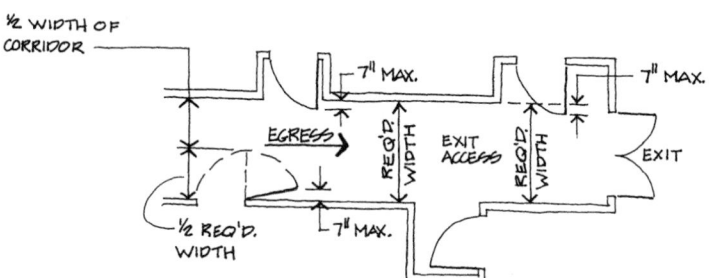

The *exit* is that portion of a means of egress that is separated from the area of the building from which escape is made, by walls, floors, doors, or other means that provide the protected path necessary for the occupants to proceed with safety to a public space. The most common form the exit takes is an enclosed stairway. In a single-story building the exit is the door opening to the outside.

After determining occupant load (Table A, p. 191) for spaces, rooms, floors, etc., use the following guidelines:

___ 1. In some cases one exit can be used (see above), but often buildings or rooms need two exits (see Table K, p. 200). In more than one story, stairs become part of an exit. Elevators are not exits. Three exits are required when the occupant load is 501 to 1000 and four when over 1000.

___ 2. In buildings 4 stories and higher and in types I and II-B construction, the exit stairs are required to have 2-hour enclosure; otherwise, 1 hour is acceptable.

___ 3. When two exits are required (for unsprinklered buildings), they have to be separated by a distance equal to half the diagonal dimension of the floor and/or room the exits are serving (measured in straight lines). See sketch below. If the building is sprinklered, the minimum separation is ⅓.

___ 4. Where more than two exits are required, two of them need to be separated by at least half the diagonal dimension. The others are spaced to provide good access from any direction.

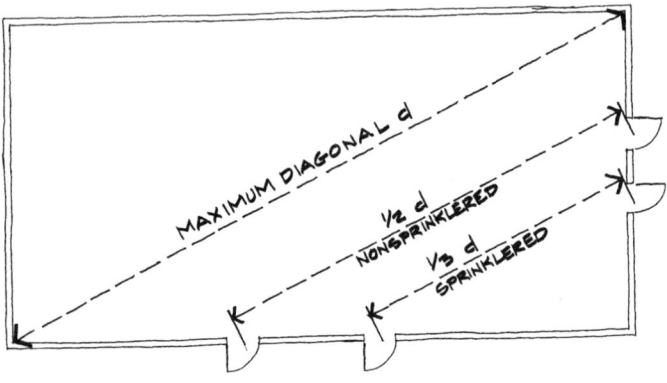

___ 5. May exit from room through adjoining rooms (except rooms that are accessory or high-hazard occupancy), provided adjoining rooms (other than DUs) are not kitchens, storerooms, toilets, closets, etc. Foyers, lobbies, and reception rooms are not considered adjoining rooms and can always be exited through.

___ 6. The total exit width required (in inches) is determined by multiplying the occupant load by the factors shown in Table J (p. 204). This width should be divided equally among the required number of exits.

___ 7. Total occupant load for calculating exit stair width is defined as the sum of the occupant load on the floor in question. The maximum exit stair width calculated is maintained for the entire exit. See sketch on p. 184.

___ 8. Minimum exit door width is 36″ with 32″ clear opening. Maximum door width is 48″.

___ 9. The width of exit stairs, and the width of landings between flights of stairs, must all be the same and must meet the minimum exit stair width requirements as calculated or:

> 44″ min. width for an occupant load of 51 or more
>
> 36″ min. width for 50 or less

___ 10. Doors must swing in the direction of travel when serving a hazardous area or when serving an occupant load of usually 50 or more.

A *horizontal exit* is a way of passage through a 2-hour fire wall into another area of the same building or into a different building that will provide refuge from smoke and fire. Horizontal exits cannot provide more than half of the required exit capacity, and any such exit must discharge into an area capable of holding the occupant capacity of the exit. The area is calculated at 3 SF/occupant. In institutional occupancies the area needed is 15 SF/ambulatory person and 30 SF/nonambulatory person.

Exit discharge is that portion of a means of egress between the termination of an exit and a public way. The most common form this takes is the door out of an exit stairway opening onto a public street. Exits can discharge through a courtyard with 1-hour walls that connect the exit with a public way. 50% of the exits can discharge through a street floor lobby area if the entire street floor is sprinklered, the path through the lobby is unobstructed and obvious, and the level of discharge is separated from floors below.

Smokeproof enclosures for exits are required in any building with floors 75′ above (or 30′ below) the lowest ground level where fire trucks have access. A smokeproof enclosure is an exit stair that is entered through a vestibule that is ventilated by either natural or mechanical means such that products of combustion from a fire will be limited in their penetration of the exit-stair enclosure. Smokeproof enclosures are required to be 2-hour construction. They must discharge directly to the outside or directly through a 2-hour exit passageway to the outside. In a *sprinklered* building, mechanically pressurized and vented stairways can be substituted for smokeproof enclosures.

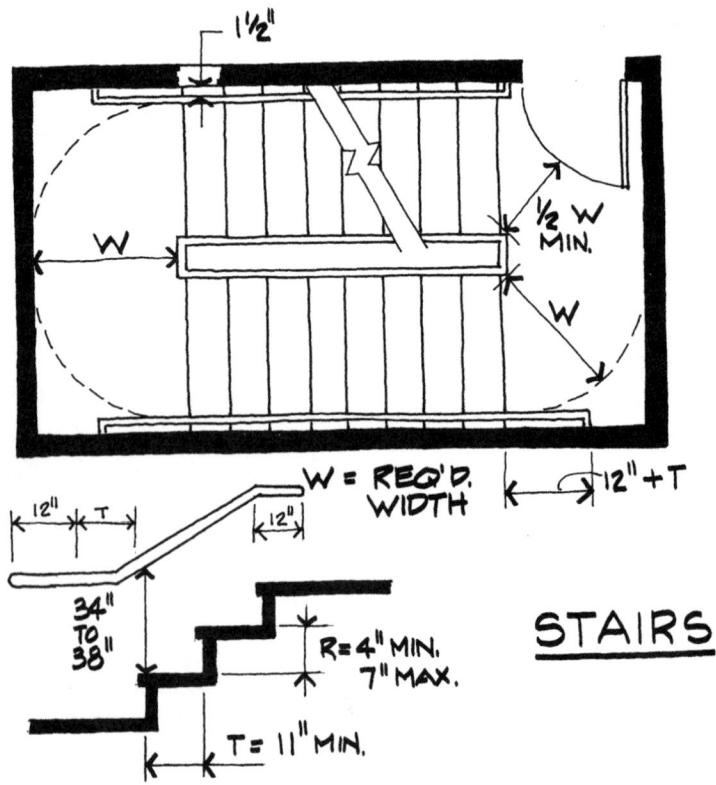

STAIRS

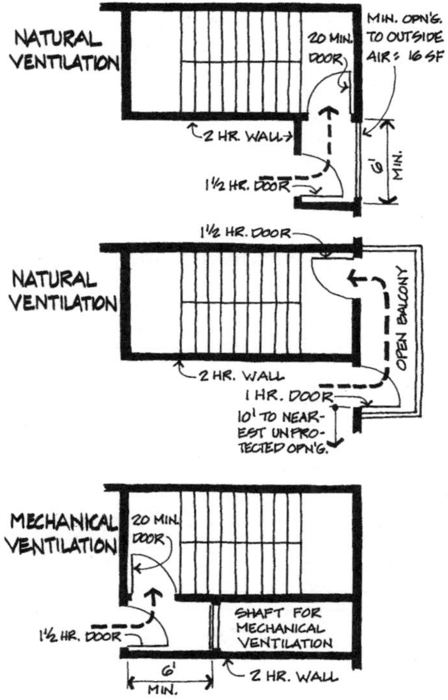

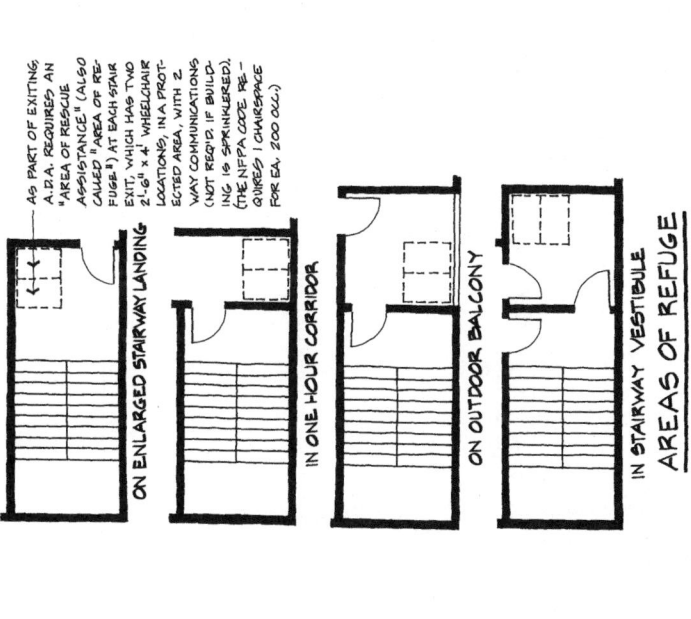

AREAS OF REFUGE

ON ENLARGED STAIRWAY LANDING

AS PART OF EXITING, A.D.A. REQUIRES AN "AREA OF RESCUE ASSISTANCE" (ALSO CALLED "AREA OF REFUGE") AT EACH STAIR EXIT, WHICH HAS TWO 2'-6" x 4' WHEELCHAIR LOCATIONS, IN A PROTECTED AREA, WITH 2-WAY COMMUNICATIONS (NOT REQ'D. IF BUILDING IS SPRINKLERED). (THE NFPA CODE, RE-QUIRES 1 CHAIRSPACE FOR EA. 200 OCC.)

IN ONE HOUR CORRIDOR

ON OUTDOOR BALCONY

IN STAIRWAY VESTIBULE

SMOKE PROOF ENCLOSURES

NATURAL VENTILATION

MIN. OPNG. TO OUTSIDE AIR = 16 SF

6' MIN.

20 MIN. DOOR

2 HR. WALL

1½ HR. DOOR

NATURAL VENTILATION

OPEN BALCONY

1½ HR. DOOR

2 HR. WALL

1 HR. DOOR

10' TO NEAREST UNPROTECTED OPNG.

MECHANICAL VENTILATION

SHAFT FOR MECHANICAL VENTILATION

2 HR. WALL

6' MIN.

20 MIN. DOOR

1½ HR. DOOR

Code Requirements for Stairs

Code requirements	Tread min.	Riser		Min. width	Headroom
		Min.	Max.		
General (including HC)	11"	4"	7"	36"	6–8"
Private stairways (occ. <10)	9"		8"	20"	6–8"
Winding—min. required T at 12" from narrow side*	6" at any pt.				
Spiral—at 12" from column*	7½"			26"	

Only permitted in R-3 dwellings and R-1 private apartments*

Circular: inside radius not less than 2× width of stair, min. T depth = 11" @ 12" from inside and 10" @ inside, 36" min. width.

Rules of thumb for general stairs:

Interior	2R + T = 25
Exterior	2R + T = 26

Open risers not permitted in most situations.

*Requires handrail.

Code Requirements for Ramp Slopes

Type	Max. slope	Max. rise	Max. run
Required for accessibility	1:12*	5′	
Others	1:8*	5′	
Assembly with fixed seats	1:5		
HC, new facilities	1:12*	2.5′	30′
HC, existing facilities	1:10*	6″	5′
	1:8*	3″	2′
HC, curb ramps	1:10	6″	5′
Historic buildings	1:6		

Any walking surface steeper than 5% is a ramp.

Landings are to be as wide as widest ramp to landing. Depth to be 5′ min. Where landing is at a corner, it shall be 5′ × 5′ min.

*Requires handrails for ramps > 1:15.

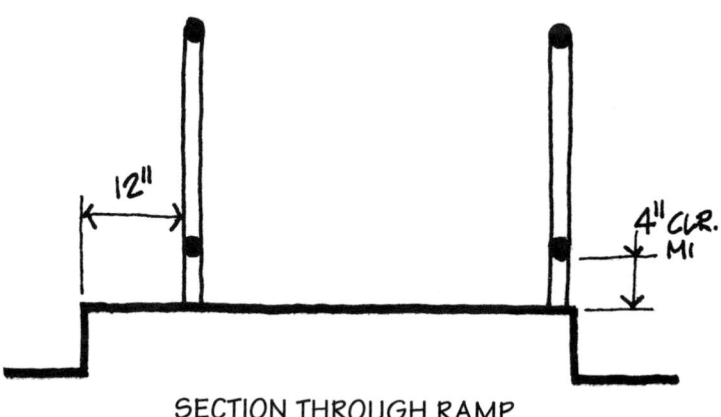

SECTION THROUGH RAMP

NOTES

NOTES

NOTES

TABLE A

TABLE 1004.1.1
MAXIMUM FLOOR AREA ALLOWANCES PER OCCUPANT

FUNCTION OF SPACE	FLOOR AREA IN SQ. FT. PER OCCUPANT
Accessory storage areas, mechanical equipment room	300 gross
Agricultural building	300 gross
Aircraft hangars	500 gross

Airport terminal
  Baggage claim
  Baggage handling
  Concourse
  Waiting areas
	
	20 gross
	300 gross
	100 gross
	15 gross

Assembly Gaming floors (keno, slots, etc.)	 11 gross
Assembly with fixed seats	See Section 1004.7

Assembly without fixed seats
  Concentrated (chairs only—not fixed)
  Standing space
  Unconcentrated (tables and chairs)
	
	7 net
	5 net
	15 net

Bowling centers, allow 5 persons for each lane including 15 feet of runway, and for additional areas	7 net
Business areas	100 gross
Courtrooms—other than fixed seating areas	40 net
Day care	35 net
Dormitories	50 gross

TABLE 1004.1.1, *(continued)*
MAXIMUM FLOOR AREA ALLOWANCES PER OCCUPANT

FUNCTION OF SPACE	FLOOR AREA IN SQ. FT. PER OCCUPANT
Educational 　Classroom area 　Shops and other vocational room areas	20 net 50 net
Exercise rooms	50 gross
H-5 Fabrication and manufacturing areas	200 gross
Industrial areas	100 gross
Institutional areas 　Inpatient treatment areas 　Outpatient areas 　Sleeping areas	240 gross 100 gross 120 gross
Kitchens, commercial	200 gross
Library 　Reading rooms 　Stack area	50 net 100 gross
Locker rooms	50 gross
Mercantile 　Areas on other floors 　Basement and grade floor areas 　Storage, stock, shipping areas	60 gross 30 gross 300 gross
Parking garages	200 gross
Residential	200 gross
Skating rinks, swimming pools 　Rink and pool 　Decks	50 gross 15 gross
Stages and platforms	15 net
Warehouses	500 gross

For SI:　1 square foot = 0.0929 m^2.

TABLE B (IBC 2009)
OCCUPANCY CLASSIFICATIONS

ASSEMBLY GROUP A:

NOTE: A room or space used for assembly purposes by less than 50 persons & accessory to another occupancy shall be included as part of that occupancy.

A-1 Assembly uses, usually with fixed seating, intended for the production and viewing of the performing arts or motion pictures including, but not limited to:

Motion picture theaters
Theaters
TV & radio studios admitting an audience

A-2 Assembly uses intended for food and/or drink consumption including but not limited to:

Banquet halls   Restaurants
Night clubs   Taverns & bars

A-3 Assembly uses intended for worship, recreation, or amusement and other assembly uses not classified elsewhere in Group A, including, but not limited to:

Amusement   Gymnasiums
arcades   Indoor swim-
Art galleries   ming pools
Auditoriums   Indoor tennis
Bowling alleys   courts
Churches   Lecture halls
Community   Libraries
halls   Museums
Courtrooms   Passenger
Dance halls   stations
Exhibition   (wait areas)
halls   Pool and
Funeral   billiard
parlors   parlors

A-4 Assembly uses intended for viewing of indoor sporting events and activities with spectator seating, including, but not limited to:

Arenas   Swimming pools
Skating rinks Tennis courts

A-5 Assembly uses intended for participation in or viewing outdoor activities including, but not limited to:

Amusement   Bleachers
park   Grandstands
structures   Stadiums

BUSINESS GROUP B:

Includes, among others, the use of a building or structure, or a portion thereof, for office, professional or service-type transactions, including storage of records and accounts. Business occupancies include, but not limited to:

Airport traffic control towers
Animal hospitals, kennels & pounds
Banks
Barber and beauty shops
Car wash
Civic administration
Clinic - outpatient
Dry cleaning & laundries (pick-up & delivery stations & self-serv.)
Educ. occupancies above 12th grade
Electronic data processing
Fire & police stations
Laboratories: testing & research
Motor vehicle showrooms
Post offices
Print shops
Professional services (architects, attorneys, dentists, physicians, engineers, etc.)
Radio & TV stations
Telephone exchanges

EDUCATIONAL GROUP E:

Includes, among others, the use of a building or structure, or a portion thereof, by six or more persons at any one time for educational purposes through the 12th grade.

Also, for educational, supervision or personal care services for more than five children older than 2½ years of age.

FACTORY GROUP F:

Includes, among others, the use of a building or structure, or a portion thereof, for assembling, dissembling, fabricating, finishing, manufacturing, packaging, repair or processing operations that are not classified as a Group H hazardous occupancy.

F-1 Moderate Hazard

Factory Industrial uses not classified as F-2, including, but not limited to:

Aircraft	Jute products
Appliances	Laundries
Athletic equip.	Leather products
Automobiles and other motor vehicles	Machinery
	Metals
Bakeries	Millwork
Beverages (alcoholic)	Motion picture &
Bicycles	TV filming
Boats: building	Musical instruments
Brooms or brushes	Optical goods
Business machines	Paper mill or prod.
Cameras & photo equip.	Photographic film
Canvas or sim. fabric	Plastic products
Carpets & rugs (includes cleaning)	Printing or publ.
	Recreational veh.
Clothing	Refuse incineration
Const. & agri machinery	Shoes
Disinfectants	Soaps & detergents
Dry cleaning & dyeing	Textiles
Elect. light plants & power houses	Tobacco
	Trailers
Electronics	Upholstering
Engines (incl. rebuilding)	Wood: distillation
Food processing	Woodworking

Furniture	(cabinet)
Hemp products	

F-2 Low Hazard

Factory Industrial uses that involve the fabrication or manufacturing of noncombustible materials which during finishing, packing or processing do not involve a significant fire hazard, including, but not limited to:

Beverages (nonalcoholic)	Glass products
	Gypsum
Brick & masonry	Ice
Ceramic products	Metal products
Foundries	(fab. & assemb.)

HIGH-HAZARD GROUP H:

The IBC provides a detailed and complicated definition of each classification. Usually the classification will have to be done by the Building Official, the Fire Department, or a special consultant. Because the H occupancies have become so confusing here is a very brief description of each:

H-1: Containing high explosion hazard materials.

H-2: Where flammable or combustible liquids or dusts are being created, mixed, or dispensed.

H-3: Use of flammable or combustible liquids including organic peroxides and oxidizers that present high fire or heat release hazards.

H-4: Containing health hazard materials such as corrosive and toxic chemicals.

H-5: Semiconductor fabrication facilities.

INSTITUTIONAL GROUP I:

This occupancy is where people with physical limitations in a medical setting or people with restricted limitations in a penal setting are housed.

I - 1: Housing more than 16 persons in a residential setting (R - 3 if 5 or less).

I - 2: Medical buildings with 24 hour care of more than 5 people, such as hospitals, nursing homes, etc. (R - 3 if less than 5 people). This also includes 24 hour child care under 2½ years old.

I - 3: Housing more than 5 people in secured conditions such as prisons and jails. The code further has 5 subconditions.

I - 4: Day care facilities (R - 3 if 5 or less persons and E under certain conditions).

MERCANTILE GROUP M:

Buildings used for the display, sale, and stocking of goods such as department stores, drug stores, markets, sales rooms, retail or wholesale stores. This also includes motor vehicle service stations.

RESIDENTIAL GROUP R:

NOTE: One and two family dwellings are covered under the International Residential Code (IRC). See page 71. Otherwise:

R - 1: Transient lodging (under 30 days), including hotels and motels.

R - 2: Three or more dwelling units (DUs) where occupancy is mainly permanent, such as apartments, dormitories, convents, fraternity, and sorority houses.

R - 3: Buildings containing one or two DUs for adult or child care of any age for less than 24 hours with not more than 5 people.

R - 4: Residential care or assisted living facilities, where number of residents is greater than 5 but not greater than 16.

STORAGE GROUP S:

Warehousing, subdivided as follows:

S - 1 MODERATE-HAZARD:

Stores flammable products that are not classified as H. This also includes some car repair garages and aircraft hangars. See code for details.

S - 2 LOW-HAZARD:

Storage of noncombustible materials. See code for details.

UTILITY GROUP U:

Building of an accessory nature, not classified elsewhere. Carports and private garages are included.

OTHER SPECIAL OCCUPANCIES:

Covered Mall Buildings
High Rise Buildings (above 75' high).
Atriums
Underground Buildings

TABLE D

TABLE 601
FIRE-RESISTANCE RATING REQUIREMENTS FOR BUILDING ELEMENTS (hours)

BUILDING ELEMENT	TYPE I A	TYPE I B	TYPE II A[d]	TYPE II B	TYPE III A[d]	TYPE III B	TYPE IV HT	TYPE V A[d]	TYPE V B
Primary structural frame[g] (see Section 202)	3[a]	2[a]	1	0	1	0	HT	1	0
Bearing walls Exterior[f, g]	3	2	1	0	2	2	2	1	0
Interior	3[a]	2[a]	1	0	1	0	1/HT	1	0
Nonbearing walls and partitions Exterior					See Table 602				
Nonbearing walls and partitions Interior[e]	0	0	0	0	0	0	See Section 602.4.6	0	0
Floor construction and secondary members (see Section 202)	2	2	1	0	1	0	HT	1	0
Roof construction and secondary members (see Section 202)	1 1/2[b]	1[b, c]	1[b, c]	0[c]	1[b, c]	0	HT	1[b, c]	0

TABLE 601, *(continued)*
FIRE-RESISTANCE RATING REQUIREMENTS FOR BUILDING ELEMENTS (hours)

For SI: 1 foot = 304.8 mm.

a. Roof supports: Fire-resistance ratings of primary structural frame and bearing walls are permitted to be reduced by 1 hour where supporting a roof only.

b. Except in Group F-1, H, M and S-1 occupancies, fire protection of structural members shall not be required, including protection of roof framing and decking where every part of the roof construction is 20 feet or more above any floor immediately below. Fire-retardant-treated wood members shall be allowed to be used for such unprotected members.

c. In all occupancies, heavy timber shall be allowed where a 1-hour or less fire-resistance rating is required.

d. An approved automatic sprinkler system in accordance with Section 903.3.1.1 shall be allowed to be substituted for 1-hour fire-resistance-rated construction, provided such system is not otherwise required by other provisions of the code or used for an allowable area increase in accordance with Section 506.3 or an allowable height increase in accordance with Section 504.2. The 1-hour substitution for the fire resistance of exterior walls shall not be permitted.

Not less than the fire-resistance rating required by other sections of this code.

f. Not less than the fire-resistance rating based on fire separation distance (see Table 602).

g. Not less than the fire-resistance rating as referenced in Section 704.10

TABLE E (handwritten annotation)

TABLE 508.4
REQUIRED SEPARATION OF OCCUPANCIES (HOURS)

OCCUPANCY	A[d], E S	A[d], E NS	I-1, I-3, I-4 S	I-1, I-3, I-4 NS	I-2 S	I-2 NS	R S	R NS	F-2, S-2[b], U S	F-2, S-2[b], U NS	B, F-1, M, S-1 S	B, F-1, M, S-1 NS	H-1 S	H-1 NS	H-2 S	H-2 NS	H-3, H-4, H-5 S	H-3, H-4, H-5 NS
A[d], E	N	N	1	2	2	NP	1	2	N	1	1	2	NP	NP	3	4	2	3[a]
I-1, I-3, I-4	—	—	N	N	2	NP	1	NP	1	2	1	2	NP	NP	3	NP	2	NP
I-2	—	—	—	—	N	N	2	NP	2	NP	2	NP	NP	NP	3	NP	2	NP
R	—	—	—	—	—	—	N	N	1[c]	2[c]	1	2	NP	NP	3	4	2	3[a]
F-2, S-2[b], U	—	—	—	—	—	—	—	—	N	N	1	2	NP	NP	3	4	2	3[a]
B, F-1, M, S-1	—	—	—	—	—	—	—	—	—	—	N	N	NP	NP	2	3	1	2[a]
H-1	—	—	—	—	—	—	—	—	—	—	—	—	N	NP	NP	NP	NP	NP
H-2	—	—	—	—	—	—	—	—	—	—	—	—	—	—	N	—	1	NP
H-3, H-4, H-5	—	—	—	—	—	—	—	—	—	—	—	—	—	—	—	—	1[e,f]	NP

TABLE 508.4, *(continued)*
REQUIRED SEPARATION OF OCCUPANCIES (HOURS)

For SI: 1 square foot = 0.0929 m².

S = Buildings equipped throughout with an automatic sprinkler system installed in accordance with Section 903.3.1.1.

NS = Buildings not equipped throughout with an automatic sprinkler system installed in accordance with Section 903.3.1.1.

N = No separation requirement.

NP = Not permitted.

a. For Group H-5 occupancies, see Section 903.2.5.2.

b. The required separation from areas used only for private or pleasure vehicles shall be reduced by 1 hour but to not less than 1 hour.

c. See Section 406.1.4.

d. Commercial kitchens need not be separated from the restaurant seating areas that they serve.

e. Separation is not required between occupancies of the same classification.

f. For H-5 occupancies, see Section 415.8.2.2.

TABLE K

TABLE 1005.2.1
MINIMUM NUMBER OF EXITS FOR OCCUPANT LOAD

OCCUPANT LOAD	MINIMUM NUMBER OF EXITS
1-500	2
501-1,000	3
More than 1,000	4

TABLE 1015.1
SPACES WITH ONE EXIT OR EXIT ACCESS DOORWAY

OCCUPANCY	MAXIMUM OCCUPANT LOAD
A, B, E[a], F, M, U	49
H-1, H-2, H-3	3
H-4, H-5, I-1, I-3, I-4, R	10
S	29

a. Day care maximum occupant load is 10.

TABLE G

TABLE 715.4
FIRE DOOR AND FIRE SHUTTER FIRE PROTECTION RATINGS

TYPE OF ASSEMBLY	REQUIRED ASSEMBLY RATING (hours)	MINIMUM FIRE DOOR AND FIRE SHUTTER ASSEMBLY RATING (hours)
Fire walls and fire barriers having a required fire-resistance rating greater than 1 hour	4 3 2 $1\frac{1}{2}$	3 3[a] $1\frac{1}{2}$ $1\frac{1}{2}$

Fire barriers having a required fire-resistance rating of 1 hour:
  Shaft, exit enclosure and exit passageway walls
  Other fire barriers
	1
	1
		1
		$\frac{3}{4}$

Fire partitions:
  Corridor walls
 	
  Other fire partitions
	1
	0.5
	1
	0.5
		$\frac{1}{3}$[b]
		$\frac{1}{3}$[b]
		$\frac{3}{4}$
		$\frac{1}{3}$

Exterior walls	3 2 1	$1\frac{1}{2}$ $1\frac{1}{2}$ $\frac{3}{4}$
Smoke barriers	1	$\frac{1}{3}$[b]

a. Two doors, each with a fire protection rating of $1\frac{1}{2}$ hours, installed on opposite sides of the same opening in a fire wall, shall be deemed equivalent in fire protection rating to one 3-hour fire door.
b. For testing requirements, see Section 715.4.3.

TABLE 1018.1
CORRIDOR FIRE-RESISTANCE RATING

OCCUPANCY	OCCUPANT LOAD SERVED BY CORRIDOR	REQUIRED FIRE-RESISTANCE RATING (hours)	
		Without sprinkler system	With sprinkler system[c]
H-1, H-2, H-3	All	Not Permitted	1
H-4, H-5	Greater than 30	Not Permitted	1
A, B, E, F, M, S, U	Greater than 30	1	0
R	Greater than 10	Not Permitted	0.5
I-2[a], I-4	All	Not Permitted	0
I-1, I-3	All	Not Permitted	1[b]

a. For requirements for occupancies in Group I-2, see Sections 407.2 and 407.3.

b. For a reduction in the fire-resistance rating for occupancies in Group I-3, see Section 408.8.

c. Buildings equipped throughout with an automatic sprinkler system in accordance with Section 903.3.1.1 or 903.3.1.2 where allowed.

TABLE 1005.1
EGRESS WIDTH PER OCCUPANT SERVED

OCCUPANCY	WITHOUT SPRINKLER SYSTEM		WITH SPRINKLER SYSTEM[a]	
	Stairways (inches per occupant)	Other egress components (inches per occupant)	Stairways (inches per occupant)	Other egress components (inches per occupant)
Occupancies other than those listed below	0.3	0.2	0.2	0.15
Hazardous: H-1, H-2, H-3 and H-4	0.7	0.4	0.3	0.2
Institutional: I-2	NA	NA	0.3	0.2

For SI: 1 inch = 25.4 mm. NA = Not applicable.

a. Buildings equipped throughout with an automatic sprinkler system in accordance with Section 903.3.1.1 or 903.3.1.2.

TABLE L

TABLE 1016.1
EXIT ACCESS TRAVEL DISTANCE[a]

OCCUPANCY	WITHOUT SPRINKLER SYSTEM (feet)	WITH SPRINKLER SYSTEM (feet)
A, E, F-1, M, R, S-1	200	250[b]
I-1	Not Permitted	250[c]
B	200	300[c]
F-2, S-2, U	300	400[c]
H-1	Not Permitted	75[c]
H-2	Not Permitted	100[c]
H-3	Not Permitted	150[c]
H-4	Not Permitted	175[c]
H-5	Not Permitted	200[c]
I-2, I-3, I-4	Not Permitted	200[c]

For SI: 1 foot = 304.8 mm.

a. See the following sections for modifications to exit access travel distance requirements:

  Section 402.4: For the distance limitation in malls.
  Section 404.9: For the distance limitation through an atrium space.
  Section 407.4: For the distance limitation in Group I-2.
  Sections 408.6.1 and 408.8.1: For the distance limitations in Group I-3.
  Section 411.4: For the distance limitation in special amusement buildings.
  Section 1014.2.2: For the distance limitation in Group I-2 hospital suites.
  Section 1015.4: For the distance limitation in refrigeration machinery rooms.
  Section 1015.5: For the distance limitation in refrigerated rooms and spaces.
  Section 1021.2: For buildings with one exit.
  Section 1028.7: For increased limitation in assembly seating.
  Section 1028.7: For increased limitation for assembly open-air seating.
  Section 3103.4: For temporary structures.
  Section 3104.9: For pedestrian walkways.

b. Buildings equipped throughout with an automatic sprinkler system in accordance with Section 903.3.1.1 or 903.3.1.2. See Section 903 for occupancies where automatic sprinkler systems are permitted in accordance with Section 903.3.1.2.

c. Buildings equipped throughout with an automatic sprinkler system in accordance with Section 903.3.1.1.

TABLE N

TABLE 1021.2
STORIES WITH ONE EXIT

STORY	OCCUPANCY	MAXIMUM OCCUPANTS (OR DWELLING UNITS) PER FLOOR AND TRAVEL DISTANCE
First story or basement	A, B[d], E[e], F[d], M, U, S[d]	49 occupants and 75 feet travel distance
	H-2, H-3	3 occupants and 25 feet travel distance
	H-4, H-5, I, R	10 occupants and 75 feet travel distance
	S[a]	29 occupants and 100 feet travel distance
Second story	B[b], F, M, S[a]	29 occupants and 75 feet travel distance
	R-2	4 dwelling units and 50 feet travel distance
Third story	R-2[c]	4 dwelling units and 50 feet travel distance

For SI: 1 foot = 304.8 mm.

a. For the required number of exits for parking structures, see Section 1021.1.2.

b. For the required number of exits for air traffic control towers, see Section 412.3.

c. Buildings classified as Group R-2 equipped throughout with an automatic sprinkler system in accordance with Section 903.3.1.1 or 903.3.1.2 and provided with emergency escape and rescue openings in accordance with Section 1029.

d. Group B, F and S occupancies in buildings equipped throughout with an automatic sprinkler system in accordance with Section 903.3.1.1 shall have a maximum travel distance of 100 feet.

e. Day care occupancies shall have a maximum occupant load of 10.

NOTES

NOTES

NOTES

EXAMPLE:

PROBLEM: A TENANT IMPROVEMENT IS TO GO INTO A NEW "SHELL" OFFICE BUILDING, ON THE GROUND FLOOR. THE BUILDING IS 12 STORIES HIGH. PLANS SHOW: 40,000 SF PER FLOOR; OCCUPANCY = B (OFFICE); CONSTRUCTION TYPE = II-A; AND BUILDING IS SPRINKLERED.
THE TENANT IMPROVEMENT IS TO BE A 10,000 SF CAFETERIA WITH 25% KITCHEN AND 75% DINING.
WHAT ARE THE BASIC CODE REQUIREMENTS?

SOLUTION: STARTING ON P.169 AND CONTINUING:

a. <u>OCCUPANT LOAD</u>:
PER IBC (SEE TABLE A, P.191):

SPACE	AREA		FACTOR		OCCUPANTS
DINING	7500 SF	÷	15	=	500
KITCHEN	2500 SF	÷	200	=	13
					513

b. <u>DETERMINE OCCUPANCY CLASSIFICATION</u>:
PER IBC (SEE TABLE B, P.193): A-2

NOTE: EVEN THOUGH THE BASIC BUILDING IS CLASSIFIED B, THIS SPECIAL OCC-UPANCY CAN BE CONSIDERED AN A. THE KITCHEN IS CLASSIFIED AS PART OF A, UNLESS NOTED OTHERWISE.

c. <u>DETERMINE CONSTRUCTION TYPE</u>:
PER PLANS, THE BUILDING IS DESIGNED AS II-A.

– CONTINUED –   1/3

d. CHECK FIREPROOFING :

1. RATED CONSTRUCTION :
PER TABLE D, P.196, THE IBC REQUIRES TYPE II-A CONST.
TO BE RATED 1-HOUR FIRE RESISTANCE FOR ALL ELEMENTS.

NOTE: ANY DUCT SHAFTS WILL HAVE TO GO THROUGH A 1-
HOUR FLOOR ABOVE, REQUIRING SPECIAL PROT-
ECTION.

2. FLAME SPREAD:
PER TABLE ON P. 173, CLASSIFICATIONS FOR A-2,
SPRINKLERED ARE :

	CLASS	FLAME SPREAD
ROOMS	C	76 TO 200
EXITWAYS	B	26 TO 75

BECAUSE THE SPACE IS TO BE SPRINKLERED, ANY
CARPET MAY BE CLASS II (SEE P. 174).
WHEN SELECTING FINISHES, THE DESIGNER WILL HAVE
TO ADHERE TO THE ABOVE BY SOMETIMES CHECKING
PRODUCT LITERATURE SPECS. SINCE CLASS C (FOR
WALLS AND CEILINGS) AND CLASS II (FOR CARPET)
IS THE LEAST RESTRICTIVE, THIS SHOULD NOT BE TOO MUCH
OF A PROBLEM.

e. OCCUPANCY SEPARATIONS:
FOR THE IBC CODE, PER TABLE E (SEE P. 198) THE SEP-
ARATION TO HAVE 1-HOUR PROTECTION. THIS MEANS
THAT THE WALLS AROUND AND THE FLOOR ABOVE WILL HAVE
TO HAVE 1 HOUR OF PROTECTION. PER P. 413, A WALL
ASSEMBLY OF METAL STUDS WITH 1 LAYER OF RATED
GYPBOARD ON EACH SIDE WILL GET A 1-HOUR WALL. THIS
WILL HAVE TO GO ALL THE WAY UP TO THE UNDERSIDE OF
THE 2ND FLOOR STRUCTURE. ASSUMING THE FLOOR CON-
STRUCTION, ABOVE IS METAL DECK ON STEEL BEAMS,
PROBABLY ADDED SPRAY-ON FIRE PROOFING WILL GET IT
FROM 1 HOUR UP TO 2 (SEE P. 172), THE DUCT SHAFT

– CONTINUED –

2/3

NOTED IN d-1, ABOVE, WILL NOW HAVE TO GO THROUGH A 2-HOUR FLOOR. ANY DUCTS GOING THROUGH THE PERIMETER (BUT NOT EXTERIOR) WILL HAVE TO HAVE FIRE DAMPERS FOR A 2-HOUR WALL.

NOTE: SEE NOTE d AT TABLE E, REGARDING KITCHENS.

f. SPRINKLER REQUIREMENTS:

THE BUILDING IS ALREADY SPRINKLERED. IF IT WERE NOT, DUE TO THE CAFETERIA, IT WOULD HAVE TO BE PER "SPRINKLER REQUIREMENTS, ASSEMBLY OCCUPANCIES" ON P 175.

g. FIRE AREAS, WALLS, BARRIERS, AND PARTITIONS:

PER TABLE D (P.196), NON-BEARING INTERIOR PARTITIONS REFER TO A CODE SECTION, NOT COVERED IN THIS BOOK. A FURTHER CHECK OF THE CODE WOULD REVEAL THAT THESE MUST BE 1-HOUR RATED. PER P. 413, METAL STUDS W/ ON LAYER OF RATED GYPBOARD ON EACH SIDE WOULD DO.

h. EXITING

PER a., THE OCCUPANT LOAD IS 513. PER ITEM 1 ON P. 181, 3 EXITS ARE REQUIRED. TYPICALLY ONE IS THE REAR SERVICE DOOR OF THE KITCHEN. ASSUMING A LARGE OPEN DINING ROOM, THE OTHER TWO EXITS WOULD BE THE MAIN ENTRY PLUS AN ADDED EXIT DOOR (WHICH WOULD NEED TO BE LOCATED 1/3RD THE DIAGONAL DIMENSION FROM THE ENTRY).

THE SIZE OF EXITS WOULD BE PER ITEM 6 ON P.182:

IBC (TABLE J, P. 204): 513 OCC. × 0.15 = 77"
77" ÷ 3 EXITS = 25½"/EXIT

NFPA (TABLE H, P. 203): 525 OCC. × 0.2 = 105"
105" ÷ 3 EXITS = 35"/EXIT

SINCE A MIN. EXIT WIDTH IS 36" (SEE P.179), NEED 3 EXIT DOORS AT LEAST 3' WIDE, EACH.

IF ANY EXITS ARE NOT DIRECTLY TO THE OUTSIDE, THEY WILL HAVE TO BE TO A RATED CORRIDOR.

WHEN A PRELIM. DESIGN & CODE STUDY DONE, THE DESIGNER SHOULD REVIEW WITH CODE OFFICIAL.

- END -

3/3

NOTES

NOTES

4. Accessibility (ADA requirements) ⑳

___ *a. General:* This section concerns accessibility for the disabled as required by *ADA,* the Americans with Disabilities Act (Title 3, the national civil rights law), in nongovernment buildings (Title 2 applies to government buildings), as amended by ICC/ANSI A 117.1—1998. Local or state laws may (in part) be more restrictive regarding alterations and new buildings. For each item under consideration, *the more restrictive law applies.*

___ *b. ADA* applies to:

   ___ (1) *Places of public accommodation* (excluding private homes and clubs, as well as churches). Often, buildings will have space both for the general public and for employees only.

   ___ (2) *Commercial facilities* (employees only) requirements are less restrictive, requiring only an accessible entry, exit, and route through each type of facility function. Only when a disabled employee is hired (under Title 1) do more restrictive standards apply.

___ *c. Existing buildings* are to comply by removing "architectural barriers," as much as possible, when this is "readily achievable" (not requiring undue expense, hardship, or loss of space). This effort, in theory, is to be ongoing until all barriers are removed. When barriers can't be readily removed, "equivalent facilitation" is allowed. Priorities of removal are:

   ___ (1) Entry to places of public accommodation

   ___ (2) Access to areas where goods and services are made available to the public

   ___ (3) Access to restroom facilities

   ___ (4) Removal of all other barriers

___ *d. Alterations* to existing buildings require a higher standard. To the maximum extent possible, the altered portions are to be made accessible. If the altered area is a "primary function" of the building, then an accessible "path of travel" must be provided from the entry to the area (including public restrooms, telephones, and drinking fountains), with exemption only possible when cost of the path exceeds 20% of the cost to alter the primary function.

___ *e. New buildings or facilities* must totally comply, with only exceptions being situations of "structural impracticability."

___ *f.* See *Index,* p. 525, for a complete list of *ADA* requirements.

ACCESSIBLE ROUTE PER A.D.A.
(INTERIOR AND EXTERIOR)

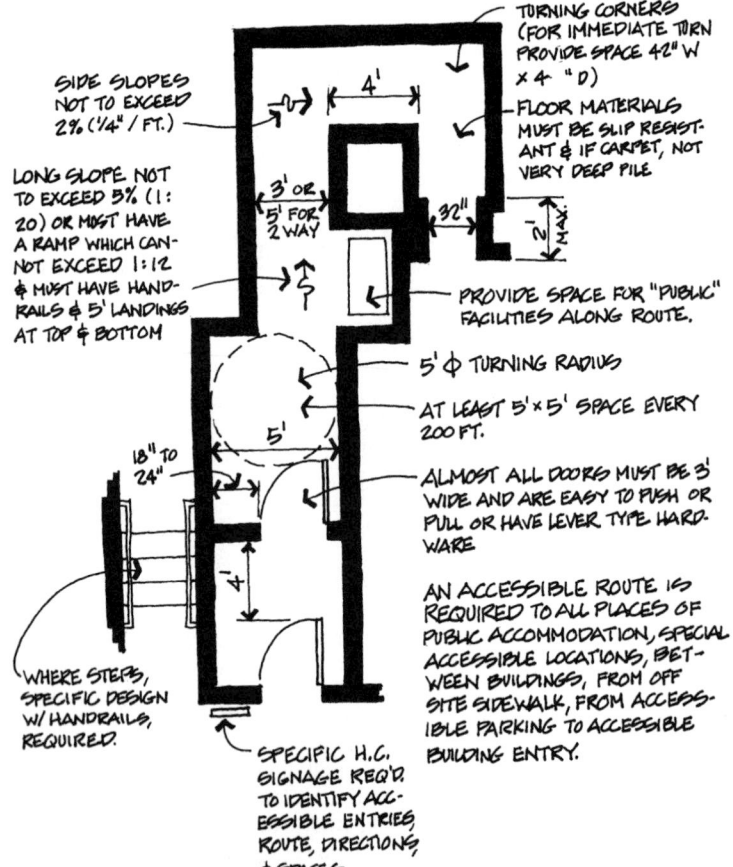

SIDE SLOPES NOT TO EXCEED 2% (¼" / FT.)

LONG SLOPE NOT TO EXCEED 5% (1: 20) OR MUST HAVE A RAMP WHICH CAN- NOT EXCEED 1:12 & MUST HAVE HAND- RAILS & 5' LANDINGS AT TOP & BOTTOM

4'

3' OR 5' FOR 2 WAY

18" TO 24"

5'

4'

TURNING CORNERS (FOR IMMEDIATE TURN PROVIDE SPACE 42" W x 4 " D)

FLOOR MATERIALS MUST BE SLIP RESIST- ANT & IF CARPET, NOT VERY DEEP PILE

32"

2' MAX.

PROVIDE SPACE FOR "PUBLIC" FACILITIES ALONG ROUTE.

5' ⌀ TURNING RADIUS

AT LEAST 5' × 5' SPACE EVERY 200 FT.

ALMOST ALL DOORS MUST BE 3' WIDE AND ARE EASY TO PUSH OR PULL OR HAVE LEVER TYPE HARD- WARE

AN ACCESSIBLE ROUTE IS REQUIRED TO ALL PLACES OF PUBLIC ACCOMMODATION, SPECIAL ACCESSIBLE LOCATIONS, BET- WEEN BUILDINGS, FROM OFF SITE SIDEWALK, FROM ACCESS- IBLE PARKING TO ACCESSIBLE BUILDING ENTRY.

WHERE STEPS, SPECIFIC DESIGN W/ HANDRAILS, REQUIRED.

SPECIFIC H.C. SIGNAGE REQ'D. TO IDENTIFY ACC- ESSIBLE ENTRIES, ROUTE, DIRECTIONS, & SPACES

NOTES

NOTES

The interior designer usually is working with an existing building which has already set energy requirements (especially for external climatic loads). However, tenant improvements or remodels do affect internal loads. Use this section as a guide to energy conservation.

___ 1. **Building Type**

All buildings produce internal heat. All buildings are affected by external loads (heating or cooling) based on the climate, and internal loads (heat from equipment, lights, people, etc.). Large commercial buildings tend to be internally dominated. Residences or small commercial buildings tend to be the exact opposite.

___ 2. **Human Comfort**

The comfort zone may be roughly defined as follows: Most people in the temperate zone, sitting indoors in the shade in light clothing, will feel tolerably comfortable at temperatures ranging from *70° to 80°F* as long as the relative humidity lies between *20 and 50%*. As humidity increases, they will begin to become uncomfortable at lower and lower temperatures until the relative humidity reaches *75 to 80%*, when discomfort at any temperature sets in. But if they are sitting in a draft, the range of tolerable temperature shifts upward, so that temperatures of *85°F* may be quite comfortable in the *20 to 50%* relative humidity range, if local air is moving at *200 ft/min*. Indoor air moving more slowly than *50 ft/min* is generally unnoticed, while flows of *50 to 100 ft/min* are pleasant and hardly noticed. Breezes from *100 to 200 ft/min* are pleasant, but one is constantly aware of them, while those from *200 to 300 ft/min* are at first slightly unpleasant, then annoying and drafty.

___ 3. **Checklist for Passive Building Design** (Strategies for Energy Conservation):

Note: Many of the following items conflict, so it is impossible to choose all.

   _X_   Cold climate or winter

     _X_   Hot climate or summer

Indoor/outdoor rooms: Courtyards, covered patios, seasonal screened and glassed-in porches, greenhouses, atriums, and sunrooms can be located in the building plan for summer cooling and winter heating benefits as with these techniques.

<u>Cold</u>   <u>Hot</u>

___   ___   *a.* Provide outdoor semiprotected areas for year-round climate moderation.

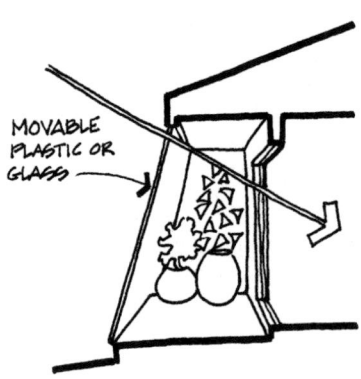

MOVABLE PLASTIC OR GLASS

___   *b.* Provide solar-oriented interior zone for maximizing heat.

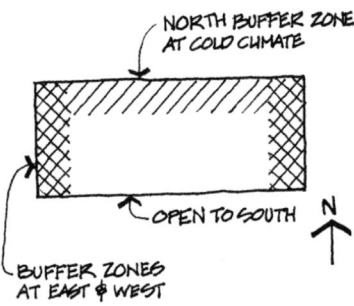

NORTH BUFFER ZONE AT COLD CLIMATE

OPEN TO SOUTH   N

BUFFER ZONES AT EAST & WEST

___   ___   *c.* Plan specific rooms or functions to coincide with solar orientation (i.e., storage on "bad" orientation such as west, living on "good" orientation such as south).

<u>Solar walls and windows:</u> Using the winter sun for heating a building through solar-oriented windows and walls is covered by these techniques.

___   *d.* Use high-capacitance materials to store solar heat gain. Best results are by distributing "mass" locations throughout interior. On average, provide 1 to 1¼ CF of concrete

Cold Hot

or masonry per each SF of south-facing glass. For an equivalent effect, 4 times more mass is needed when not exposed to sun. Do not place carpeting on these floor areas.

___   *e.* This same "mass effect" can be used in reverse in hot, dry (clear sky) climates. "Flush" building during cool night to pre-cool for next day. Be sure to shade the mass. A maximum area of up to 2"-thick clay, concrete, or plaster finishes works best. The night average wind speed is generally about 75% of the average 24-hour wind speed reported by weather bureaus. About 30 air changes per hour is an adequate rate to cool the building.

___   *f.* Maximize south-facing glazing (with overhangs as needed). On average, south-facing glass should be 10 to 25% of floor area. For north latitude/cold climates this can go up to 50%. For south latitude/hot climates this strategy may not be appropriate.

Thermal envelope: Many climatic design techniques to save energy are based on insulating the interior space from the exterior environment.

___   ___   *g.* Provide air shafts for natural or mechanically assisted house-heat recovery. This can be recirculated warm air at high ceilings or recovered heat from chimneys.

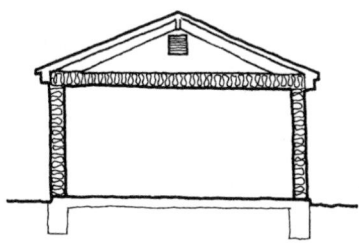

___   *h.* Centralize heat sources within building interior (fireplaces, furnaces, hot water heater, cooking, laundry, etc.). Lower-level positioning for these is most desirable.

<u>Cold</u> <u>Hot</u>

____   *i.* Put heat sources (HW, laundry, etc.) outside
              building.

____ ____   *j.* Use vestibule or exterior "wind shield" at
              entryways. Orient away from undesirable
              winds.

____ ____   *k.* Locate low-use spaces, storage, utility, and
              garage areas to provide buffers. Locate at
              "bad" orientations (i.e., on north side in
              cold climate or west side in hot climate).

____ ____   *l.* Subdivide interior to create separate heat-
              ing and cooling zones. One example is sep-
              arate living and sleeping zones.

____ ____   *m.* Minimize window and door openings (usu-
              ally N, E, and W).

____ ____   *n.* Provide ventilation openings for air flow to
              and from specific spaces and appliances.
              See p. 361 for fireplaces.

<u>Natural ventilation:</u> A simple concept by which to cool a building.
<u>Cold</u> <u>Hot</u>

___   *o.* Use "open plan" interior to promote natural ventilation air flow.

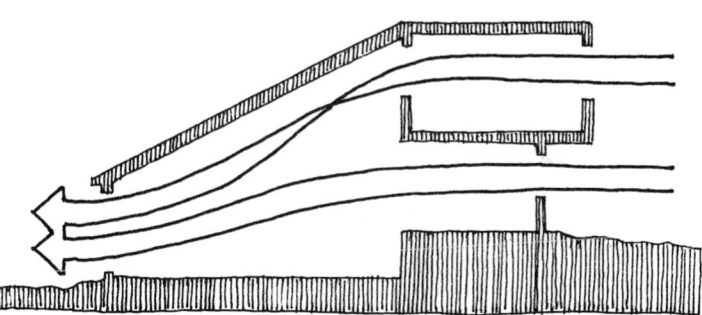

___   *p.* Provide vertical air shafts to promote interior air flow.

___   *q.* Use double roof and wall construction for ventilation within the building shell.

___   *r.* Orient door and window openings to facilitate natural ventilation from prevailing breezes. For best results:

> Windows on opposite sides of rooms.
> Inlets and outlets of equal size giving maximum air change.
> A smaller inlet increases air speed.

___ **4. Checklist for Active Building Design** (Strategies for Energy Conservation)

___ *a.* Whenever possible, use fans in lieu of compressors, as they use about 80% less energy. In residential construction this may take the form of "whole-house" fans. See *o*, below.

___ *b.* Design for natural lighting in lieu of artificial lighting. In hot climates or summers, avoid direct sun.

___ *c.* Lighting consumes about 8% of the energy used in residences and 27% in commercial buildings nationwide. Use high-efficiency lighting (50–100 lumens per watt). See Part 16A. Provide switches or controls to light only areas needed and to take advantage of daylighting.

___ *d.* Use gas rather than electric when possible, as this can be up to 75% less expensive.

___ *e.* Use efficient equipment and appliances
   ___ (1) Microwave rather than convection ovens.
   ___ (2) Refrigerators rated 5–10 kBtu/day or 535–1070 kwh/yr.

___ *f.* If fireplaces, use high-efficiency type with tight-fitting high-temperature glass, insulated, and radiant-inducing boxed with outside combustion air. See p. 361.

___ *g.* Use night setback and load-control devices.

___ *h.* Use multizone HVAC.

___ *i.* Locate ducts in conditioned space or tightly seal and insulate.

___ *j.* Insulate hot and cold water pipes (R = 1 to 3).

___ *k.* Locate air handlers in conditioned space.

___ *l.* Install thermostats away from direct sun and supply grilles.

___ *m.* Use heating equipment with efficiencies of 70% for gas and 175% for electric, or higher.

___ *n.* Use cooling equipment with efficiencies of SEER = 10, COP = 2.5 or higher.

___ *o.* Use "economizers" on commercial HVAC to take advantage of good outside temperatures.

___ *p.* In dry, hot climates, use evaporative cooling.

___ *q.* Use gas or solar in lieu of electric hot water heating. Insulate hot water heaters.

___ *r.* For some building types and at some locations, utilities have peak load rates, such as on summer afternoons. These peak rates should be identified and designed for. Therefore, designing for peak loads may be more important than yearly energy savings. In some cases saving energy and saving energy cost may not be the same. Large buildings sometimes use thermal storage systems, such as producing ice in off-peak hours to use in the heat of the day. These systems work best in areas with large daily and yearly temperature swings, high electricity costs, and big cost differences between on- and off-peak rates.

___ *s.* HVAC system as a % of total energy use:

Residential

Cold climate	70%
Hot climate	40%

Office

Cold climate	40%
Hot climate	34%

___ *t.* TBSs (total building management systems) conserve considerable energy.

NOTES

__ K. SUSTAINABLE DESIGN

$\textcircled{4}$ $\boxed{10A}$

Green design is a whole-systems approach, utilizing design and building techniques to minimize environmental impact and reduce a building's energy consumption while contributing to the health of its occupants. "If brute force doesn't work, you're not using enough of it" has been the typical American approach. The green movement takes the opposite approach.

___ 1. *General*

    ___ *a.* *Save energy:* Ongoing energy use is the single greatest source of environmental impact from a building. See "Energy," p. 219.

    ___ *b.* *Recycle buildings:* Reuse existing buildings and infrastructure instead of developing open space.

        Note: Unfortunately, it is often less expensive to *remove* existing old buildings and build anew than to preserve but refurbish, in order to bring the old buildings up to the latest codes.

    ___ *c.* *Create community:* Design communities to reduce dependence on the automobile and encourage alternative transportation, including walking.

    ___ *d.* *Reduce material use:* Optimize design to make use of smaller spaces and utilize materials efficiently.

    ___ *e.* *Protect and enhance the site:* Try to work with existing grades and landscaping rather than bulldozing the site.

    ___ *f.* *Select low-impact materials:* Buildings account for about 40% of the material resources flowing through the economy. Specify low-environmental-impact resources and efficient materials. Most environmental impacts associated with building materials occur before installation. Raw materials have been extracted from the ground or harvested from forests. Pollutants have been emitted during manufacture, and energy has been invested during production. Avoid materials that generate pollution, are not salvaged from other uses, deplete limited natural resources, or are made from toxic or hazardous constituents. Specify materials with low embodied energy used in extracting, manufacturing, and shipping.

MATERIALS LIFE CYCLE

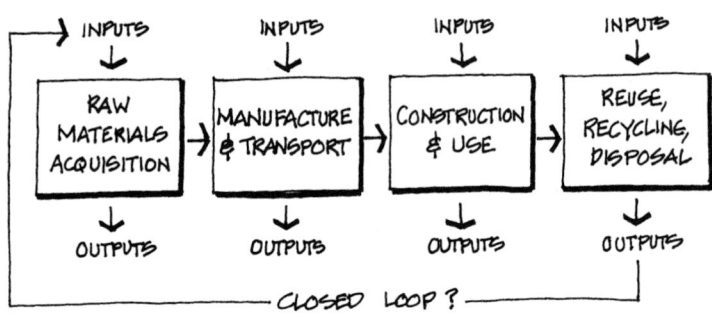

___ g. *Maximize longevity:* Design for durability and adaptability. The longer a building lasts, the longer the period over which to amortize its environmental impacts. Specify durable, nondecaying, easily maintained, or replaceable materials. Design for adaptability (new building uses).

___ h. *Save water:* Design buildings and landscape that use water efficiently.

___ i. *Make buildings healthy:* Provide a safe, comfortable indoor environment. Over 60,000 commercial chemicals are in use that were unknown 40 years ago. Many common building materials and products produce harmful off-gases.

___ j. *Minimize construction and demolition waste:* Return, reuse, and recycle job site waste.

___ 2. **Finishes**

___ • *Gypsum board:* Utilize recycled materials, synthetic gypsum. Collect scrap for recycling use as soil amendment.

___ • *Tile work:* Select recycled content material or recycle as mosaic or aggregate.

___ • *Acoustic ceiling panels:* Utilize high-recycled-content materials. Utilize perlite or mineralized wood fiber to avoid fiber shedding.

___ • *Wood flooring:* Select FSC-certified products. Select salvaged flooring. Select engineered flooring with thick-faced veneers that are able to be refinished. Select nontoxic, low-VOC finishes and additives. Consider bamboo in lieu of wood.

___ • *Resilient flooring:* Select materials with low-VOC

adhesives. Select recycled-content materials. Vinyl and rubber are the best green materials.

___ • *Carpet:* Select materials with low VOC content. Avoid synthetics. Select 100% recycled-content face fiber and backing. Avoid synthetic latex backing.

___ • *Painting:* Select low- or no-VOC products. Select the least toxic alternatives. Select recycled content.

___ 3. *Equipment*

___ • *Residential appliances:* Select water-conserving appliances with "green seal." Select horizontal-axis clothes washers. Select energy-star-rated appliances.

___ 4. *Furnishings*

___ • Select materials with organically grown plant fibers.

___ • Select natural dyes or undyed fabrics.

___ • Use products with FSC-certified wood.

___ • Use refurbished furniture.

___ • Select fibers that are naturally pest resistant.

___ • Select recycled synthetic fibers.

___ 5. *Mechanical*

___ • Select low-flow water-efficient plumbing fixtures (1.6 GPF, or below, toilets).

___ • Select compost toilet systems.

___ • Choose cooling equipment that does not utilize ozone-depleting refrigerants.

___ • Design graywater systems.

___ • Avoid PVC piping (recycle problems).

___ • Use domestic water-heat exchangers.

NOTES

___ L. ACOUSTICS ② ③ ④ ⑩ ⑪ₐ

Sound is a series of pressure vibrations that move through an elastic medium. Its alternating compressions and rarefactions may be far apart (low-pitched), close together (high-pitched), wide (loud), or narrow (soft).

All perceived sound has a *source, path,* and *receiver.* Each source has a size, direction, and duration. Paths can be air-borne or structure-borne.

Sound has four quantifiable properties: *velocity, frequency, intensity,* and *diffuseness.* Regarding *velocity*, sound travels much faster through solids than air (and faster through warm air than cool air). *Frequency* is sound's vibrations per second, or hertz (Hz). This varies according to its purity and pitch. The average human pitch for hearing is about 1000 Hz. *Intensity* is the power level (or loudness) measured in decibels (dB). Attenuation is the loss of a sound's intensity as it travels outward from a source. *Diffuse* noise (blanket or background noise level) is sound emanating from a multiple of similar sound sources.

There is both a positive function and a negative function to consider in acoustic design.

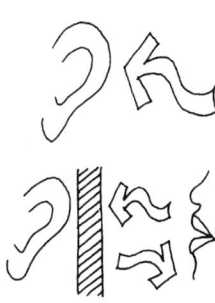

The positive function is to ensure that the reverberation characteristics of a building are appropriate to their function. See 1 below.

On the negative side, the task is to make certain that unwanted outside noises are kept out of quiet areas of the building. See 2 below.

___ 1. *Room Acoustics*

Sound can be likened to light. *Sound control* uses reflection and diffusion to enhance acoustics in such spaces as auditoriums and sound studios, and absorption for noise control in more typical spaces such as offices.

___ *a.* *Reflection:* The geometry of the room is important in effective sound control. Large concave surfaces concentrate sound and should usually be avoided, while convex surfaces disperse sound.

___ *b.* *Diffusion* promotes uniform distribution of continuous sound and improves "liveness" (very important in performing arts). It is increased by objects and surface irregularities. Ideal diffusing surfaces neither absorb nor reflect sound but scatter it.

    c. Absorption (see table on p. 241) provides the most effective form of noise control. Sound pressure waves travel at the speed of sound (1100 fps), which is a slow enough velocity that reflections of the original sound-wave form can interfere with perception of the original, intended signal. *Reverberation time* is the measure of this problem.

    Sound of any kind emitted in a room will be absorbed or reflected off the room surfaces. Soft materials absorb sound energy. Hard materials reflect sound energy back into the space. The reflected sound can reinforce the direct sound and enhance communication if the room size and room surfaces are configured appropriately. Annoying reverberation (echoes) occur in rooms more than *30 feet* long. Echoes are stronger when the reflection surface is highly reflective and is concave toward the listener.

    The room volume and surface characteristics will determine the reverberation time for the room. <u>Reverberation time</u> is the time in seconds that it takes for a sound to decay through 60 decibels. It is calculated as follows:

$$RT = \frac{0.05 \times \text{Room Volume (cf)}}{\text{Average Absorption of Room}}$$

Desirable room reverberation times are:

Office and commercial spaces	0.5 seconds
Rooms for speech	1.0 seconds
Rooms for music	1.5 seconds
Sports arenas	2.0 seconds

    The <u>*absorption,*</u> also called <u>noise reduction coefficient (NRC)</u>, of a surface is the product of the acoustic coefficient for the surface multiplied by the area of the surface. The sound absorption of a room is the sum of the sound absorptions of all the surfaces in the room. <u>The higher the coefficient, the more sound absorbed, with *1.0* (complete absorption) being the highest possible.</u> Generally, a material with a coefficient below *0.2* is considered to be reflective and above *0.2* to be absorbing. Some common acoustic coefficients are:

1½" glass fiber clg. panels	1.0
Carpet and pad	0.6
Acoustic tile (no paint)	0.8
Cloth-upholstered seats	0.6
An audience	0.8
Concrete	0.02
Gypsum board	0.05
Glass	0.09
Tile	0.01
Fabric	0.30

The average absorption coefficient of a room should be at least *0.2*. Average absorption above *0.5* is usually not desirable, nor is it economically justified. A lower value is suitable for large rooms; and larger values for controlling sound in small or noisy rooms. Although absorptive materials can be placed anywhere, ceiling treatment is more effective in large rooms, while wall treatment is more effective in small rooms. If additional absorptive material is being added to a room to improve it, the total absorption should be increased at least *3 times* to bring absorption to between *0.2 and 0.5*. An increase of *10 times* is about the practical limit. Each doubling of the absorption in a room reduces RT by ½.

EXAMPLE:

WHAT IS THE ABSORPTION COEFFICIENT AND REVERBERATION TIME FOR A 20' × 10' × 9' H OFFICE WITH CARPET FLOOR, ACOUSTIC TILE CEILING AND GYPB'D. WALL (BUT ⅓ OF WHICH HAS SOUND ABSORPTION MATERIAL)?

ABSORPTION COEFFICIENT:

FLOOR	0.6 × 200 =	120
⅔ WALL	0.05 × 356 =	18
⅓ WALL	0.8 × 178 =	142
CEILING	0.8 × 200 =	160
		440

$$\text{AVER. COEF. OF ABSORPTION} = \frac{\text{TOTAL ABSORB.}}{\text{TOTAL RM. SURFACE}} = \frac{440}{20 \times 9 \times 2 + 10 \times 9 \times 2 + 10 \times 20 \times 2}$$

$$= \frac{440}{940 \text{ SF}} = 0.47 \quad \text{O.K.}$$

$$\text{REVERBERATION TIME} = \frac{0.05 \times (10 \times 20 \times 9)}{440} = 0.2 < 0.5 \quad \text{OK.}$$

___ *d.* Other factors affecting acoustics:

___ If a corridor is appreciably higher than it is wide, some absorptive material should be placed on the walls as well as the ceiling, especially if the floor is hard. If the corridor is wider than its height, ceiling treatment is usually enough.

___ Acoustically critical rooms require an appropriate volume of space. Rooms for speech require 120 CF per audience seat. Rooms for music require 270 CF per audience seat.

___ "Ray diagramming" can be a useful tool in sound control. As with light, the reflective angle of a sound wave equals its incident angle. In like manner, concave shapes focus sound and convex shapes disperse sound.

___ Amplification: If listeners are within 110′ of the sound source, a cluster of loudspeakers is usually located well above and slightly in front of the sound source. If some listeners are more than 110′ away, an overhead network of small speakers, located no more than 35′ apart, should be used.

___ **2. *Sound Isolation***

Sound travels through walls and floors by causing building materials to vibrate and then broadcast the noise into the quiet space. There are two methods of setting up the vibration: through structure-borne sound, and air-borne sound.

Structure-borne sound is the vibration of building materials by vibrating pieces of equipment, or caused by walking on hard floors.

Air-borne sound is a pressure vibration in the air. When it hits a wall, the wall materials are forced to vibrate. The vibration passes through the materials of the wall. The far side of the wall then passes the vibration back into the air.

Noise Reduction and Sound Isolation Guidelines

___ *a.* Choose a quiet, protected site. Orient building with doors and windows away from noise.

___ *b.* Use site barriers such as walls or landscape (dense tree lines or hedges, a combination of deciduous and evergreen shrubs, reduce sound more efficiently).

___ *c.* Avoid placing noisy areas near quiet areas. Areas with similar noise characteristics should be placed next to each other. Place bedrooms next to bedrooms and living rooms next to living rooms.

___ *d.* As the distance from the sound source increases, pressure at the listener's ear will decrease by the inverse square law (as with light). Therefore, separate sound sources by distance.

___ *e.* Orient spaces to minimize transmission problems. Space windows of adjoining apartments max. distance apart. Place noisy areas back to back. Place closets between noisy and quiet areas.

___ *f.* Massive materials (concrete or masonry) are the best noise-isolation materials.

___ *g.* Choose quiet mechanical equipment. Use vibration isolation, sound-absorbing duct lining, resilient pipe connections. Design for low flow velocities in pipes and ducts.

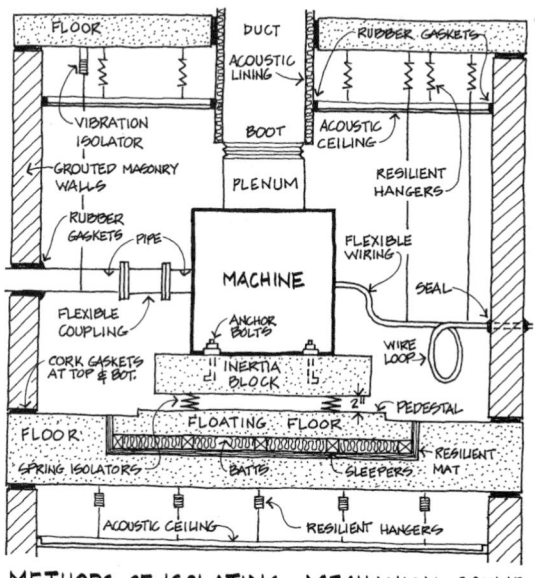

METHODS OF ISOLATING MECHANICAL SOUND

_____ *h.* Reducing structure-borne sound from walking on floors is achieved by carpet (with padding, improves greatly).

_____ *i.* Avoid flanking of sound over ceilings.

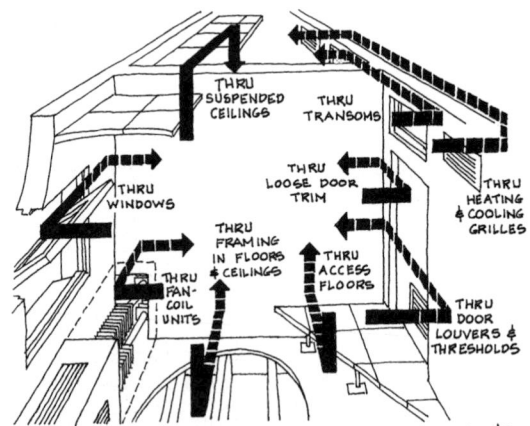

THRU SUSPENDED CEILINGS

THRU TRANSOMS

THRU LOOSE DOOR TRIM

THRU WINDOWS

THRU HEATING & COOLING GRILLES

THRU FRAMING IN FLOORS & CEILINGS

THRU ACCESS FLOORS

THRU FAN-COIL UNITS

THRU DOOR LOUVERS & THRESHOLDS

(11A)   **FLANKING SOUND PATHS BETW'N. RM'S.**

_____ *j.* Avoid flanking of sound at wall and floor intersections.

_____ *k.* Wall and floor penetrations (such as elect. boxes) can be a source of sound leakage. A 1-square-inch wall opening in a 100-SF gypsum-board partition can transmit as much sound as the entire partition.

___ *l.* Many sound leaks can be plugged in the same manner as is done for air leaks, by caulking.

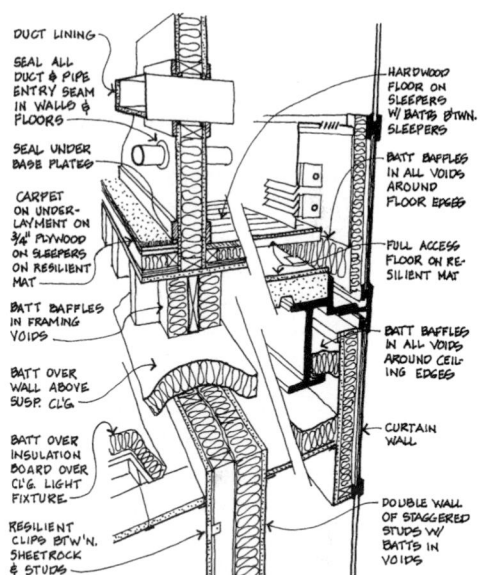

REMEDIES FOR REDUCING NOISE TRANSMISSION

(11A)

___ *m.* Walls and floors are classified by Sound Transmission Class (*STC*), which is a measure of the reduction of loudness provided by various barriers. The higher the number, the better. In determining the required STC rating of a barrier, the following rough guidelines may be used:

___ *n.* The best remedy for reducing impact noise is to cushion the noise at its source.

STC Effect on Hearing

STC	Effect on Hearing
25	Normal speech clearly heard through barrier.
30	Loud speech can be heard and understood fairly well. Normal speech can be heard but barely understood.
35	Loud speech is unintelligible but can be heard.
42–45	Loud speech can be heard only faintly. Normal speech cannot be heard.

46–50   Loud speech not audible. Loud sounds other than speech can be heard only faintly, if at all.

See p. 165 for recommended STC room barriers.

Rough Estimating of STC Ratings

When the wall or floor assembly is less than that desired, the following modifications can be made. Select the appropriate wall or floor assembly. To improve the rating, select modifications (largest number, + ½ next largest, + ½ next largest, etc):

a. Light frame walls

Base design	*STC Rating*
Wood studs W/ ½″ gyp'bd.	32
Metal studs W/ ⅝″ gyp'bd.	39

Modification	*Added STC*
Staggered Studs	+9
Double surface skin	+3 to +5
Absorption insulation	+5

b. Heavy walls
The greater the density, the higher the rating. Density goes up in the following order: CMU, brick, concrete.

Base Design	*STC Rating*
4-inch CMU, brick, concrete	37–41, 42
6-inch	42, 46
8-inch	47, 49, 51
12-inch	52, 54, 56

Modification	*Added STC*
Furred-out surface	+7 to +10
Add plaster, ½″	+2 to +4
Sand-filled cores	+3

c. Wood floors

Base Design	*STC Rating*
½-in plyw'd, subfloor with oak floor, no ceiling	25

Modification	*Added STC*
Add carpet	+10
⅝-inch gyp'bd. ceiling	+10
Add resilient damping board	+7
Add absorbtion insul.	+3

d. Concrete floors

Base Design	*STC Rating*
4-, 6-, 8-inch thick concrete	41, 46, 51
Modification	*Added STC*
Resil. Susp. Ceiling	+12
Add sleepers	+7
Add absorption insul.	+3

e. Glass

¼″ float	26
double glaze	32

f. Doors

wood HC	26
SC	29
metal	30
special acoustical	35 to 38

Costs: **Sound attenuation blankets: 2″ thick = $1.00/SF (10% L and 90% M). Add $0.20 per added inch up to 4″.**

EXAMPLE:

ROUGHLY ESTIMATE HOW TO GET S.T.C. = 45 FOR AN OFFICE WALL PARTITION MADE OF WOOD STUDS AND GYPB'D.

FROM ABOVE, A WOOD STUD PARTITION W/ 1/2" GYPB'D. IS ...... S.T.C.= 32

ADD STAGGERED STUDS FOR FULL CREDIT	+9
ADD DOUBLE GYPPB'D. FOR 1/2 CREDIT, BOTH SIDES: 1/2 × 5 =	+2.5
ADD ABSORPTION BATTS BETWEEN STUDS: 1/2 × 5 =	+2.5
TOTAL =	46.0

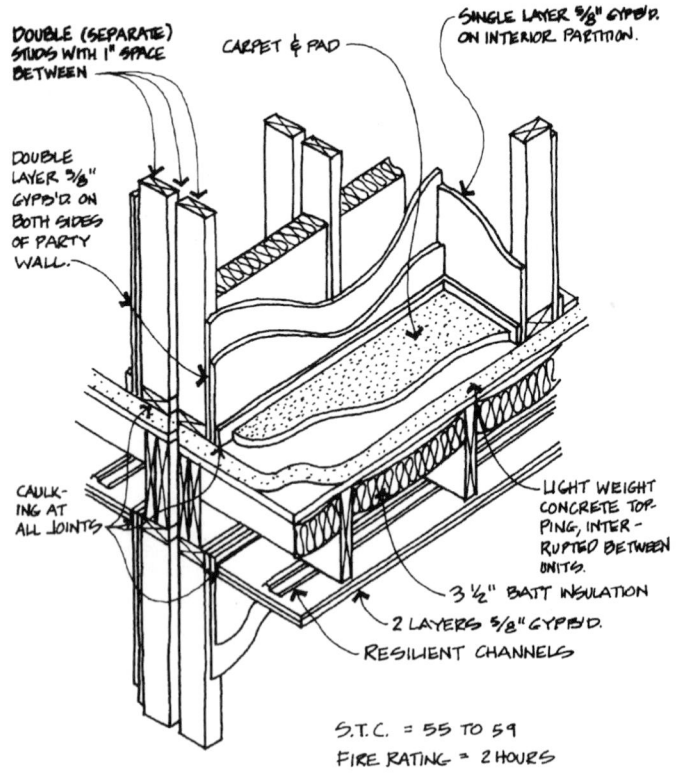

DOUBLE (SEPARATE) STUDS WITH 1" SPACE BETWEEN

CARPET & PAD

SINGLE LAYER 5/8" GYPB'D. ON INTERIOR PARTITION.

DOUBLE LAYER 3/8" GYPB'D ON BOTH SIDES OF PARTY WALL.

CAULKING AT ALL JOINTS

LIGHT WEIGHT CONCRETE TOPPING, INTERRUPTED BETWEEN UNITS.

3 1/2" BATT INSULATION

2 LAYERS 5/8" GYPB'D.

RESILIENT CHANNELS

S.T.C. = 55 TO 59
FIRE RATING = 2 HOURS

PARTY WALL DETAIL

(5)

USE OF ABSORPTION IN COMMON OCCUPANCIES

ROOM OCCUPANCY	CEILING TREATMENT	WALL TREATMENT	SPECIAL
AUDITORIUMS, CHURCHES, THEATERS, CONCERT HALLS, RADIO, RECORDING & T.V. STUDIOS, SPEECH & MUSIC RMS.			●
BOARDROOMS, TELECONFERENCING	●	●	
CLASSROOMS	●	○	
COMMERCIAL KITCHENS	●		
COMPUTER AND BUSINESS MACHINE ROOMS	●		
CORRIDORS AND LOBBIES	○		
GYMNASIUMS, ARENAS, & RECREATIONAL SPACES	●	●	
HEALTH CARE PATIENT ROOMS	●		
LABORATORIES	●		
LIBRARIES	●		
MECHANICAL EQUIPMENT ROOMS			●
MEETING AND CONFERENCE ROOMS	●	○	
OPEN OFFICE PLAN	●	●	
PRIVATE OFFICES	●		
RESTAURANTS	●	○	
SCHOOLS & INDUSTRIAL SHOPS, FACTORIES	●	●	
STORES AND COMMERCIAL SHOPS	●		

● STRONGLY RECOMMENDED
○ ADVISABLE

SOUND ISOLATION CRITERIA ④

SOURCE ROOM OCCUPANCY	RECEIVER ROOM ADJACENT	SOUND ISOLATION REQUIREMENT (MINIMUM) FOR ALL PATHS BE-TWEEN SOURCE AND RECEIVER
EXECUTIVE AREAS, DOCTOR'S SUITES, PERSONNEL OFFICERS, LARGE CONFERENCE ROOMS, CONFIDENTIAL PRIVACY REQUIREMENTS	ADJACENT OFFICES AND RELATED SPACES	STC 50-55
NORMAL OFFICES, REGULAR, CONFERENCE ROOMS FOR GROUP MEETINGS, NORMAL PRIVACY REQM'TS	ADJACENT OFFICES & SIMILAR ACTIVITIES	STC 45-50
LARGE GENERAL BUSINESS OFFICES, DRAFTING AREAS, BANKING FLOORS	CORRIDORS, LOBBIES, DATA PROCESSING, SIMILAR ACTIVITIES	STC 40-45
SHOP AND LABORATORY OFFICE IN MANUFACTURING LABORATORY OR TEST AREAS, NORM. PRIVACY	ADJACENT OFFICES TEST AREAS, CORRID	STC 40-45
MECHANICAL EQUIPMENT ROOMS	ANY SPACE	STC 50-60 +
MULTIFAMILY DWELLINGS	NEIGHBORS (SEPARATE OCCUPANCY)	
(a.) BEDROOMS	BEDROOMS	STC 48-55
	BATHROOMS	STC 52-58
	KITCHENS	STC 52-58
	LIVING ROOMS	STC 50-57
	CORRIDORS	STC 52-58
(b.) LIVING ROOMS	LIVING ROOMS	STC 48-55
	BATHROOMS	STC 50-57
	KITCHENS	STC 48-50
SCHOOL BUILDINGS (a.) CLASSROOMS	ADJ. CLASSROOMS	STC 50
	LABORATORIES	STC 50
	CORRIDORS	STC 45
(b.) LARGE MUSIC OR DRAMA AREA	ADJ. MUSIC OR DRAMA AREA	STC 60
(c.) MUSIC PRACTICE ROOMS	MUSIC PRACTICE RMS	STC 55
INTERIOR OCCUPIED SPACES	EXTERIOR OF BLDG.	STC 35-60
THEATERS, CONCERT HALLS, LECTURE HALLS, RADIO, AND T.V. STUDIOS	ANY AND ALL ADJACENT	USE QUALIFIED ACOUSTICAL CONSULTANT.

NOTES

NOTES

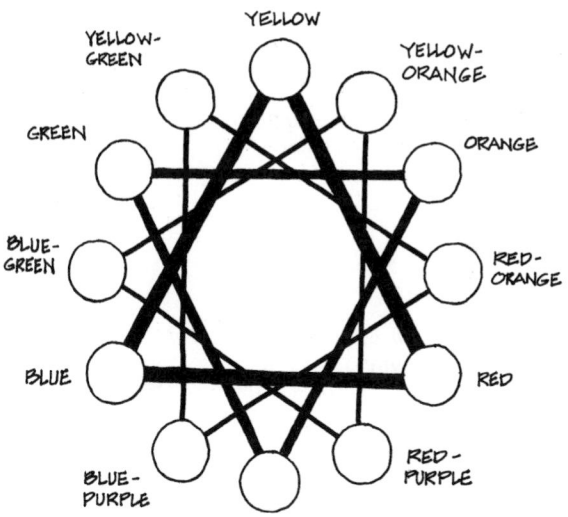

YELLOW

YELLOW-GREEN

YELLOW-ORANGE

GREEN

ORANGE

BLUE-GREEN

RED-ORANGE

BLUE

RED

BLUE-PURPLE

RED-PURPLE

PURPLE

THE COLOR WHEEL
(FOR PIGMENTS)

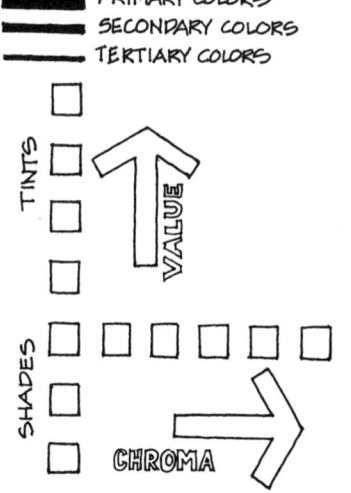

▬▬▬ PRIMARY COLORS
▬▬▬ SECONDARY COLORS
▬▬▬ TERTIARY COLORS

TINTS

VALUE

SHADES

CHROMA

THINK OF COLOR IN THREE DIMENSIONS:

1. HUE ("COLOR")
2. VALUE (LIGHT TO DARK)
3. CHROMA (SATURATION-INTENSITY)

COLOR CONTRASTS

1. ONE COLOR VS. ANOTHER
2. DARK VS. LIGHT
3. COMPLEMENTARY (RED VS. GREEN, YELLOW VS VIOLET, ORANGE VS BLUE)
4. WARM VS COOL (HOT RED-ORANGE - YELLOW VS COOL BLUE - GREEN)
5. SMALL VS LARGE

___ 1. <u>Basic Color Schemes</u>

___ *a.* *Triadic schemes.* Made from any three hues that are equidistant on the color wheel.

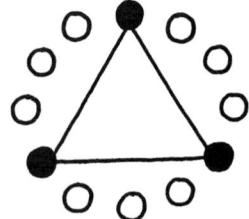

___ *b.* *Analogous or related schemes.* Consist of hues that are side by side.

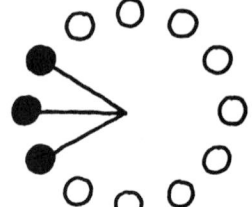

___ *c.* *Monochromatic schemes.* Use only one color (hue) in a range of values and intensities, coupled with neutral blacks or whites.

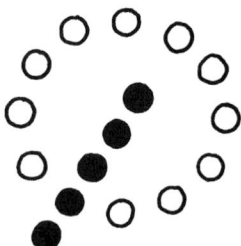

___ *d.* *Complementary schemes.* Use contrast by drawing from exact opposites on the color wheel. Usually, one of the colors is dominant while the other is used as an accent. Usually vary the amount and brightness of contrasting colors.

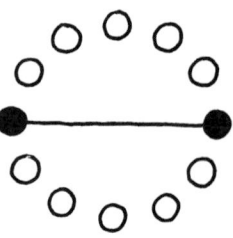

___ *e.* *Split complemen-*
tary schemes. Con-
sist of one hue and the
two hues on each side
of its compliment.

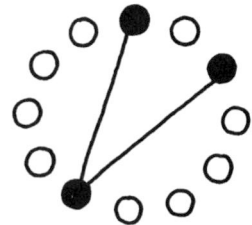

___ *f.* *Double complemen-*
tary schemes. Com-
posed of two adjacent
hues and their respec-
tive hues, directly
opposite on the color
wheel.

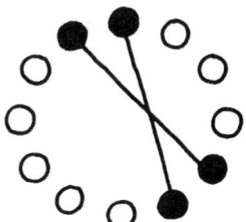

___ *g.* *Many-hued schemes.*
Those with more than
three hues. These usu-
ally need a strong
dose of one color as a
base with added col-
ors that are closely
matched in value.

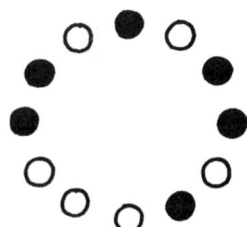

___ 2. <u>Rules of Thumb</u>

   ___ *a.* Your dominant color should cover about ⅔ of the
room's area. Equal areas of color are usually less
pleasing. Typical areas to be covered by the main
color are the walls, ceiling, and part of the floor.

   ___ *b.* The next most important color usually is in the floor
covering, the furniture, or the draperies.

   ___ *c.* The accent colors act as the "spice" for the scheme.

   ___ *d.* Study the proposed colors in the lighting conditions
of where they will be used (natural light, type of
artificial light).

   ___ *e.* The larger the area, the brighter a color will seem.
Usually duller tones are used for large areas.

___ *f.* Contrast is greater from light to dark than it is from hue to hue or dull to bright saturation.

___ *g.* Colors that seem identical but are slightly different will seem more divergent when placed together.

___ *h.* Bold, warm (red/orange) and dark colors will "advance." These can be used to bring in end walls, to lower ceilings, or to create a feeling of closeness in a room.

___ *i.* Cool (blue/green), dull and light colors "recede." These can be used to heighten ceilings or to widen a room.

___ *j.* Related colors tend to blend into "harmony."

___ *k.* From an economic point of view, dark colors absorb more light (and heat) and will require more lighting. Light colors reflect light, requiring less lighting.

___ *l.* Colors will appear darker and more saturated when reflected from a glossy surface than when reflected from a matte surface.

___ *m.* A color on a textured surface will appear darker than on a smooth surface.

___ *n.* Bright colors increase in brilliance when increased in area, and pale colors fade when increased in area.

___ *o.* Incandescent (warm) lighting normally adds a warming glow to colors. Under this light, consider "cooling" down or graying bright reds, oranges, or yellows.

___ *p.* Low atmospheric lighting tends to gray down colors.

___ *q.* Fluorescent lighting changes the hue of colors in varying ways depending on the type used. In some instances it will accent blue tones and make reds look colder. It may make many colors look harsher.

___ *r.* Southern exposures will bring in warm tones of sunlight.

___ *s.* Northern exposures will bring in cool light.

___ *t.* Cool pale colors tend to promote relaxation and shorten the passing of time. Therefore, they are good for repetitive work. Warm bright colors tend to promote activity and heighten awareness of time. Therefore, they are better for entertainment and romantic settings

___ *u.* Cool colors tend to make warm conditions more tolerable. Warm colors do the same for cold conditions.

___ *v.* Advancing colors (red-yellow) usually make objects larger. Receding colors (green-violet) usually make things look smaller.

___ 3. <u>**Percent Light Reflected from Typical Walls and Ceilings**</u>

Class	Surface	Color	% Light reflected
Light	Paint	white	81
		ivory	79
		cream	74
	Stone	cream	69
Medium	Paint	buff	63
		lt. green	63
		lt. grey	58
	Stone	grey	56
Dark	Paint	tan	48
		dk. grey	26
		olive green	17
		lt. oak	32
		mohogany	8
	Cement	natural	25
	Brick	red	13

___ 4. <u>**Typical Reflectance %**</u>
   ___ *a.* Commercial
         Ceiling   80%
         Walls   50%
         Floors   20%
   ___ *b.* Industrial
         Ceiling   50%
         Walls   50%
         Floor   20%
   ___ *c.* Classrooms
         Ceiling   70–90%
         Walls   40–60%
         Floor   30–50%
         Desk top 35–50%
         Blackboard 20%

___ 5. <u>**Surfaces:**</u>
   ___ *a.* Specular: A smooth, shiny surface that casts a mirrorlike image of the arriving light.
   ___ *b.* Matte: A smooth, dull surface that emits an inarticulate shine.
   ___ *c.* Diffuse: A rough, dull surface that widely scatters the arriving light.

NOTES

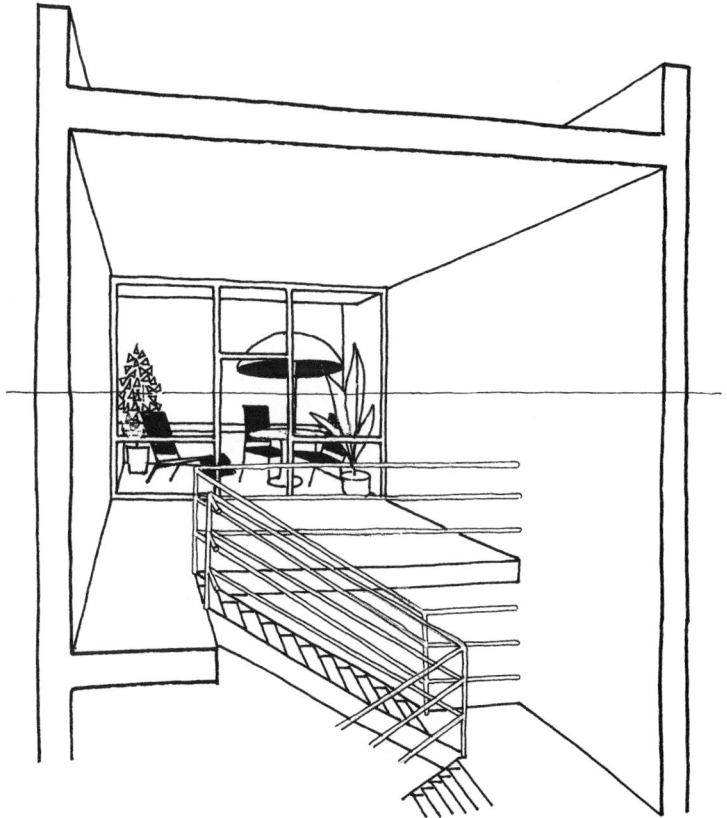

2

SITE
FURNISHINGS

NOTES

CAST ALUM. SIDE CHAIR W/ CUSHION

COST : $1200

OVAL ALUM. SIDE CHAIR W/ MESH SLING

COST : $510

TABLE OF METAL STAND W/ GLASS TOP   (42"φ)

COST: $870

UMBRELLA FOR TABLE TOP

COST : $140

METAL CHAIR
COST: $755

METAL CHAIR
COST: $445

METAL BENCH

COST: $2230

METAL BENCH

COST: $1870

WOOD ARMCHAIR

COST: $575

WOOD BENCH

5' LONG: $875
8' LONG: $1440

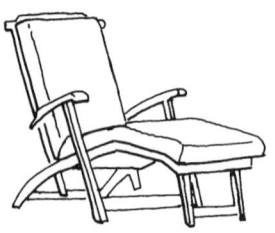

STEAMER CHAIR, WOOD
W/CUSHION
COST: $780 - $1020

RATTAN CHAIR W/CUSHION

COST: $575-$660

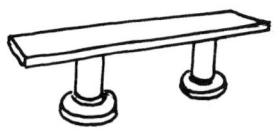

CONCRETE BENCH
COST: $190

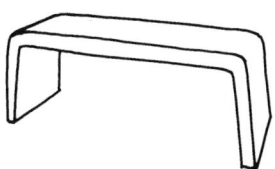

CONCRETE BENCH
COST: $245

CONC. ASH RECEPTACLE,
16" x 24"
COST: $115

CONC. WASTE RECEPTACLE,
30" x 32", 30 GAL.
COST: $460

CONC. PARK WASTE REC.
COST: $340

NOTES

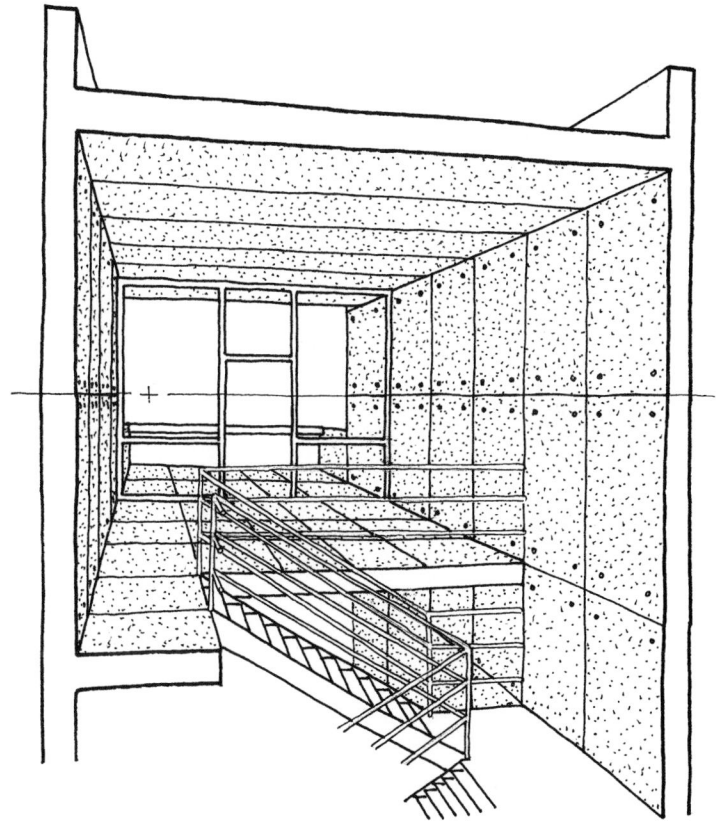

3 CONCRETE FINISHES

NOTES

___ A. GENERAL

Since concrete is usually used for the structural frame and floors of buildings, the interior designer is seldom concerned with this material. However, there may be occasions where the interior designer may have need to specify concrete finishes. This may occur where working with an architect for exterior or interior walls or possibly remodeling a one-story interior where a new special slab on grade is to be done.

Concrete walls and floors are usually covered. When they are left exposed, and with special finish treatments, this can have a desirable visual effect.

___ B. FLOORS (SLAB ON GRADE)

Exposed concrete horizontal surfaces are usually left to the exterior of the building, but they can also be used inside for special effects. Possible finishes are:

___ 1. **Steel Trowel**
Normal floor finish prior to a floor covering being installed. Slippery when wet. May be left exposed inside industrial spaces, with a hardener.

___ 2. **Broom Finish**
Roughens surface to increase slip resistance. Usually used on outside walks.

___ 3. **Exposed Aggregate**
Although usually used for outside walks and patios, can be used inside for special effect. Aggregate is usually ¾″ to 1½″ river run smooth rock of different colors.
Costs: Add $1.00 to $2.20/SF to cost of slab.

___ 4. **Colored Concrete**
Can be an integral concrete mix or hand-cast on surface.
Costs: Add $1.00 to $4.30/SF to cost of slab.

___ 5. **Joint Patterns**
For visual effect by sawing joints.
Costs: Cutting 1″ deep joints: add $1.60/LF to cost of slab.

___ 6. **Simulated Colored Concrete Patterns**
Brick, stone, or other decorative textures and materials can be replicated on surface.
Costs: Stamp patterns: Add $12.30 to $61.20/SF to cost of slab.

__ C. WALLS

__ 1. Colored concrete, chemical retardant, or exposed aggregate
   Costs: Chemical retardation (exposed aggregate): Add $0.90/SF to cost of wall.

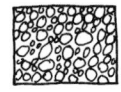

EXP. AGG.

__ 2. Cast with Different Moldings
   Costs: Add $3.25 to $7.80/SF to cost of wall.

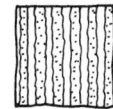

CAST

__ 3. Treating or Tooling the Concrete Surface
   Done in final stages of hardening.
   Costs: Add $1.70 to $4.20/SF to cost of wall.

BUSH HAMMERED

__ D. CEILINGS

When ceilings are of exposed concrete, they are usually the exposed underside of the floor or roof, above.

__ 1. Exposed Plank
   Sometimes used in hotels, these ceilings are often too low to put a drop ceiling under. The surface often has acoustical spray applied.

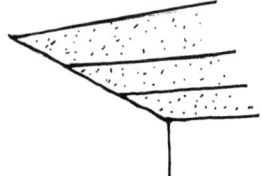

__ 2. Waffle Slabs
   Sometimes used in public buildings, these ceilings are often high and are intended to be left exposed as a design element.

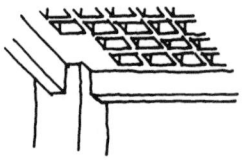

NOTES

<u>NOTES</u>

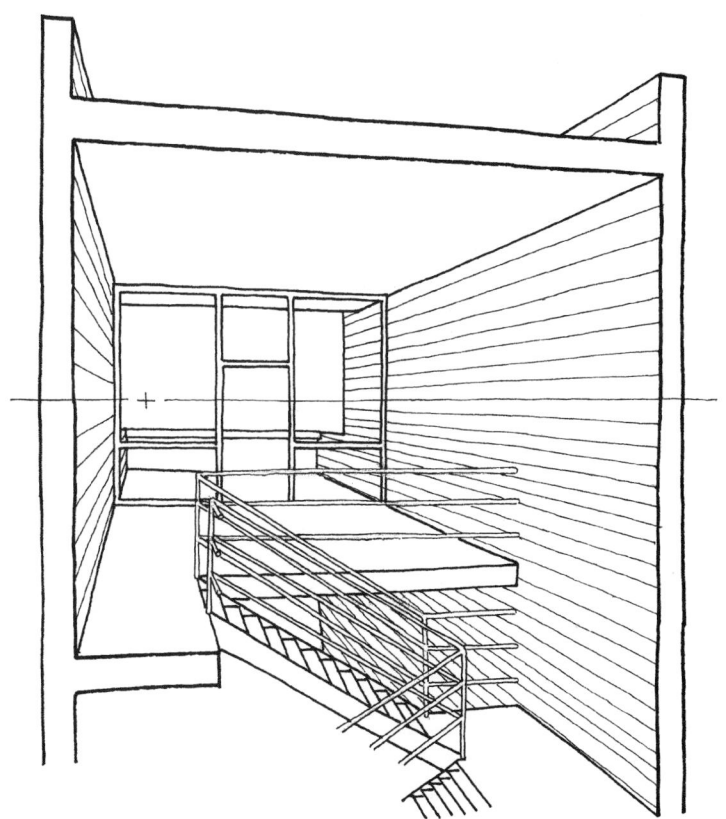

4 MASONRY

NOTES

__ A. MASONRY MATERIALS AND FINISHES

(4) (45A) (46) (49)

___ 1. __General:__ Masonry, like concrete, is seldom used by the
   interior designer. Nevertheless, there may be occasions
   where masonry walls may be needed in a remodel. For
   masonry floors, see p. 342.
   Masonry consists of:
   ___ *a.* Brick
      ___ (1) Fired
      ___ (2) Unfired ("adobe")
   ___ *b.* Concrete block (concrete masonry units or CMU)
   ___ *c.* Stone
   ___ *d.* Glass block
___ 2. __Bond Joints__
___ 3. __Bond Patterns__

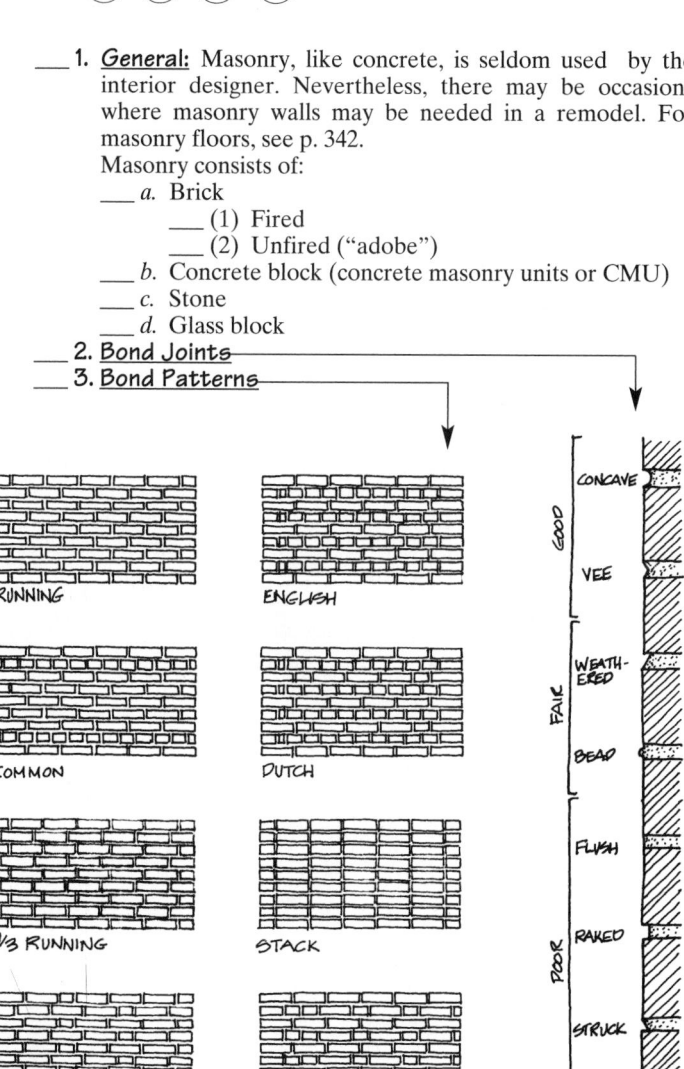

RUNNING

ENGLISH

COMMON

DUTCH

⅓ RUNNING

STACK

GARDEN WALL

FLEMISH

GOOD
CONCAVE
VEE

FAIR
WEATH-ERED
BEAD

POOR
FLUSH
RAKED
STRUCK
EXT-RUDED

___ **4. <u>Coatings</u>:** Must be (see p. 348)
        ___ (1) "Bridgeable" (seal cracks)
        ___ (2) Breathable (do not trap vapor)

___ **5. <u>Brick</u>**
    ___ *a.* <u>*Types*</u>
        ___ (1) Common (building)
        ___ (2) Face
            ___ (*a*) FBX   Select
            ___ (*b*) FBS   Standard
            ___ (*c*) FBA   Architectural
        ___ (3) Clinker
        ___ (4) Glazed
        ___ (5) Fire
        ___ (6) Cored
        ___ (7) Sand-lime (white, yellow)
        ___ (8) Pavers
    ___ *b.* <u>*Weatherability*</u>
        ___ (1) NW—Negligible weathering; for indoor or sheltered locations.
        ___ (2) MW—Moderate weather locations.
        ___ (3) SW—Severe weather locations and/or earth contact.
    ___ *c.* <u>*Positions*</u>

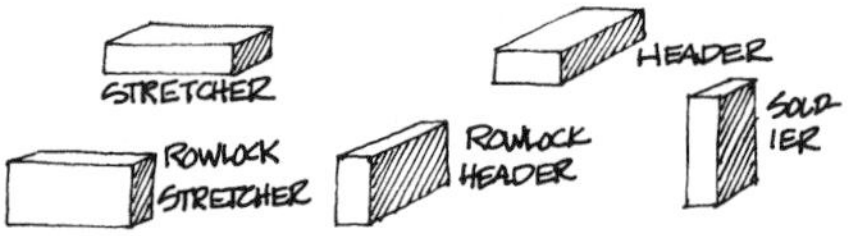

___ *d.* *Sizes:* (modular brick based on 4″ module with ⅜″ joint)

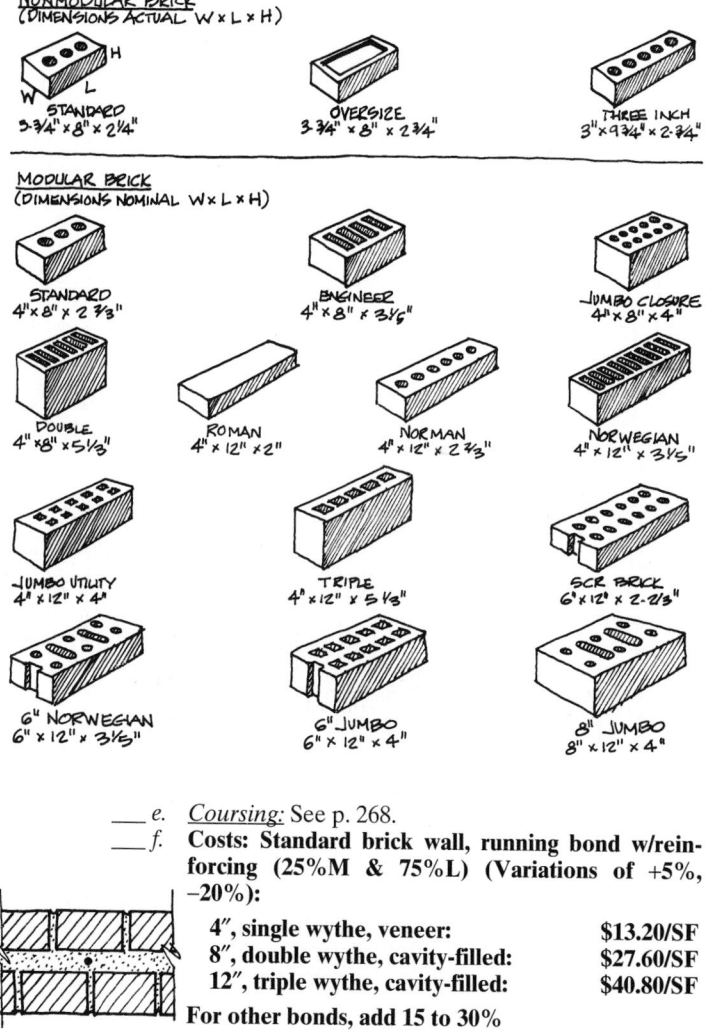

NONMODULAR BRICK
(DIMENSIONS ACTUAL W × L × H)

STANDARD
3-¾" × 8" × 2¼"

OVERSIZE
3-¾" × 8" × 2¾"

THREE INCH
3" × 9¾" × 2-¾"

MODULAR BRICK
(DIMENSIONS NOMINAL W × L × H)

STANDARD
4" × 8" × 2⅔"

ENGINEER
4" × 8" × 3⅕"

JUMBO CLOSURE
4" × 8" × 4"

DOUBLE
4" × 8" × 5⅓"

ROMAN
4" × 12" × 2"

NORMAN
4" × 12" × 2⅔"

NORWEGIAN
4" × 12" × 3⅕"

JUMBO UTILITY
4" × 12" × 4"

TRIPLE
4" × 12" × 5⅓"

SCR BRICK
6" × 12" × 2-⅔"

6" NORWEGIAN
6" × 12" × 3⅕"

6" JUMBO
6" × 12" × 4"

8" JUMBO
8" × 12" × 4"

___ *e.* *Coursing:* See p. 268.
___ *f.* **Costs: Standard brick wall, running bond w/reinforcing (25%M & 75%L) (Variations of +5%, −20%):**

4″, single wythe, veneer:	**$13.20/SF**
8″, double wythe, cavity-filled:	**$27.60/SF**
12″, triple wythe, cavity-filled:	**$40.80/SF**

For other bonds, add 15 to 30%

BRICK COURSING

COURSE	NONMODULAR						MODULAR				
	2 1/4" THICK BRICKS		2 5/8" THICK BRICKS		2 3/4" THICK BRICKS		NOMINAL THICKNESS (HEIGHT) OF BRICK				
	3/8" JOINT	1/2" JOINTS	3/8" JOINT	1/2" JOINT	3/8" JOINT	1/2" JOINT	2"	2 2/3"	3 1/5"	4"	5 1/3"
1	2 5/8"	2 3/4"	3"	3 1/8"	3 1/8"	3 1/4"	2"	2 11/16"	3 3/16"	4"	5 5/16"
2	5 1/4"	5 1/2"	6"	6 1/4"	6 1/4"	6 1/2"	4"	5 5/16"	6 3/8"	8"	10 11/16"
3	7 7/8"	8 1/4"	9"	9 3/8"	9 3/8"	9 3/4"	6"	8"	9 5/8"	1'-0"	1'-4"
4	10 1/2"	11"	1'-0"	1'-0 1/2"	1'-0 1/2"	1'-1"	8"	10 11/16"	1'-0 13/16"	1'-4"	1'-9 5/16"
5	1'-1 1/8"	1'-1 3/4"	1'-3"	1'-3 5/8"	1'-3 5/8"	1'-4 1/4"	10"	1'-1 5/16"	1'-4"	1'-8"	2'-2 11/16"
6	1'-3 3/4"	1'-4 1/2"	1'-6"	1'-6 3/4"	1'-6 3/4"	1'-7 1/2"	1'-0"	1'-4"	1'-7 3/16"	2'-0"	2'-8"
7	1'-6 3/8"	1'-7 1/4"	1'-9"	1'-9 7/8"	1'-9 7/8"	1'-10 3/4"	1'-2"	1'-6 11/16"	1'-10 3/8"	2'-4"	3'-1 5/16"
8	1'-9"	1'-10"	2'-0"	2'-1"	2'-1"	2'-2"	1'-4"	1'-9 5/16"	2'-1 5/8"	2'-8"	3'-6 11/16"
9	1'-11 5/8"	2'-0 3/4"	2'-3"	2'-4 1/8"	2'-4 1/8"	2'-5 1/4"	1'-6"	2'-0"	2'-4 13/16"	3'-0"	4'-0"
10	2'-2 1/4"	2'-3 1/2"	2'-6"	2'-7 1/4"	2'-7 1/4"	2'-8 1/2"	1'-8"	2'-2 11/16"	2'-8"	3'-4"	4'-5 5/16"
11	2'-4 7/8"	2'-6 1/4"	2'-9"	2'-10 3/8"	2'-10 3/8"	2'-11 3/4"	1'-10"	2'-5 5/16"	2'-11 3/16"	3'-8"	4'-10 11/16"
12	2'-7 1/2"	2'-9"	3'-0"	3'-1 1/2"	3'-1 1/2"	3'-3"	2'-0"	2'-8"	3'-2 3/8"	4'-0"	5'-4"
13	2'-10 1/8"	2'-11 3/4"	3'-3"	3'-4 5/8"	3'-4 5/8"	3'-6 1/4"	2'-2"	2'-10 11/16"	3'-5 5/8"	4'-4"	5'-9 5/16"
14	3'-0 3/4"	3'-2 1/2"	3'-6"	3'-7 3/4"	3'-7 3/4"	3'-9 1/2"	2'-4"	3'-1 5/16"	3'-8 13/16"	4'-8"	6'-2 11/16"
15	3'-3 3/8"	3'-5 1/4"	3'-9"	3'-10 7/8"	3'-10 7/8"	4'-0 3/4"	2'-6"	3'-4"	4'-0"	5'-0"	6'-8"
16	3'-6"	3'-8"	4'-0"	4'-2"	4'-2"	4'-4"	2'-8"	3'-6 11/16"	4'-4 3/16"	5'-4"	7'-1 5/16"
17	3'-8 5/8"	3'-10 3/4"	4'-3"	4'-5 1/8"	4'-5 1/8"	4'-7 1/4"	2'-10"	3'-9 5/16"	4'-6 3/8"	5'-8"	7'-6 11/16"
18	3'-11 1/4"	4'-1 1/2"	4'-6"	4'-8 1/4"	4'-8 1/4"	4'-10 1/2"	3'-0"	4'-0"	4'-9 5/8"	6'-0"	8'-0"
19	4'-1 7/8"	4'-4 1/4"	4'-9"	4'-11 3/8"	4'-11 3/8"	5'-1 3/4"	3'-2"	4'-2 11/16"	5'-0 13/16"	6'-4"	8'-5 5/16"
20	4'-4 1/2"	4'-7"	5'-0"	5'-2 1/2"	5'-2 1/2"	5'-5"	3'-4"	4'-5 5/16"	5'-4"	6'-8"	8'-10 11/16"
21	4'-7 1/8"	4'-9 3/4"	5'-3"	5'-5 5/8"	5'-5 5/8"	5'-8 1/4"	3'-6"	4'-8"	5'-7 3/16"	7'-0"	9'-4"
22	4'-9 3/4"	5'-0 1/2"	5'-6"	5'-8 3/4"	5'-8 3/4"	5'-11 1/2"	3'-8"	4'-10 11/16"	5'-10 3/8"	7'-4"	9'-9 5/16"
23	5'-0 3/8"	5'-3 1/4"	5'-9"	5'-11 7/8"	5'-11 7/8"	6'-2 3/4"	3'-10"	5'-1 5/16"	6'-1 5/8"	7'-8"	10'-2 11/16"
24	5'-3"	5'-6"	6'-0"	6'-3"	6'-3"	6'-6"	4'-0"	5'-4"	6'-4 13/16"	8'-0"	10'-8"

CONCRETE BLOCK TYPES & SIZES

NOMINAL DIMENSIONS W × L × H (ACTUAL DIMENSIONS ARE 3/8" LESS)

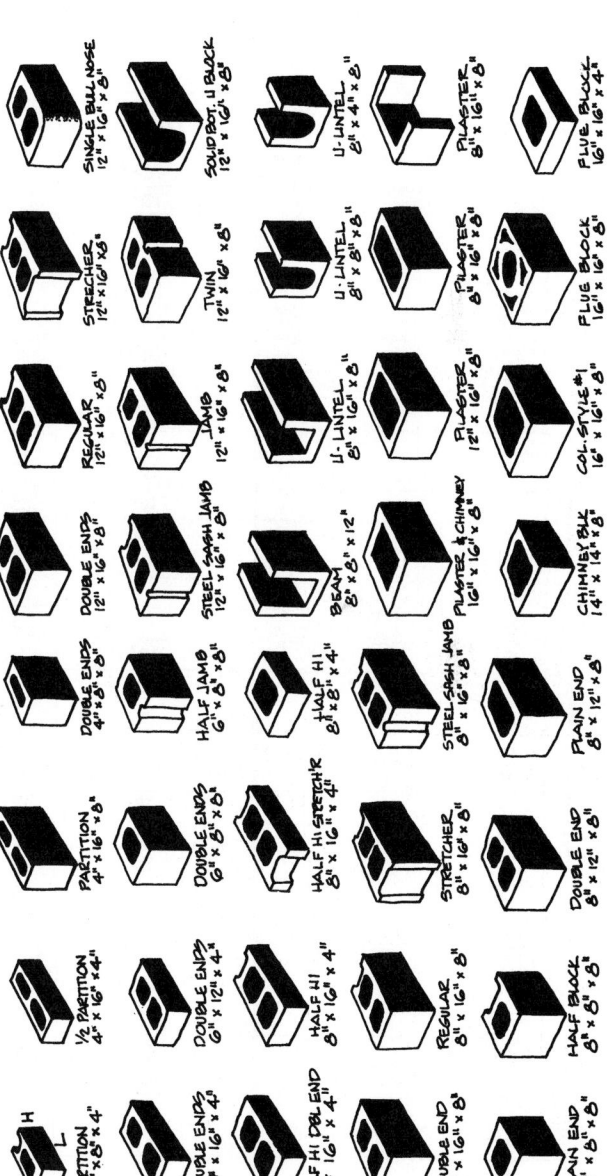

SINGLE BULL NOSE 12" × 16" × 8"

SOLID BOT. U BLOCK 12" × 16" × 8"

U-LINTEL 8" × 4" × 8"

PILASTER 8" × 16" × 8"

PLUG BLOCK 16" × 16" × 4"

STRETCHER 12" × 16" × 8"

TWIN 12" × 16" × 8"

U-LINTEL 8" × 16" × 8"

PILASTER 8" × 16" × 8"

PLUG BLOCK 16" × 16" × 8"

REGULAR 12" × 16" × 8"

JAMB 12" × 16" × 8"

U-LINTEL 8" × 16" × 8"

PILASTER 12" × 16" × 8"

COL. STYLE #1 16" × 16" × 8"

DOUBLE ENDS 12" × 16" × 8"

STEEL SASH JAMB 12" × 16" × 8"

BEAM 8" × 8" × 12"

PLASTER & CHIMNEY 16" × 16" × 8"

CHIMNEY BLK 14" × 14" × 8"

DOUBLE ENDS 4" × 8" × 8"

HALF JAMB 6" × 8" × 8"

HALF HI 8" × 8" × 4"

STEEL SASH JAMB 8" × 16" × 8"

PLAIN END 8" × 12" × 8"

PARTITION 4" × 16" × 8"

DOUBLE ENDS 6" × 8" × 8"

HALF HI STRETCH'R 8" × 16" × 4"

STRETCHER 8" × 16" × 8"

DOUBLE END 8" × 12" × 8"

1/2 PARTITION 4" × 16" × 4"

DOUBLE ENDS 6" × 12" × 4"

HALF HI 8" × 16" × 4"

REGULAR 8" × 16" × 8"

HALF BLOCK 8" × 8" × 8"

PARTITION 4" × 8" × 4"

DOUBLE ENDS 6" × 16" × 4"

HALF HI DBL END 8" × 16" × 4"

DOUBLE END 8" × 16" × 8"

PLAIN END 8" × 8" × 8"

CONCRETE BLOCK COURSING

CSC	4" HIGH BLK.	8" HIGH BLK.	CSC	4" HIGH BLK.	8" HIGH BLK.
1	4"	8"	38	12'-8"	25'-4"
2	8"	1'-4"	39	13'-0"	26'-0"
3	1'-0"	2'-0"	40	13'-4"	26'-8"
4	1'-4"	2'-8"	41	13'-8"	27'-4"
5	1'-8"	3'-4"	42	14'-0"	28'-0"
6	2'-0"	4'-0"	43	14'-4"	28'-8"
7	2'-4"	4'-8"	44	14'-8"	29'-4"
8	2'-8"	5'-4"	45	15'-0"	30'-0"
9	3'-0"	6'-0"	46	15'-4"	30'-8"
10	3'-4"	6'-8"	47	15'-8"	31'-4"
11	3'-8"	7'-4"	48	16'-0"	32'-0"
12	4'-0"	8'-0"	49	16'-4"	32'-8"
13	4'-4"	8'-8"	50	16'-8"	33'-4"
14	4'-8"	9'-4"	51	17'-0"	34'-0"
15	5'-0"	10'-0"	52	17'-4"	34'-8"
16	5'-4"	10'-8"	53	17'-8"	35'-4"
17	5'-8"	11'-4"	54	18'-0"	36'-0"
18	6'-0"	12'-0"	55	18'-4"	36'-8"
19	6'-4"	12'-8"	56	18'-8"	37'-4"
20	6'-8"	13'-4"	57	19'-0"	38'-0"
21	7'-0"	14'-0"	58	19'-4"	38'-8"
22	7'-4"	14'-8"	59	19'-8"	39'-4"
23	7'-8"	15'-4"	60	20'-0"	40'-0"
24	8'-0"	16'-0"	61	20'-4"	40'-8"
25	8'-4"	16'-8"	62	20'-8"	41'-4"
26	8'-8"	17'-4"	63	21'-0"	42'-0"
27	9'-0"	18'-0"	64	21'-4"	42'-8"
28	9'-4"	18'-8"	65	21'-8"	43'-4"
29	9'-8"	19'-4"	66	22'-0"	44'-0"
30	10'-0"	20'-0"	67	22'-4"	44'-8"
31	10'-4"	20'-8"	68	22'-8"	45'-4"
32	10'-8"	21'-4"	69	23'-0"	46'-0"
33	11'-0"	22'-0"	70	23'-4"	46'-8"
34	11'-4"	22'-8"	71	23'-8"	47'-4"
35	11'-8"	23'-4"	72	24'-0"	48'-0"
36	12'-0"	24'-0"	73	24'-4"	48'-8"
37	12'-4"	24'-8"	74	24'-8"	49'-4"

___ 6. <u>Concrete Block (CMU)</u>
___ *a.* Types: Plain (gray), colored, pavers, special shapes (such as "slump"), and special surfaces (split faced, scored, etc.).
___ *b.* Size: See p. 219.
___ *c.* Coursing: See p. 220.
___ *d.* **Costs: CMU (Regular weight, gray, running bond, typical reinforcing and grout)**

4″ walls:	**$7.20/SF**	**(Typical 25 to 30%M and 75 to**
6″ walls:	**$7.80/SF**	**70%L)**
8″ walls:	**$9.00/SF**	**(Variations for special block, such as**
12″ walls:	**$11.00/SF**	**glazed, decorative, screen, etc. + 15 to 150%)**

Deduct 30 to 40% for residential work.

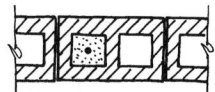

___ 7. <u>Stone</u>
___ *a.* <u>*Type unit*</u>
___ (1) *Ashlar:* Best for strength and stability; is square-cut on level beds. Joints of ½″ to ¾″.

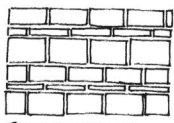 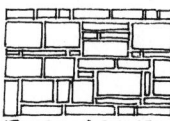 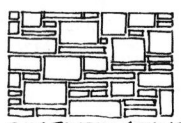

COURSED ASHLAR　　RANDOM ASHLAR　　3 HT. RANDOM ASHLAR

___ (2) *Squared stone* (course rubble): Next-best for strength and stability; is fitted less carefully than ashlar, but more carefully than rubble.

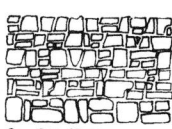 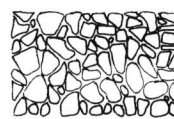

COURSE RUBBLE　　MOSAIC

___ (3) *Rubble:* Built with a minimum of dressing, with joints unevenly coursed, or in a completely irregular pattern. Stones are lapped for bond and many stones extend through wall (when full-width wall) to bond it transversely. If built carefully, with all interstices completely filled with good cement mortar, has ample durability for ordinary structures.

___ *b.* *Typical materials*
    ___ (1) Limestone
    ___ (2) Sandstone
    ___ (3) Quartzite
    ___ (4) Granite

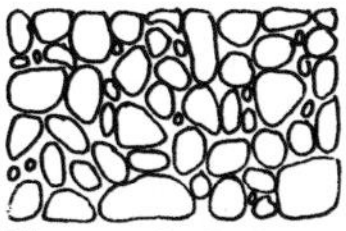

RANDOM RUBBLE

___ *c.* *Wall types*
    ___ (1) Full width
    ___ (2) Solid veneer (metal ties to structural wall)
    ___ (3) Thin veneer (set against mortar bed against structural wall)

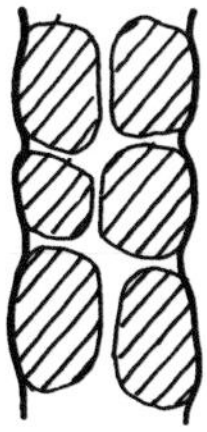

___ *d.* **Costs: 4″ veneer (most common): $14.00 to $17.50/SF (40%M and 60%L) (Variation: + 50%) 18″ rough stone wall (dry): $52/CF (40%M and 60%L)**

___ **8. Glass Block**
    ___ *a.* *Thickness* 3″ and 4″.
    ___ *b.* *Size,* 4½″, 6″, 7½″, 8″, 9½″, and 12″ square.
    ___ *c.* *Reinforcing* at 16″ oc.
    ___ *d.* *Interior Panels:* 25 LF max. and 250 SF max.
    ___ *e.* **Costs: 4″ thick, 6, 8, or 12″ sq = $32 to $40.50/SF (55% M and 45% L).**

NOTES

NOTES

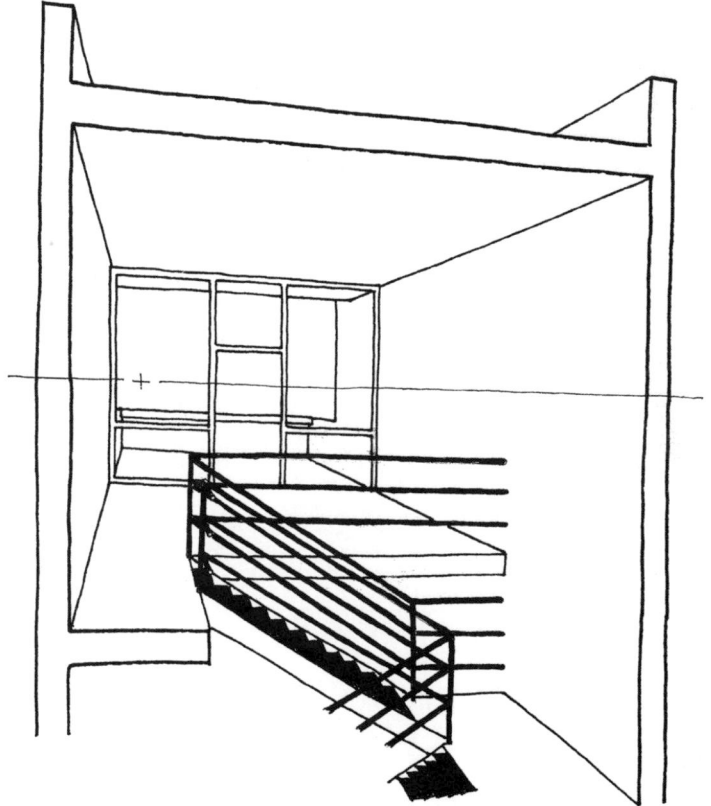

5 METALS

NOTES

___ A. METALS

(4) (27) (32) (45A) (49)

___ 1. General
    ___ *a.* *Ferrous metals* (contain iron)
        ___ (1) *Iron:* Soft, easily worked, oxidizes rapidly, susceptible to acid.
        ___ (2) *Cast-iron:* Brittle, corrosion-resistant, high compressive strength. Used for gratings, stairs, etc.
        ___ (3) *Malleable iron:* Same as cast-iron, but better workability.
        ___ (4) *Wrought iron:* Soft, corrosion- and fatigue-resistant, machinable. Used for railings, grilles, screws, and ornamental items.
        ___ (5) *Steel:* Iron with carbon. Strongest metal. Used for structural purposes.
        ___ (6) *Stainless steel:* An alloy for maximum corrosion resistance. Used for flashing, handrails, hardware, connections, and equipment.
    ___ *b.* *Nonferrous metals* (not containing iron)
        ___ (1) *Aluminum:* Soft, ductile, high corrosion resistance, low strength.
        ___ (2) *Lead:* Dense, workable, toxic, corrosion-resistant. Improved with alloys for hardness and strength. Used as waterproofing, sound isolation, and radiation shielding.
        ___ (3) *Zinc:* Corrosion-resistant, brittle, low-strength. Used in "galvanizing" of other metals for corrosion resistance for roofing, flashing, hardware, connections, etc.
        ___ (4) *Chromium and nickel:* Used as alloy for corrosion-resistant bright "plating."
        ___ (5) *Monel:* High corrosion resistance. Used for fasteners and anchors.
        ___ (6) *Copper:* Resistant to corrosion, impact, and fatigue. Ductile. Used for wiring, roofing, flashing, and piping.
        ___ (7) *Bronze:* An alloy for "plating."
        ___ (8) *Brass:* Copper with zinc. Used for hardware, handrails, grilles, etc.
___ 2. Metal Corrosion
    ___ *a.* Galvanic action, or corrosion, occurs between dissimilar metals or metals and other metals when sufficient moisture is present to carry an electric

current. The farther apart two metals are on the following list, the greater the corrosion of the more susceptible one:

Anodic (+): Most susceptible to corrosion
Magnesium
Zinc
Aluminum
Cadmium
Iron/steel
Stainless steel (active)
Soft solders
Tin
Lead
Nickel
Brass
Bronzes
Nickel-copper alloys
Copper
Stainless steel (passive)
Silver solder
Cathodic (−): Least susceptible to corrosion

___ *b.* Metals deteriorate also when in contact with chemically active materials, particularly when water is present. Examples include aluminum in contact with concrete or mortar, and steel in contact with treated wood.

___ **3. Gauges:** See pp. 279.

___ **4. Light Metal Framing**

   ___ *a.* *Joists*

      ___ (1) Makes an economical floor system for light loading and spans up to 32′

      ___ (2) Depths: 6″, 8″, 9″, 10″, 12″

      ___ (3) Spacings: 16″, 24″, 48″ oc

      ___ (4) Gauges: 12 through 18 (light = 20–25 GA; structural = 18–12 GA)

      ___ (5) Bridging, usually 5′ to 8′ oc

   ___ *b.* *Studs*

      ___ (1) Sizes

         ___ (*a*) Widths: ¾″, 1″, 1⅜″, 1⅝″, 2″

         ___ (*b*) Depths: 2½″, 3⅝″, 4″, 6″, 8″

METAL GAUGES

GAUGE NO.	GRAPHIC SIZES	U.S. STD. REVISED		GRAPHIC SIZES
		DECIMAL	FRACTION	
000	■	.3750"	3/8"	●
00	■	.3437"	11/32"	●
0	■	.3125"	5/16"	●
1.	■	.2812"	9/32"	●
2.	■	.2656"	17/64"	●
3.	■	.2391"	15/64"	●
4.	■	.2242"	7/32°	●
5.	■	.2092"	13/64"	●
6.	■	.1943"	3/18"	●

7	■	.1793"	11/64" +	●
8	■	.1644"	11/64" −	●
9	■	.1495"	6/32" −	●
10	■	.1345"	9/64" −	●
11	■	.1196"	1/8" −	●
12	■	.1046"	7/64" −	●
13	■	.0897"	3/32" −	●
14	■	.0747"	5/64" −	●
15	■	.0673"	1/16" +	●
16	■	.0598"	1/16" −	●
17	■	.0538"	3/64" +	●
18	■	.0478"	3/64" +	●
19	■	.0418"	3/64" −	●
20	■	.0359"	1/32" +	●
21	■	.0329"	1/32" +	●
22	■	.0299"	1/32" −	●
23	■	.0269"	1/32" −	●
24	■	.0239"	1/32" −	●
25	●	.0209"	1/64" +	●
26	●	.0179"	1/64" +	●
27	●	.0164"	1/64" +	●
28	●	.0149"	1/64" −	●
29	●	.0135"	1/64" −	●
30	●	.0120"	1/64" −	●

___ (2) Gauges: 14, 15, 16, 18, 20

___ (3) Spacings: 12″, 16″, 24″ oc

___ **5.** <u>Miscellaneous Metals</u>

    ___ *a.* <u>*Nails*</u>

        ___ (1) Size: Penny designated as d. A 2-penny nail is 1″ long. Each additional "penny" adds ¼″ length, to:

            12-penny = 3¼″ long

            16-penny = 3½″

            20-penny = 4″

            30-penny = 4½″

            40-penny = 5″

            50-penny = 5½″

            60-penny = 6″

        Rule of thumb: Use nail with length 3× thickness of board being secured.

        ___ (2) Types

<u>COMMON NAIL</u> GREAT HOLDING POWER DUE TO LARGE HEAD. USE WHERE NOT SEEN.

<u>BOX NAIL</u> SIMILAR TO COMMON BUT SLIMMER SHANK. EASILY BENT BUT LESS LIKLEY TO SPLIT WOOD.

<u>FINISHING NAIL</u> SMALL HEAD PERMITS IT TO BE COUNTER-SUNK TO CREATE FINISHED APPEARANCE.

<u>CASING NAIL</u> A FINISH NAIL BUT W/ THICKER SHANK & MORE ANGULAR HEAD FOR FLOORING & DOOR CASINGS.

<u>ROOFING NAIL</u> MADE WITH A WIDE HEAD AND USUALLY GALVANIZED. DIFFERENT TYPES BASED ON TYPE ROOFING.

<u>MASONRY NAIL</u> CASE-HARDENED W/ A FLUTED SHANK FOR USE IN FASTENING WOOD TO CONC. OR MASONRY.

<u>DRYWALL NAIL</u> COATED W/ CEMENT W/ A SPECIAL HEAD FOR RECESSING BELOW DRYWALL SURFACE.

<u>BRAD</u> A SMALL FINISHING NAIL FOR SMALL TRIM.

___ *b.*   Screws and bolts

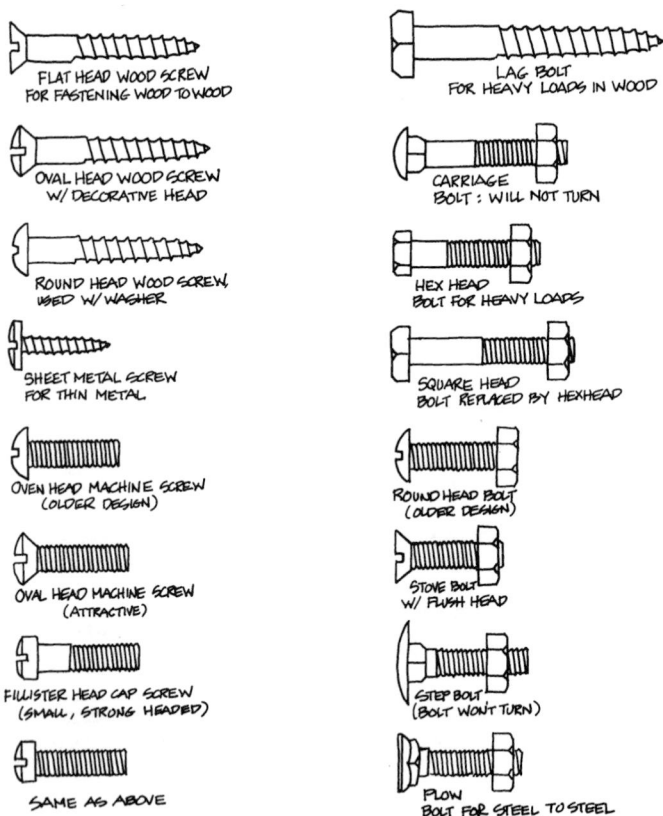

FLAT HEAD WOOD SCREW
FOR FASTENING WOOD TO WOOD

OVAL HEAD WOOD SCREW
W/ DECORATIVE HEAD

ROUND HEAD WOOD SCREW,
USED W/ WASHER

SHEET METAL SCREW
FOR THIN METAL

OVEN HEAD MACHINE SCREW
(OLDER DESIGN)

OVAL HEAD MACHINE SCREW
(ATTRACTIVE)

FILLISTER HEAD CAP SCREW
(SMALL, STRONG HEADED)

SAME AS ABOVE

LAG BOLT
FOR HEAVY LOADS IN WOOD

CARRIAGE
BOLT : WILL NOT TURN

HEX HEAD
BOLT FOR HEAVY LOADS

SQUARE HEAD
BOLT REPLACED BY HEXHEAD

ROUND HEAD BOLT
(OLDER DESIGN)

STONE BOLT
W/ FLUSH HEAD

STEP BOLT
(BOLT WON'T TURN)

PLOW
BOLT FOR STEEL TO STEEL

___ 6. <u>Stairs</u>
   ___ *a.* See p. 179 for code requirements.
   ___ *b.* **Costs: steel pan w/cement fill, 4′ wide w/landing, w/o railing: cost per flight (12 risers) = \$4920 (80% M & 20% L). Add \$295 for each additional riser.**

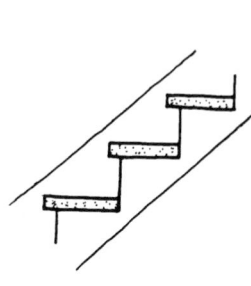 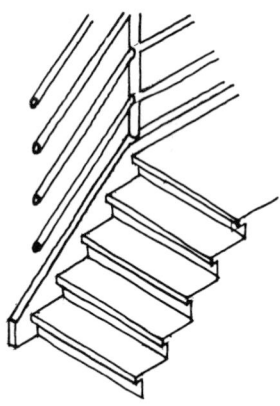

___ 7. <u>Railings</u>
   ___ *a.* Code requirements (IBC)
      ___ (1) Guardrails are required where walking surface ends and there is a drop of more than 30″. This includes landings at stairs.
         ___ (*a*) Heights required:
              36″ residential
              42″ other
         ___ (*b*) If open rails, openings between rails must be no more than 4″ at intermediates and no more than 6″ at bottom.
   ___ *b.* Requirements for handrails at stairs:
      ___ (1) No handrail required in private garages. Not required in private residences where stairway is less than 4 risers high.
      ___ (2) Handrail required on one side when stairway is less than 44″ wide, or in private residence.
      ___ (3) Handrail required on both sides when stairway is more than 44″ in width, and additional railings are required with each additional 88″ in width.

___ (4) See below for additional requirements.

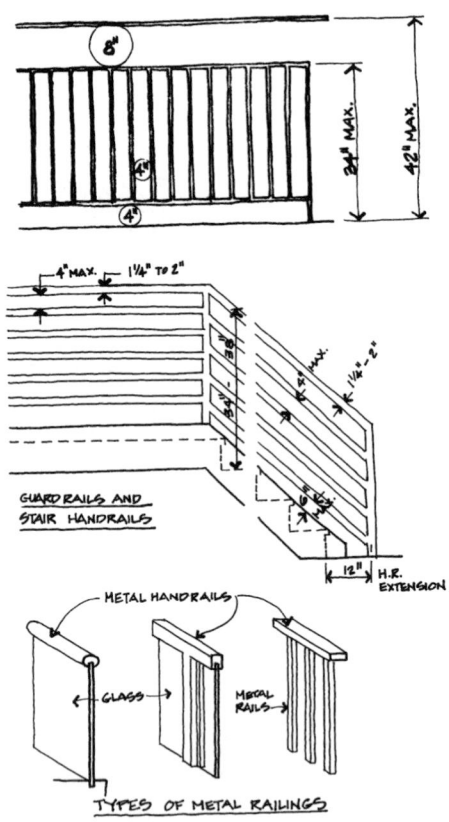

GUARD RAILS AND
STAIR HANDRAILS

H.R.
EXTENSION

METAL HANDRAILS

GLASS   METAL RAILS

TYPES OF METAL RAILINGS

___ *c.* **Costs:**
   Metal railings:
      Stair: steel pipe, 3-rail on stair, 1¼″ diameter:
         $37/LF (50% M and 50% L)
         Stainless steel: triple cost
      Wall rails: Steel or Alum: $25/LF
         Stainless steel: double cost
      Freestanding, ornamental: $180 to $720/LF

NOTES

NOTES

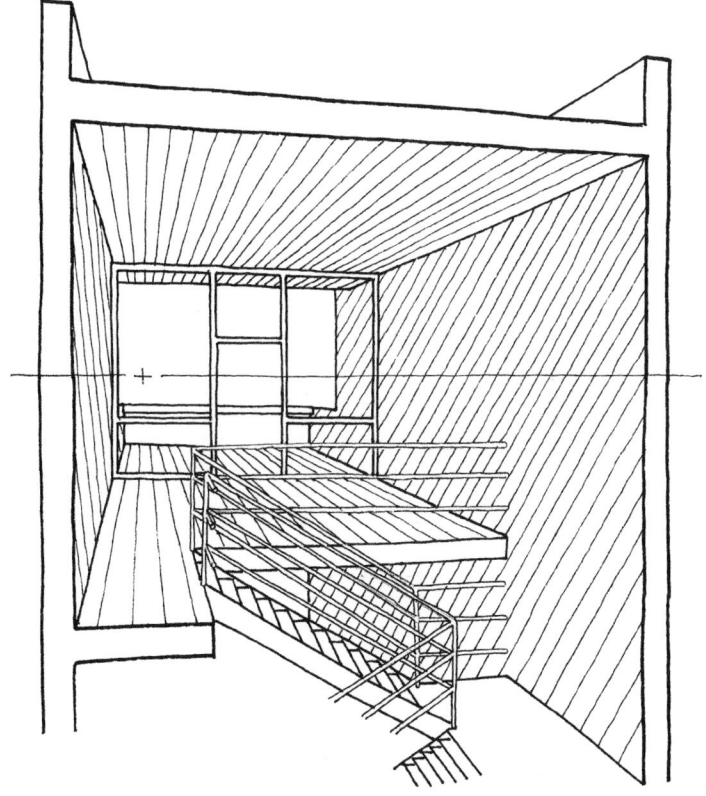

6 WOOD

NOTES

___ A. WOOD

④ ⑪ ⑫ ㉛ ㊺A ㊻ ㊾

___ **1. General** (*Note:* See p. 300 for species table. See p. 342 for wood flooring.)

    ___ *a.* *Two general types of wood* are used in buildings:

        ___ (1) *Softwood* (from evergreen trees) for general construction

        ___ (2) *Hardwood* (from deciduous trees) for furnishings and finishes

    ___ *b.* *Moisture and shrinkage:* The amount of water in wood is expressed as a percentage of its oven-dry (dry as possible) weight. As wood dries, it first loses moisture from within the cells without shrinking; after reaching the fiber saturation point (dry cell), further drying results in shrinkage. Eventually wood comes to dynamic equilibrium with the relative humidity of the surrounding air. Interior wood typically shrinks in winter and swells in summer. Average equilibrium moisture content ranges from 6 to 11%, but wood is considered dry enough for use at 12 to 15%. The loss of moisture during seasoning causes wood to become harder, stronger, stiffer, and lighter in weight. Wood is most decay-resistant when moisture content is under 20%.

___ **2. Lumber**

    ___ *a.* *Sizes*

        ___ (1) Sectional

Nominal sizes	To get actual sizes
2×'s up to 8×'s	deduct ½″
8×'s and larger	deduct ¾″

        ___ (2) Lengths

            ___ (*a*) Softwoods: cut to lengths of 6′ to 24′, in 2′ increments

            ___ (*b*) Hardwoods: cut to 1′-long increments

    ___ *b.* *Economy:* best achieved when layouts are within a 2′- or 4′-module, with subdivisions of 4″, 16″, 24″, and 48″

    ___ *c.* *Defects:* See diagram below and on following page.

DEFECT   END VIEW   LONG VIEW

BOW      

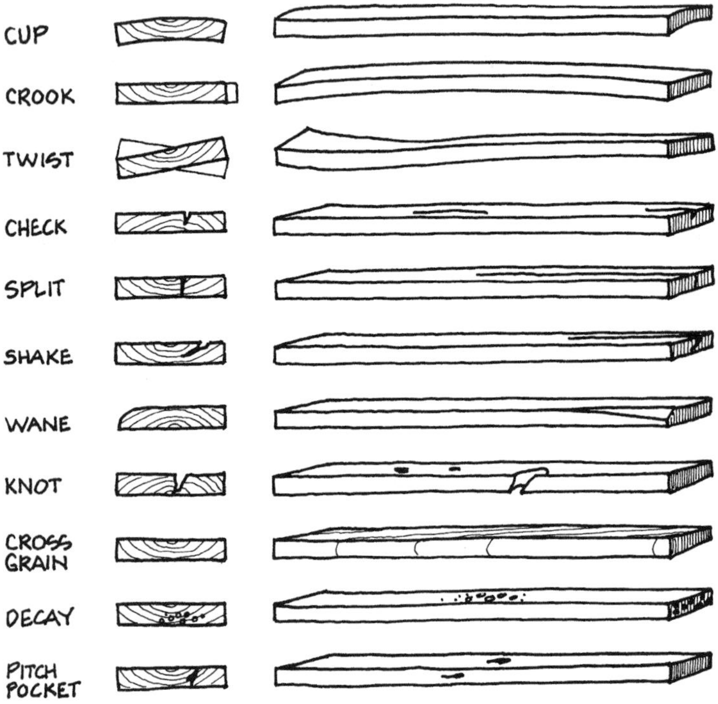

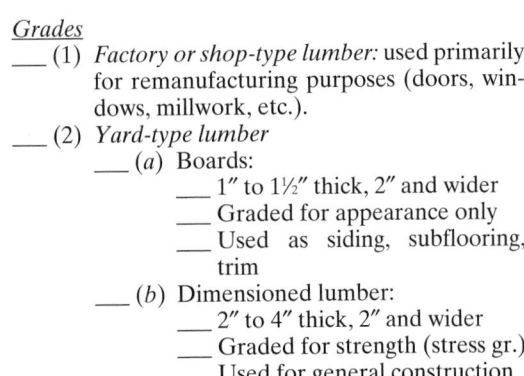

CUP

CROOK

TWIST

CHECK

SPLIT

SHAKE

WANE

KNOT

CROSS
GRAIN

DECAY

PITCH
POCKET

___ d. <u>Grades</u>
    ___ (1) *Factory or shop-type lumber:* used primarily
           for remanufacturing purposes (doors, win-
           dows, millwork, etc.).
    ___ (2) *Yard-type lumber*
        ___ (*a*) Boards:
             ___ 1″ to 1½″ thick, 2″ and wider
             ___ Graded for appearance only
             ___ Used as siding, subflooring,
                  trim
        ___ (*b*) Dimensioned lumber:
             ___ 2″ to 4″ thick, 2″ and wider
             ___ Graded for strength (stress gr.)
             ___ Used for general construction
             ___ Light framing: 2″ to 4″ wide

___ Joists and planks: 6″ and wider

___ Decking: 4″ and wider (*select and commercial*).

___ (*c*) Timbers:

___ 5″ × 5″ and larger

___ Graded for strength and serviceability

___ May be classified as "structural."

___ (3) *Structural grades* (in descending order, according to stress grade):

___ (*a*) Light framing: *Construction, Standard,* and *Utility*

___ (*b*) Structural light framing (joists, planks): *Select Structural, No. 1, 2, or 3* (some species may also be appearance-graded for exposed work).

___ (*c*) Timber: *Select Structural No. 1.*

Note: Working stress values can be assigned to each of the grades according to the species of wood.

___ (4) *Appearance grades*

___ (*a*) For natural finishes: *Select A or B.*

___ (*b*) For paint finishes: *Select C or D.*

___ (*c*) For general construction and utility: *Common, Nos. 1 thru 5.*

___ *e.* *Pressure-treated wood:* Softwood lumber treated by a process that forces preservative chemicals into the cells of the wood. The result is a material that is immune to decay. This should not generally be used for interiors. Where required:

___ (1) In direct contact with earth

___ (2) Floor joists less than 18″ (or girders less than 12″) from the ground

___ (3) Plates, sills, or sleepers in contact with concrete or masonry

___ (4) Posts exposed to weather or in basements

___ (5) Ends of beams entering concrete or masonry, without ½″ air space

___ (6) Wood located less than 6″ from earth

___ (7) Wood structural members supporting moisture-permeable floors or roofs, exposed to weather, unless separated by an impervious moisture barrier

___ (8) Wood retaining walls or crib walls
___ (9) For exterior construction such as stairs and railings, in geographic areas where experience has demonstrated the need

___ f. *Framing-estimating rules of thumb:* For 16-inch oc stud partitions, estimate one stud for every LF of wall, then add for top and bottom plates. For any type of framing, the quantity of basic framing members (in LF) can be determined based on spacing and surface area (SF):

12 inches oc	1.2 LF/SF
16 inches oc	1.0 LF/SF
24 inches oc	0.8 LF/SF

(Doubled-up members, bands, plates, framed openings, etc., must be added.) Framing accessories, nails, joist hangers, connectors, etc., may be roughly estimated by adding *0.5 to 1.5% of the cost of lumber.* Estimating lumber can be done in *board feet* where one BF is the amount of lumber in a rough-sawed board one foot long, one foot wide, and one inch thick (144 cubic inches) or the equivalent volume in any other shape. As an example, one hundred one-inch by 12-inch dressed boards, 16 feet long, contain:

$$100 \times 1 \times 12 \times 16/12 = 1600 \text{ BF}$$

Use the following table to help estimate board feet:

BF per SF of surface

	12-inch oc	16-inch oc	24-inch oc
2 × 4s	0.8	0.67	0.54
2 × 6s	1.2	1.0	0.8
2 × 8s	1.6	1.33	1.06
2 × 10s	2.0	1.67	1.34
2 × 12s	2.4	2.0	1.6

Costs: Rough lumber costs by board feet.

Studs	$0.65 / BF
Posts	$0.75 / BF
Joist	$0.75 / BF
Beams (Douglas fir)	$0.85 / BF

The above are materials only. Total in-place cost
may be estimated by doubling the above numbers.
Stud walls: 2 × 4s @ 24″ oc: $1.20 (50% M and 50%
      L) with variation of ± 10%. Add 30% for
      each jump (i.e. 16″ and 12″ o.c.) 2 × 6s
      @ 24″ oc: $1.60/SF (M, L variation, and
      spacing the same as above)

___ 3. Details

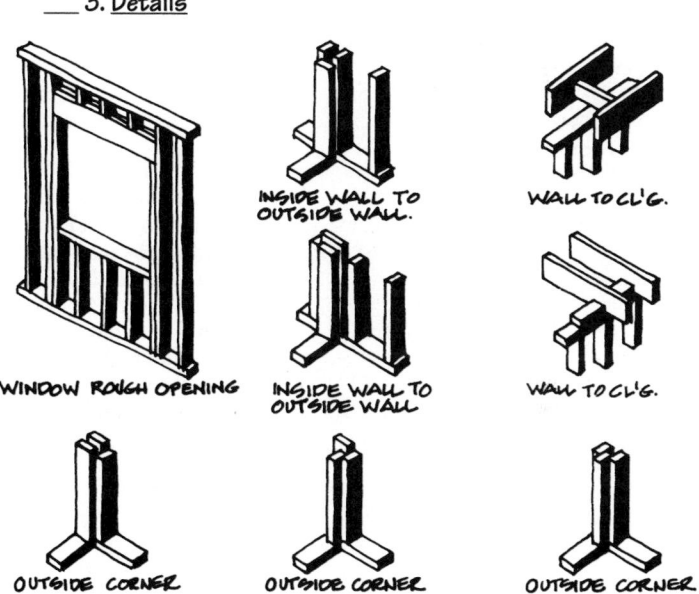

WINDOW ROUGH OPENING

INSIDE WALL TO OUTSIDE WALL.

WALL TO CL'G.

INSIDE WALL TO OUTSIDE WALL

WALL TO CL'G.

OUTSIDE CORNER

OUTSIDE CORNER

OUTSIDE CORNER

___ 4. Laminated Lumber

    ___ a. *Laminated timber* (glu-lam
      beams): For large structural
      members, these are
      preferable to solid
      timber in terms of
      finished dressed
      appearance, weather
      resistance, controlled
      moisture content, and
      size availability.

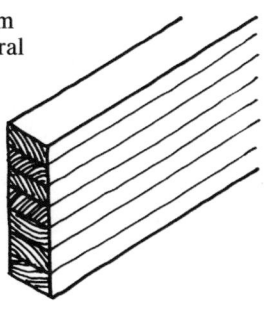

___ *b.* *Sheathing Panels*
___ (1) *Composites:* Veneer faces bonded to recon-
stituted wood cores
___ (2) *Nonveneered panels:*
___ *(a)* Oriented strand board (OSB)
___ *(b)* Particle board
___ (3) Plywood
_ *(a)* Two main types

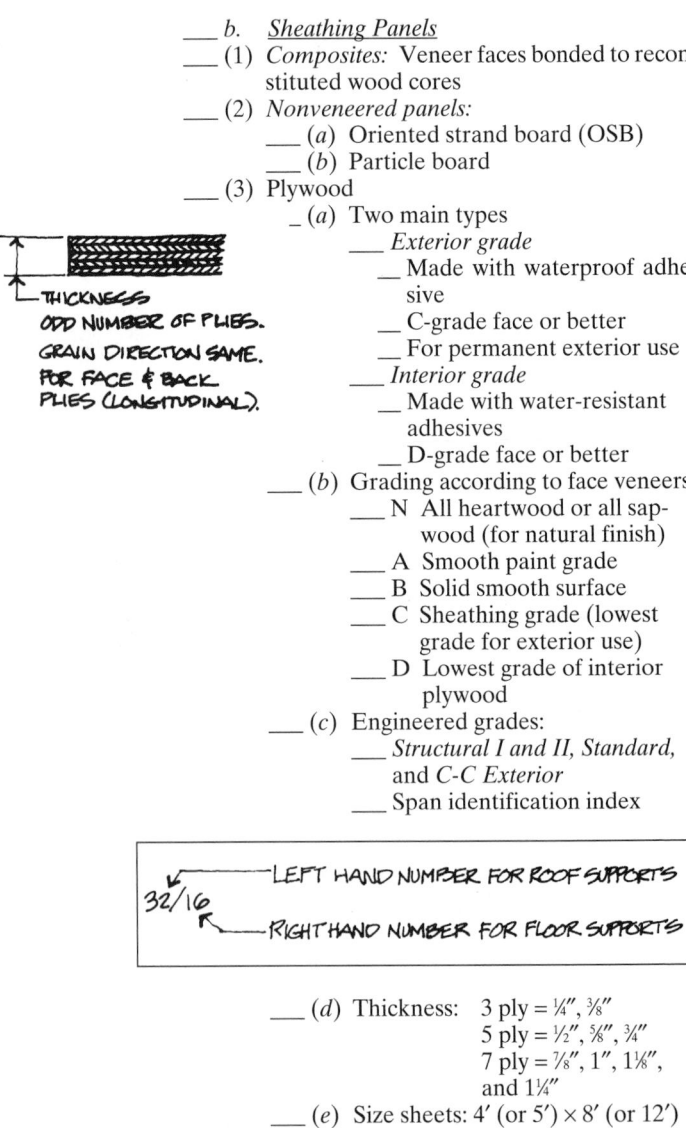

THICKNESS
ODD NUMBER OF PLIES.
GRAIN DIRECTION SAME.
FOR FACE & BACK
PLIES (LONGITUDINAL).

___ *Exterior grade*
__ Made with waterproof adhe-
sive
__ C-grade face or better
__ For permanent exterior use
___ *Interior grade*
__ Made with water-resistant
adhesives
__ D-grade face or better
___ *(b)* Grading according to face veneers
___ N All heartwood or all sap-
wood (for natural finish)
___ A Smooth paint grade
___ B Solid smooth surface
___ C Sheathing grade (lowest
grade for exterior use)
___ D Lowest grade of interior
plywood
___ *(c)* Engineered grades:
___ *Structural I and II, Standard,*
and *C-C Exterior*
___ Span identification index

LEFT HAND NUMBER FOR ROOF SUPPORTS
32/16
RIGHT HAND NUMBER FOR FLOOR SUPPORTS

___ *(d)* Thickness: 3 ply = ¼″, ⅜″
5 ply = ½″, ⅝″, ¾″
7 ply = ⅞″, 1″, 1⅛″,
and 1¼″
___ *(e)* Size sheets: 4′ (or 5′) × 8′ (or 12′)

___ 5. <u>Finish Wood</u> (Interior Hardwood Plywoods)
    ___ *a.*   *Sizes*
        ___ (1) Thicknesses: ⅛″ to 1″ in ¹⁄₁₆″ and ⅛″ increments
        ___ (2) Widths: 18″, 24″, 32″, 36″, and 48″
        ___ (3) Lengths: 4′, 5′, 6′, 7′, 8′, and 10′
    ___ *b.*   *Types*
        ___ (1) Technical: fully waterproof bond
        ___ (2) Type I (exterior): fully waterproof bond/ weather- and fungus-resistant
        ___ (3) Type II (interior): water-resistant bond
        ___ (4) Type III (interior): moisture-resistant bond
    ___ *c.*   *Grades*
        ___ (1) Premium 1: very slight imperfections
        ___ (2) Good 1: suitable for natural finishes
        ___ (3) Sound 2: suitable for painted finishes
        ___ (4) Utility 3: may have open defects
        ___ (5) Backing 4: may have many flaws
    ___ *d.*   *Grains and patterns*

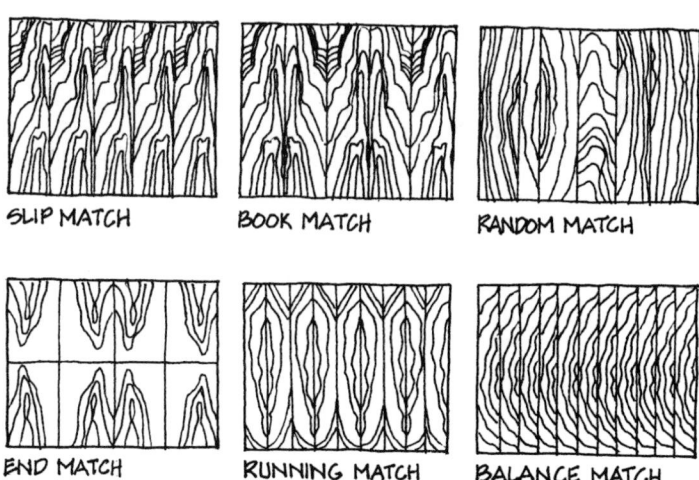

SLIP MATCH   BOOK MATCH   RANDOM MATCH

END MATCH   RUNNING MATCH   BALANCE MATCH

VENEER MATCH TYPES

Costs: Prefinished plywood paneling: $2.40 to $7.20/SF
     Trim: $4.20 to $7.80/LF
     Cabinetry: See p. *297*

SUNBURST BOX MATCH PARQUET MATCH

END GRAIN BOX HERRINGBONE SWING MATCH

DIAMOND REVERSE DIAMOND SKETCH FACE

SPECIAL WOOD VENEER MATCHING OPTIONS

___ 6. Cabinetry

Grades are often a function of surface treatment.

___ *a.* *Economy* (lowest grade)

Usually not shop-built. Has no back. Usually has a lipped door. The underside of counter is not specifically treated, thus some warpage may occur. Divisions between sections are open frame. Shelves are usually adjustable with clips.

___ *b.* *Custom* (average grade)

Usually shop-built. Has a back. The edges of all exposed plywood or particle board are covered. The divisions between one area and another are solid. The drawers have hardwood guides for better wear. Adjustable shelves are usually attached with recessed standards. Finish is usually plastic laminate.

___ *c.* *Premium* (best grade)

Shop-built. Best construction procedures and materials. The corners are mitered. There are solid panels between drawers to prevent dust travel. Drawers are made completely of hardwood and use high-quality hardware. The countertops are attached with hidden clips or screwed, not nailed down. All joints are screwed or glued with blocks. Adjustable shelves are usually attached with recessed standards. Finishes are often high-gloss plastic laminates or laquered wood.

___ *d.* *Costs:*

For standard grade (custom):

Wall cabinet = $85/LF
Base cabinet = $155/LF
Countertop = $35 to $85/LF

$$\overline{\text{\$275 to \$325/LF (90\%M and 10\%L)}}$$

For premium grade add 65%.
For economy grade deduct 45%.

___ 7. Plastic Laminates

___ *a.* High-pressure laminates are a sandwich of 8 to 10 layers of resin-impregnated papers converted by heat and pressure into a plasticlike material.

___ *b.* Thickness: $1/16''$ general purpose; $1/32''$ for vert. surfaces.

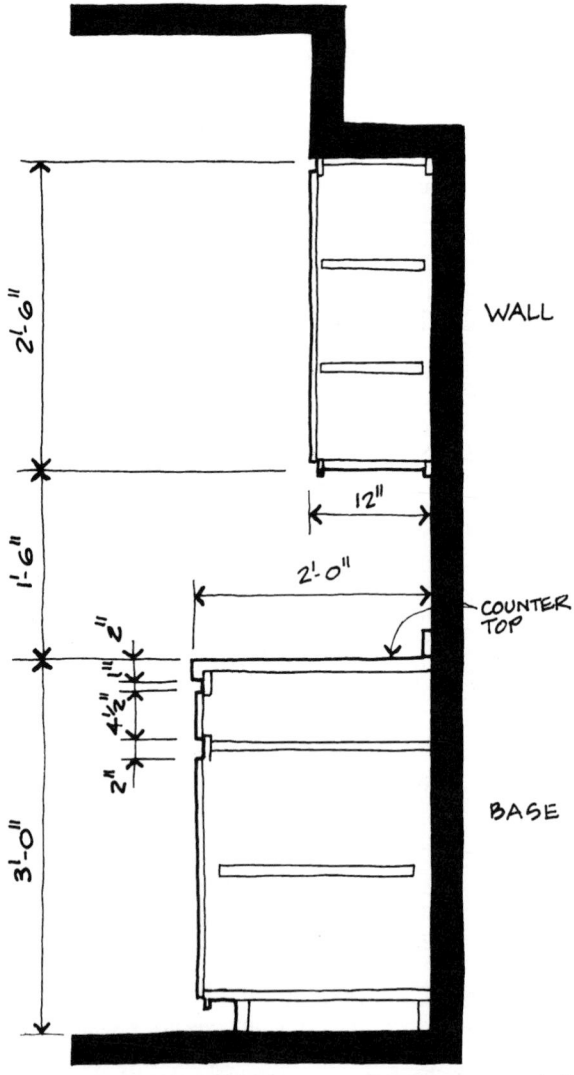

2'-6"

WALL

12"

2'-0"

COUNTER
TOP

1'-6"

2"
4½"
2"

BASE

3'-0"

SECTION THRU BASE AND WALL CABINET

___ *c.* Finishes:
  ___ gloss
  ___ satin
  ___ textured
  ___ low-glare
  ___ oil rub
___ *d.* Edge treatments

METAL METAL SELF BAND

WOOD WOOD ROLLED ROLLED

___ *e.* **Costs: materials only: $1.20 to $1.60/SF**

___ 8. <u>Solid Surfaces</u> (Solid Polymer Fabrications)
  ___ *a.* A modern synthetic with the appearance of marble or stone.
  ___ *b.* Leading manufacturer is "Corian," which is often used as a term for "solid surfaces."
  ___ *c.* Used for countertop vanities and work surfaces. Sometimes used for wall cladding and toilet partitions.
  ___ *d.* Size: ½″ typical for horizontal surface
    ¼″ for wall cladding
    Use on substrat (but
    some products can
    span up to 12′)
  ___ *e.* Made of
    ___ (1) Filler of minerals
    ___ (2) Binder of polymers
      (acrylic, polester, or
      plastic)

  ___ *f.* Colors and finishes vary from plain white and tans to grained, marbleized appearance. Surfaces come in matt, semigloss, and high gloss.
  ___ *g.* **Costs:**
    Average, ½″ thick, solid color, matt: $38 to $65/SF (50% M and 50% L)
    Top-end, with nonstandard color, semi- or high-gloss finish: $52 to $80/SF

— 7. SPECIES

● DENOTES COMMON USES & PROPERTIES
○ POSSIBLE OR LIMITED USE
□ TREATED WOOD ONLY
+ FLAME SPREAD RATING
SCALE OF 1 TO 10 WHERE 1 IS LOWEST & 10 IS HIGHEST.

SPECIES	COLOR	VENEERS	BOARDS/PLANK	DIMENSION	STRIPS/BLOCK	BLD'G. POSTS	FRAMING	SHEATHING	SIDING	PART'N FRAMING	PANELING
SOFTWOODS											
1 CEDAR, WESTERN RED	RED BROWN TO WHITE SAPW	○	●	●		●	●	●	●	●	●
2 CYPRESS, BALD	YELLOWISH BROWN		●	●		●	●	●	●	●	●
3 FIR, DOUGLAS (COAST)	REDDISH TAN		●	●	●	●	●	●	●	●	○
4 HEMLOCK, WESTERN	PALE BROWN		●	●		●	●	●	●		
5 LARCH, WESTERN	BROWN		●	●		●	●	●	●		
6 PINE – LODGEPOLE			●	●		○	●	○	○	○	
7 – PONDEROSA	WHITE TO PALE YELLOW		●	●		○	●	○	●	●	●
8 – RED	LIGHT BROWN		●	●		○	●	○	●	●	●
9 – SOUTHERN	WHITE TO PALE YELLOW		●	●	●	○	●	●	●	●	●
10 – SUGAR	CREAMY WHITE		●	●		●	●	●	●	●	●
11 REDWOOD – OLD GROWTH	DEEP RED TO DARK BROWN	○	●	●		●	○	○	○	○	●
12 SPRUCE – BLACK						○	○	○	○	○	
13 – ENGLEMAN	CREAMY WHITE		●	●		○	○	○	○	○	
14 – RED						○	●	○	○	○	
15 – SITKA	LIGHT YELLOWISH TAN		●	●		○	●	○	○	○	
HARDWOODS											
1 ASH, WHITE	CREAMY WHITE TO LT. BROWN	●			○						○
2 BEECH	WHITE TO REDDISH BROWN	●			●						●
3 BIRCH, YELLOW	LIGHT BROWN	●			●						●
4 CHERRY	REDDISH BROWN	●			●						●
5 ELM, AMERICAN	BROWN	●									●
6 LOCUST, BLACK	GOLDEN BROWN					○					○
7 MAHOGANY	REDDISH BROWN	●			●						●
8 MAPLE (HARD) SUGAR	WHITE TO REDDISH BROWN	●			●						●
9 OAK, RED	REDDISH TAN TO BROWN	●			●						●
10 POPLAR, YELLOW	WHT. TO BRN W/ GR. CAST	●								○	●
11 ROSE WOOD	MIXED REDS, BROWN, BLACK	●									●
12 TEAK	TAWNY YELLOW TO DRK BRN	●			●						●
13 WALNUT, BLACK	DARK BROWN	●	○								●

USES													PROPERTIES									NOTES	
FLOORS			ROOF		FOUND./OUTDOOR					EQUIPT													
JOISTS	ROUGH	FINISH	RAFTERS	DECKING	PILES	WD. FOUND	RET. WALLS	POSTS	DECKS	FURNITURE	CABINETS	FURNITURE	SHRINKAGE	BENDG. STRG	COMPRESSION =	COMPRESSION ⊥	HARDNESS SIDE	IMPACT BENDG	RESIST TO DECAY	WEATHERING	PAINTABILITY		
																						SOFTWOOD	
○			●	●				●	●	●	○	○	2	4	5	4	3	4	8	7	7	+78	1
○			●	●				●	●	●	○	○	5	6	6	6	6	6	8	7	7	+115	2
●	●	●	●	●	■	■	■	■	■				7	7	7	6	7	6	6	5	4	+145-130	3
●			●	●		■	■	■	■				7	6	6	5	6	7	5	5	5	+140-155	4
●			●	●	□				●				8	3	7	6	7	6	6	5	4		5
○			●						■				5	4	5	4	4	5	5	5	5	+60-75	6
○			●	●		■	■		■		○	○	4	3	4	5	4	5	5	5	6	+120-245	7
○			●	●		■			●		○	○	5	5	5	5	5	6	5	5	4	+105-200	8
○	●	●	●	●	■	■	■	■	■				7	7	7	6	7	6	5	5	5	+142	9
○			●	●					■		○	○	3	3	4	3	3	4	5	5	6		10
○			○	●				●	■	●	○	○	2	5	6	5	5	4	8	7	7	+75	11
○			○	○					□				5	4	5	3	5	5	5	5	5		12
○			○	○					□				5	3	3	3	3	4	5	5	5		13
○			○	○					□				6	5	5	5	5	4	5	5	5		14
○			●	●					□				6	5	5	5	5	5	5	5	5		15
																						HARDWOOD	
	○											●	5	6	6	6	6	6	4	5	5		1
											●	●	8	5	5	5	5	5	5	5	6		2
	○										●	●	7	5	4	4	5	6	5	5	6	+105-110	3
	○										●	●	3	4	5	3	4	3	6	5	5		4
												●	6	3	3	8	4	4	5	5	5		5
					○	○	○						2	8	8	8	8	5	8	5	5		6
		●								○	○												7
		●									○	●	6	6	6	6	6	5	5	5	6	+104	8
		●								○	○		7	3	3	5	5	3	5	5	5	+100	9
											○	○	4	3	3	2	3	3	5	5	7	+170-185	10
										●	●												11
	○		○							○	●	●											12
	○										●	●	4	6	6	4	6	4	7	6	5	+130-140	13

NOTES

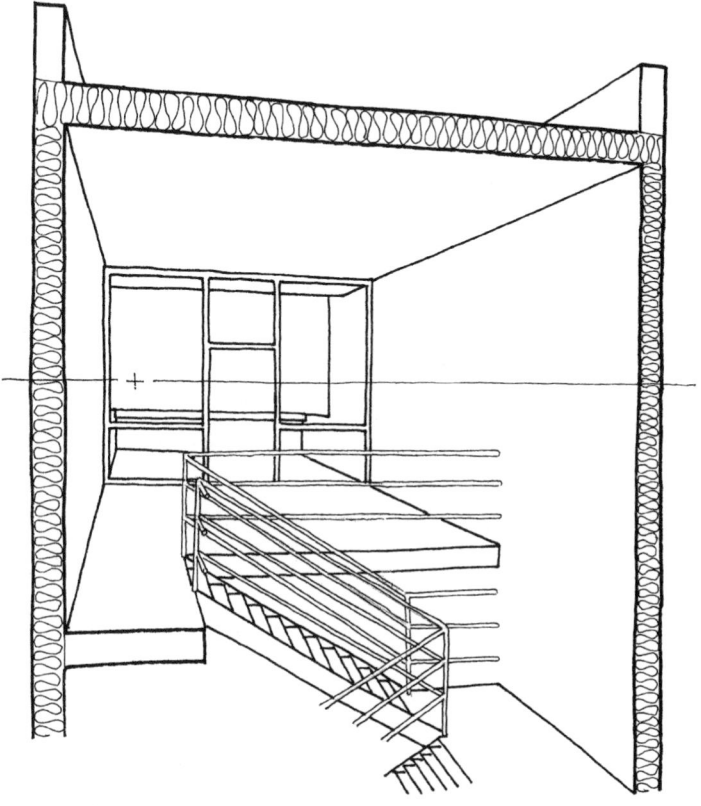

7
THERMAL
AND MOISTURE
PROTECTION

NOTES

___ A. WATER AND DAMPPROOFING

The interior designer may be involved with waterproofing where interior water is concerned. Examples might be showers and interior planters.

   ___ 1. Waterproofing

      Waterproofing is the prevention of water flow (usually under hydrostatic pressure such as saturated soil) into the building. This usually involves basement walls or decks, and can be achieved by:

      ___ a. Membranes: Layers of asphalt with plies of saturated felt or woven fabric

      ___ b. Hydrolithic: Coatings of asphalt or plastics (elastomeric)

      ___ c. Admixtures: To concrete

Typical costs:
      Elastomeric, ½₂″ neoprene: $3.95/SF (50% M and 50% L)
      Bit. membrane, 2-ply felt: $1.70/SF (35% M and 65% L)

   ___ 2. Dampproofing

      Dampproofing is preventing dampness (from earth or surface water without hydrostatic pressure) from penetrating into the building. This can be:

      ___ a. Below grade: 2 coats asphalt paint, dense cement plaster, silicons, and plastics.

      ___ b. Above grade: See paints and coatings, p. 348.

Typical costs:
      Asphalt paint, per coat: $0.25/SF (50% M and 50% L)

___ B. VAPOR BARRIERS

This section will give the interior designer an idea of vapor mitigation, if involved with roofs and exterior walls.

   ___ 1. **General**

      ___ *a.* Vapor can penetrate walls and roof by:

         ___ (1) Diffusion—vapor passes through materials due to:

            ___ (*a*) Difference in vapor pressure between inside and outside.

            ___ (*b*) Permeability of construction materials.

         ___ (2) Air leakage by:

            ___ (*a*) Stack effect

            ___ (*b*) Wind pressure

            ___ (*c*) Building pressure

      ___ *b.* Vapor is not a problem until it reaches its *dew point* and condenses into moisture, causing deterioration in the building materials of wall, roof, and floor assemblies.

   ___ 2. **Vapor Barriers**

Vapor barriers should be placed on the warm or humid side of the assembly. For *cold* climates this will be toward the inside. For *warm, humid* climates, this will be toward the outside. Barriers are also often put under slabs-on-grade to protect flooring from ground moisture.

Vapor barriers are measured by *perms* (grains/SF/hr/inch mercury vapor pressure difference). One grain equals about one drop of water. For a material to qualify as a vapor barrier, its perm rate must be *1.0* or less. A good perm rate for foil laminates, polyethylene sheets, etc. equals *0.1* or less (avoid aluminum foil against mortar). See p. 309 for perms of various materials. Care must be taken to avoid puncturing the barrier.

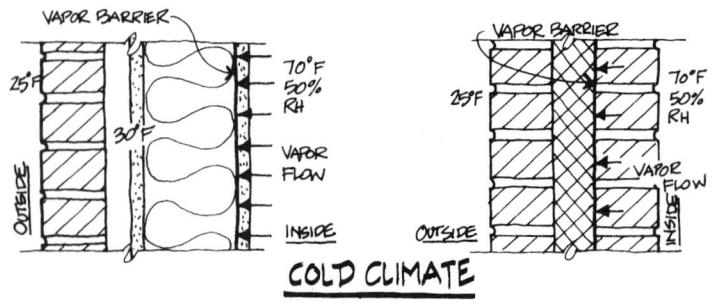

COLD CLIMATE

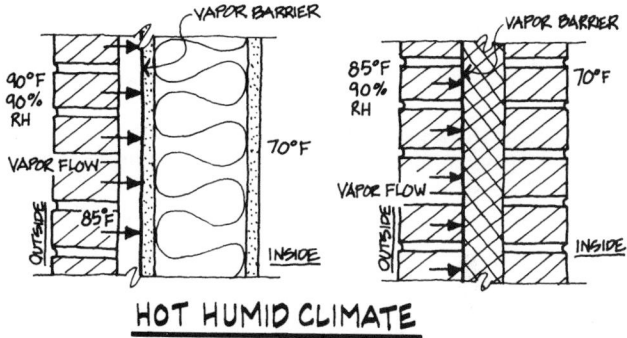

HOT HUMID CLIMATE

Other methods for sealing out moisture are elastomeric coatings on interior wall-board in cold climates and at exterior masonry or stucco walls in hot, wet climates. See p. 348 for coatings. Care must be taken to caulk all joints and cracks.

Typical costs: Polyethylene sheets, 2–10 mill. $.20 to $.30/SF

___ C. INSULATION

This section will give the interior designer typical insulation requirements and solutions for use if involved with roofs and exterior walls.

___ 1. **Insulation** is the entrapment of air within modern lightweight materials, to resist heat flow.

___ 2. **In the design of a building,** design the different elements (roof, wall, floor) to be at the minimum ΣR. Each piece of construction has some resistance to heat flow, with lightweight insulations contributing the bulk of the resistance.

$$\Sigma R = R1 + R2 + R3 + R4 + R5, \text{etc.} \qquad *(\text{air films})$$

See p. 309 for resistance (r) of elements to be added.

Another common term is U Value, the coefficient of heat transmission.

$$U = \text{Btuh/ft}^2/°F = \frac{1}{\Sigma R}$$

___ 3. **Other factors in control of heat flow:**

___ *a.* The *mass* (density or weight) of building elements (such as walls) will delay and store heat. Time lag in hours is related to thermal conductivity, heat capacity, and thickness. This increases as weight of construction goes up with about *½% per lb/CF.* Desirable time lags in temperate climates are: Roof—12 hrs; north and east walls—0 hrs; west and south walls—8 hrs. This effect can also be used to increase R values, at the approximate rate of *+0.4%* for every added lb/CF of weight.

___ *b.* *Light colors* will reflect and *dark colors* will absorb the sun's heat. Cold climates will favor dark surfaces, and the opposite for hot climates. For summer roofs, the overall effect can be 20% between light and dark.

___ 4. **Typical Batts:**
    R = 11   3½" thick
    R = 19   6"
    R = 22   6½"
    R = 26   8¼"
    R = 30   9"

Typical Costs:
    Ceiling batt, 6″ R = 19: $1.20/SF (60% M and 40% L)
                9″ R = 30: $1.30/SF
    Wall batt,   4″ R = 11: $.65/SF (50% M and 50% L)
                6″ R = 19: $.85/SF
    Add $.05/SF for foil backs.
    Rigid: $.70 to $1.80/SF, 1″, R = 4 to $1.20 to $4.20/SF, 3″,
    R = 12.5.

___ 5. Insulating Properties of Building Materials:

Material	lbs. weight	r value (per in)	Perm
Water	60		
Earth dry	75 to 95	0.33	
saturated		0.05	
Sand/gravel dry	100–120		
wet			
Concrete req.	150	0.11	
lt. wt.	120	0.59	
Masonry			
Mortar	130	0.2	
Brick, common	120	0.2	1 (4″)
8″ CMU, reg. wt.	85	1.11	0.4
lt. wt.	55	2	
Stone	±170	0.08	
Metals			
Aluminum	165	0.0007	0 (1 mil)
Steel	490	0.0032	
Copper	555	0.0004	
Wood			
Plywood	36	1.25	½″ = .4 to 1
Hardwood	40	0.91	
Softwood	30	1.25	2.9 (¾″)
Waterproofing			0.05
Vapor barrier			0.05
Insulations			
Min. wool batt	4	±3.2	>50
Fill		3.7	>50
Perlite	11	2.78	
Board polystyrene		4	1–6
fiber		2.94	
glass fiber		4.17	
urethane		8.5	

Material	lbs. weight	r value (per in)	Perm
Air			
Betwn. nonrefl.		1.34	
One side refl.		4.64	
Two sides refl.			
Inside film		0.77 (av)	
Outside film			
winter		0.17	
summer		0.25	
Doors			
Metal			
Fiber core		1.69	
Urethane core		5.56	
Wood, solid 1¾″		3.13	
HC 1⅜″		2.22	
Glass, single	160	0.90	
Plaster (stucco)	110	0.2	
Gypsum	48	0.6	
CT	145		
Terrazzo			
Acoustical CLGs			
Resilient flooring		0.05	
Carpet and pad		2.08	
Paint			0.3 to 1 (see p. 348)

NOTES

NOTES

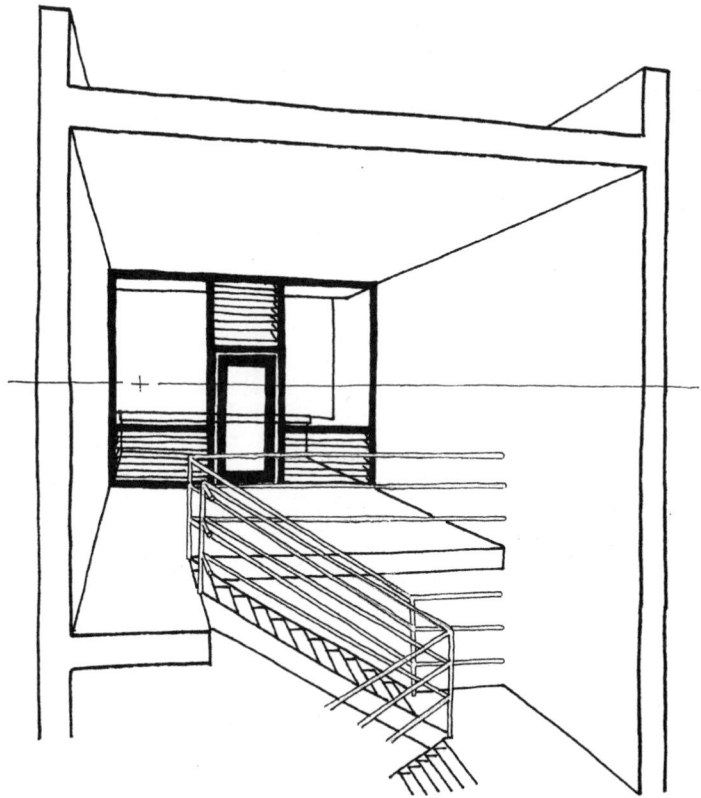

8

DOORS, WINDOWS, AND GLASS

NOTES

__ A. DOORS ④ ⑦ ⑫ ㊺A ㊾

__ 1. Accessible Door Approach (ADA) ⑳

X = 12" IF DOOR HAS BOTH A CLOSER AND LATCH, OTHERWISE X = 0"

X = 3' MIN. IF Y = 5'
X = 3'-6" MIN. IF Y = 4'-6"

Y = 4' MIN. IF DOOR HAS BOTH A CLOSER & LATCH, OTHERWISE Y = 3'-6"

Y = 4'-6" MIN. IF DOOR HAS A CLOSER, OTHERWISE Y = 4' MIN.

Y = 4' MIN. IF DOOR HAS A CLOSER, OTHERWISE, Y = 3'-6" MIN.

SLIDING & FOLDING DOORS

NOTE: ALL DOORS IN ALCOVES SHALL COMPLY W/ FRONT APPROACHES.

___ 2. *General*

   ___ *a.* Types by operation
        ___ (1) Swinging
        ___ (2) Bypass sliding
        ___ (3) Surface sliding
        ___ (4) Pocket sliding
        ___ (5) Folding

   ___ *b.* Physical types
        (1) Flush   (2) Panelled   (3) French   (4) Glass

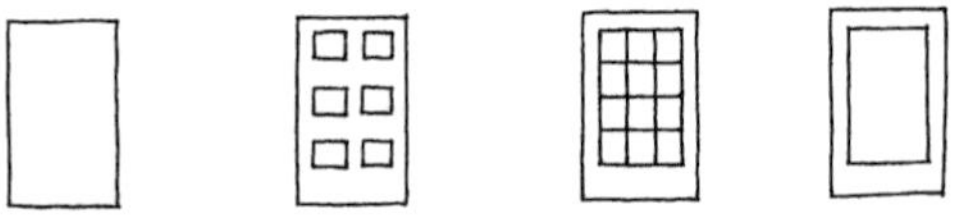

        (5) Sash   (6) Jalousie   (7) Louver

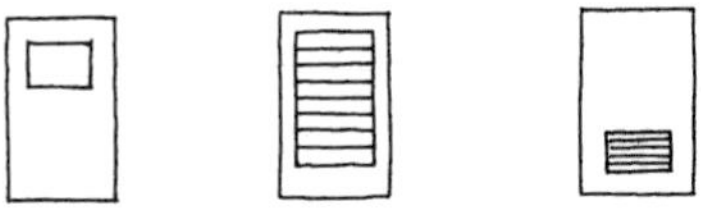

        (8) Shutter   (9) Screen   (10) Dutch

   ___ *c.* Rough openings (door dimensions +)

	Width	Height
In wood stud walls (r.o.)	+3½″	+3½″
In masonry walls (m.o)	+4″	+2″ to 4″

   ___ *d.* Fire door classifications: see p. 202.

   ___ *e.* Energy conservation: Specify exterior doors not to exceed:
        ___ (1) Residential: 0.5 CFM/SF infiltration
        ___ (2) Nonresidential: 11.0 CFM/LF crack infiltration
        ___ (3) Insulated to R-2.

___ 3. <u>Hollow Metal Doors and Frames</u>

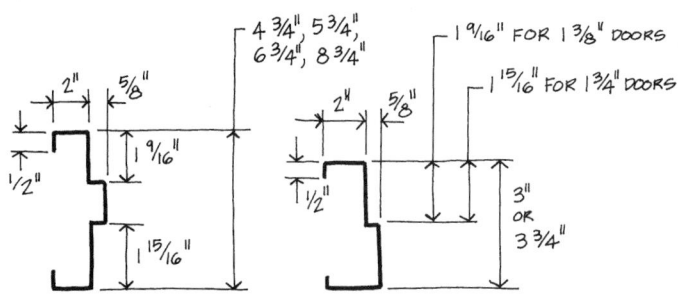

DOUBLE RABBET   SINGLE RABBET

___ *a.* Material (for gauges, see p. 278). Typical gauges of
doors (16, 18, 20) and frames (12, 14, 16, 18)

Use	Frame	Door face
Heavy (entries, stairs, public toilets, mech. rms.)	12, 14	16
Medium to low (rooms, closets, etc.)	14, 16, 20	18

___ *b.* Doors (total door construction of 16 to 22 GA)
  Thickness   1¾″ and 1⅜″
  Widths   2′ to 4′ in 2″ increments
  Heights   6′8″, 7′, 7′2″, 7′10″, 8′, 10′

**Costs: Frames: 3′ × 7′, 18 GA $8.10/SF (of opening) or 16 GA at
$7.65/SF (60% M and 40% L), can vary ±40%.**

**Doors: 3′ × 7′, 20 GA, 1¾″: $17.50/SF (85% M and 3′ × 6′8″,
20 GA, 1⅜″: $16.80/SF 15% L).**

**Add: lead lining: $720/ea.,   8″ × 8″ glass, $120/ea.,
soundproofing $35/ea.,   3-hour $145/ea.,   ¾-hour $30/ea.**

___ **4.** <u>Wood Doors</u> 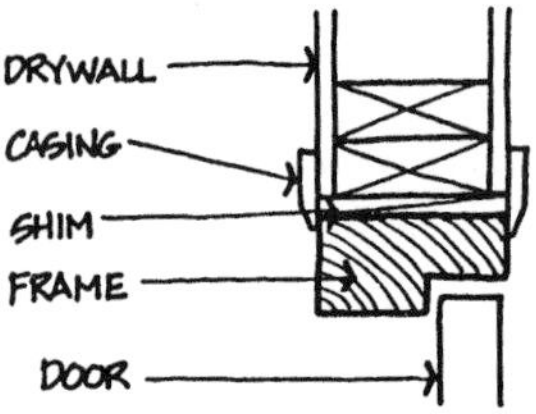
   ___ *a.* Types
      ___ (1) Flush
      ___ (2) Hollow core
      ___ (3) Solid core
      ___ (4) Panel (rail
          and stile)

DRYWALL
CASING
SHIM
FRAME
DOOR

   ___ *b.* Sizes
      Thickness: 1¾″ (solid core), 1⅜″ (hollow core)
      Widths: 1′6″ to 3′6″ in 2″ increments
      Heights: 6′, 6′6″, 6′8″, 6′10″, 7′
   ___ *c.* Materials (birch, lavan, tempered hardboard)

Flush	Panel
Hardwood veneer	#1: hardwood or pine for transp. finish
Premium: for transp. finish	#2: Doug fir plywood for paint
Good #3: For paint.	
Sound: (for paint only)	

   ___ *d.* Fire doors (with mineral composition cores) B and
      C labels.

Typical costs:
     Wood frame: interior, pine: $4.80/SF (of opening)
               exterior, pine: $9.35/SF
               (triple costs for hardwoods)
     Door: H.C. 1⅜″, hardboard $6.35/SF
          S.C. 1¾″, hardboard $13.00/SF (75% M and 25% L)
          Hardwood veneers about same costs.
     For carved solid exterior doors, multiply costs by 4 to 6.

___ 5. <u>Other Doors</u>
    ___ *a.* Sliding glass doors

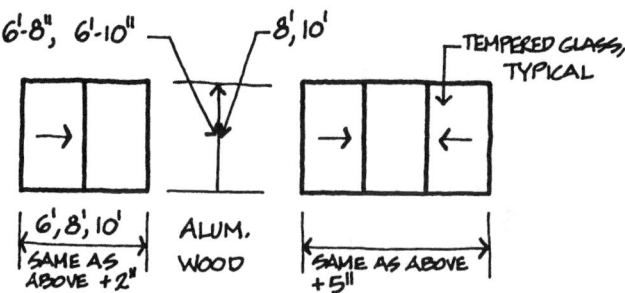

Typical costs (aluminum with ¼″ tempered glass):
    6′ wide: $925 to $1080/ea. (85% M and 15% L)
    12′ wide: $1465 to $2160/ea.
    Add 10% for insulated glass.

    ___ *b.* Aluminum "storefront" (7′ height typical)
**Typical cost with glass: $40/SF (85% M and 15% L). Variation of
–25% to +55%.**

    ___ *c.* Residential garage doors
        8′ min. width/car (9′ recommended)
        6′6″ min. height (7′2″ min. ceiling).
Costs: $34.50/SF (75% M and 25% L)

    ___ *d.* Folding doors
        2 panels: 1′6″, 2′0″, 2′6″, 3′0″ openings
        4 panels: 3′0″, 4′0″, 5′0″, 6′0″ openings
        6 panels: 7′6″ opening
        8 panels: 8′0″, 10′0″, 12′0″ openings
Costs: Accordion-folding closet doors with frame and trim: $28.00/SF

NOTES

__ B. WINDOWS ④ ⑫ ㉜ ⒋₅ₐ ㊾

For costs, see p. 324.

__ 1. _General_

__ *a.* In common with walls, windows are expected to keep out:

(1) Winter wind
(2) Rain in all seasons
(3) Noise
(4) Winter cold
(5) Winter snow
(6) Bugs and other flying objects
(7) Summer heat

They are expected, at the same time, to let in:

(1) Outside views
(2) Ventilating air
(3) Natural light
(4) Winter solar gain

__ *b.* Size designations: 3′ W × 6′ H = 3060

__ *c.* For types by operation, see p. 324.

__ *d.* For aid to selection of type, see p. 322.

__ *e.* Windows come in aluminum, steel, and wood. See pp. 326 and 327 for typical sizes.

__ *f.* Energy conservation: Specify windows to not exceed 0.34 CFM per LF of operable sash crack for infiltration (or 0.30 CFM/SF). NFRC ratings: U factor of .50 (0.25 better), SHGC of .40.

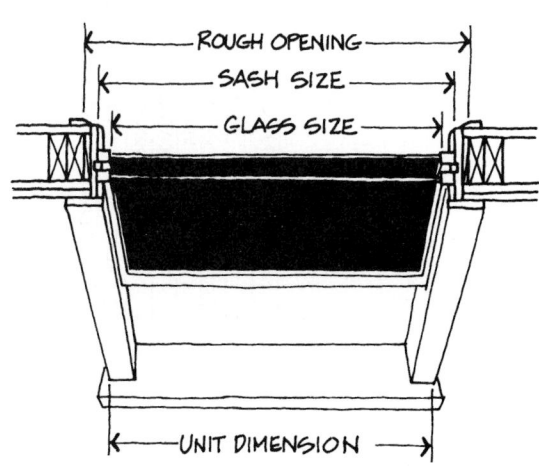

WINDOW TYPES — DISADVANTAGES

● INDICATES CHARACTERISTICS

WINDOW TYPE	ONLY 50% OF AREA OPENABLE	DOESN'T PROTECT FROM RAIN, WHEN OPEN	INCONVENIENT OPER. IF OVER OBSTRUCTION	HAZ'D. IF LOW VENT NEXT TO WALK	REQUIRES WEATHER STRIPPING	HORZ. MEMBERS OBSTRUCT VIEW	VERT. MEMBERS OBSTRUCT VIEW	WILL SAG IF NOT STRUCTURALLY STRONG	GLASS QUICKLY SOILS WHEN VENT OPEN	INFLOWING AIR CANNOT BE DIVERTED DOWN	EXCESSIVE AIR LEAKAGE	HARD TO WASH	INTERFERES W/ FURNITURE, DRAPES, ETC.	SCREENS – STORM SASH DIFFICULT TO PROV'D	SASH HAS TO BE REMOVED FOR WASHING
HORIZONTAL SLIDING	●	●	●		●		●			●			●	●	
PROJECTED					●		●			●	●				
MONITOR, CONTINUOUS					●		●			●				●	
JALOUSIE							●			●	●	●	●		
FIXED SASH														●	
BOTTOM HINGED, IN				●						●	●		●		
TOP HINGED, OUT			●	●	●					●	●				●
PIVOTED, HORIZONTAL			●	●	●					●			●	●	
PIVOTED, VERTICAL		●		●	●					●			●	●	
AWNING, CANOPY				●		●				●					
CASEMENT, IN				●			●	●					●		
CASEMENT, OUT		●		●	●		●	●							
DOUBLE HUNG, REVERSED	●	●	●		●	●				●					
DOUBLE HUNG	●	●	●		●	●				●				●	

WINDOW TYPES

O INDICATES CHARACTERISTICS

ADVANTAGES	NOT APT TO SAG	SCREEN & STORM SASH EASY TO INSTALL	PROVIDES 100% VENT OPENING	EASY TO WASH W/ PROPER HARDWARE	WILL DEFLECT DRAFTS	OFFERS RAIN PROTECTION, PARTLY OPEN	DIVERTS INFLOWING AIR UPWARD	ODD SIZES ECONOMICALLY AVAILABLE	LARGE SIZES PRACTICAL
DOUBLE HUNG	●	●							
DOUBLE HUNG, REVERSED	●	●		●					
CASEMENT, OUT			●		●				
CASEMENT, IN			●	●	●				
AWNING, CANOPY	●	●	●			●	●		
PIVOTED, VERTICAL			●	●	●				
PIVOTED, HORIZONTAL	●		●	●	●	●	●		
TOP HINGED, OUT	●	●	●			●			
BOTTOM HINGED, IN	●		●	●	●	●	●		
FIXED SASH	●								●
JALOUSIE			●		●	●	●	●	
MONITOR, CONTINUOUS	●		●		●	●	●		
PROJECTED						●	●		
HORIZONTAL SLIDING	●	●					●	●	

WINDOW TYPES BY OPERATION AND MATERIAL & COSTS

NOTE: GLASS EXCLUDED IN COSTS   * (90%M & 10%L)

TYPE		VENT	ALUMINUM	STEEL	WOOD
FIXED		0%		$25/SF AVE.	$24.50/SF AVE. * VARIATION -10% +20% PICTURE WINDOW
CASEMENT		100%		$25 TO $35/SF AVE. (85%M & 15%L)	$50/SF AVE ** VARIATION +70%, -40%
PROJECTED	AWNING	50 TO 100%	$33.50 TO $40/SF AVE. (75%M & 25%L) 7	$35 TO $40/SF AVE. *	$55/SF M AVE. (85%M & 15%L) VARIATION +60%, -40%
	HOPPER				
SLIDING		50 TO 100%	$25 TO $27/SF AVE. (80%M & 20%L)		$35/SF AVE. VARIATION ± 60%

DOUBLE-HUNG	50%	$25.50 TO $28.00/ SF AVE. *	$47/SF AVE *	$41.50/SF AVE. (85%M & 15%L) VARIATION +70%, -45%
JALOUSIE	100%	$27.50/SF AVE. (80%M & 20%L)		
PIVOTING	100%		$27.50/SF AVE. (85%M & 15%L)	

TYPICAL WOOD WINDOW SASH SIZES

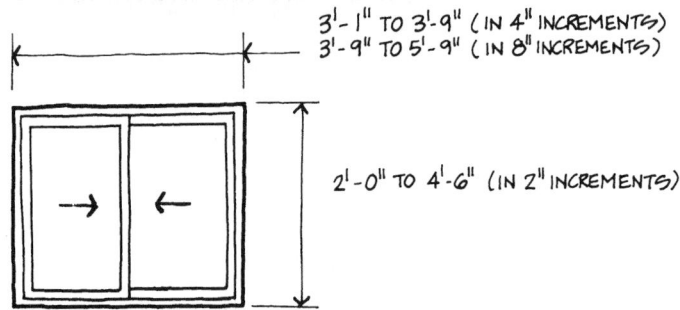

3'-1" TO 3'-9" (IN 4" INCREMENTS)
3'-9" TO 5'-9" (IN 8" INCREMENTS)

2'-0" TO 4'-6" (IN 2" INCREMENTS)

HORIZONTAL SLIDING WINDOWS

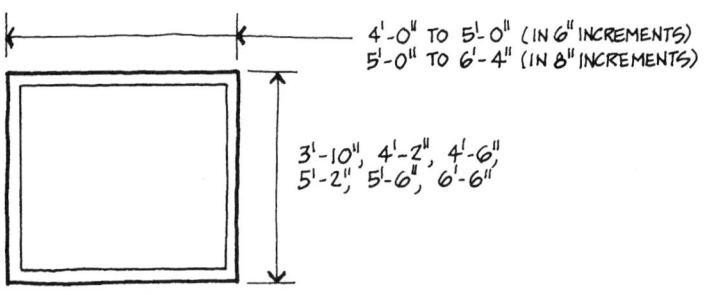

4'-0" TO 5'-0" (IN 6" INCREMENTS)
5'-0" TO 6'-4" (IN 8" INCREMENTS)

3'-10", 4'-2", 4'-6",
5'-2", 5'-6", 6'-6"

PICTURE WINDOWS

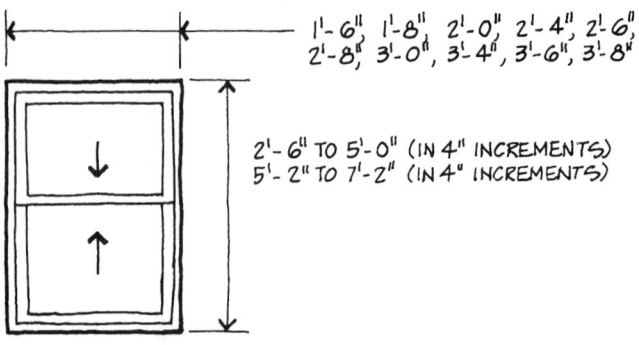

1'-6", 1'-8", 2'-0", 2'-4", 2'-6",
2'-8", 3'-0", 3'-4", 3'-6", 3'-8"

2'-6" TO 5'-0" (IN 4" INCREMENTS)
5'-2" TO 7'-2" (IN 4" INCREMENTS)

DOUBLE HUNG WINDOWS

ALUM. : RESIDENTIAL SIZES
STEEL : NO STD SIZES

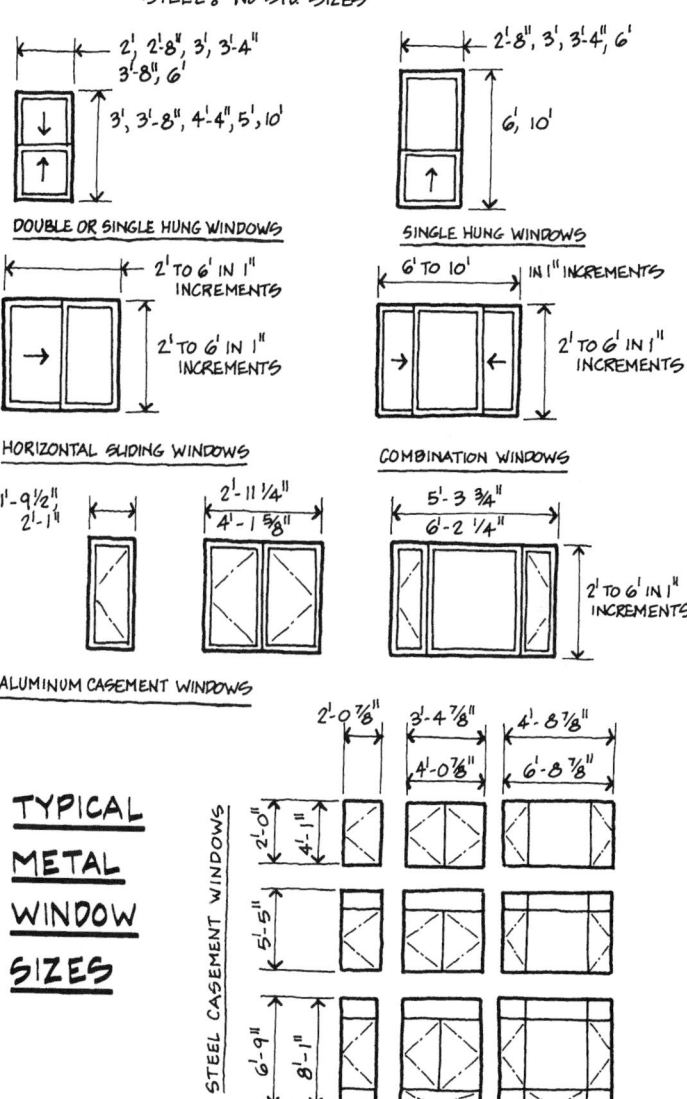

2', 2'-8", 3', 3'-4"
3'-8", 6'

3', 3'-8", 4'-4", 5', 10'

DOUBLE OR SINGLE HUNG WINDOWS

2'-8", 3', 3'-4", 6'

6', 10'

SINGLE HUNG WINDOWS

2' TO 6' IN 1"
INCREMENTS

2' TO 6' IN 1"
INCREMENTS

HORIZONTAL SLIDING WINDOWS

6' TO 10'   IN 1" INCREMENTS

2' TO 6' IN 1"
INCREMENTS

COMBINATION WINDOWS

1'-9½"
2'-1"

2'-11¼"
4'-1⅝"

5'-3¾"
6'-2¼"

2' TO 6' IN 1"
INCREMENTS

ALUMINUM CASEMENT WINDOWS

TYPICAL

METAL

WINDOW

SIZES

STEEL CASEMENT WINDOWS

2'-0⅞"   3'-4⅞"   4'-8⅞"

4'-0⅞"   6'-8⅞"

2'-0"
4'-1"

5'-5"

6'-9"
8'-1"

NOTES

___ **C. HARDWARE** (12) (20) (45A)

___ 1. <u>General Considerations</u>: How to . . .
   ___ *a.* Hang the door
   ___ *b.* Lock the door
   ___ *c.* Close the door
   ___ *d.* Protect the door
   ___ *e.* Stop the door
   ___ *f.* Seal the door
   ___ *g.* Misc. the door
   ___ *h.* Electrify the door

___ 2. <u>Recommended Locations</u>   ___ 3. <u>Door Hand Conventions</u>

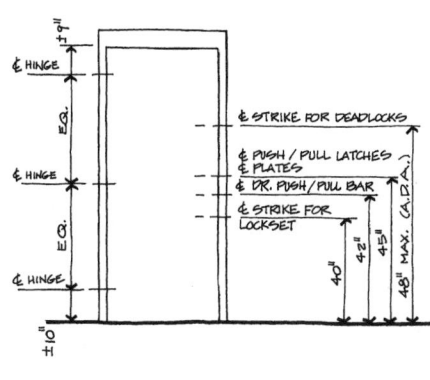

DIRECTION OF TRAVEL ASSUMED TO BE FROM OUTSIDE IN OR FROM KEYED SIDE FOR INTERIOR DOORS

___ 4. <u>Specific Considerations</u>
   ___ *a.* Function and ease of operation
   ___ *b.* Durability in terms of:
      ___ (1) Frequency of use
         ___ (*a*) Heavy
         ___ (*b*) Medium
         ___ (*c*) Light
      ___ (2) Exposure to weather and climate (aluminum and stainless steel good for humid or coastal conditions)
   ___ *c.* Material, form, surface texture, finish, and color.

___ 5. <u>Typical Hardware</u>
   ___ *a.* Locksets (locks, latches, bolts)
   ___ *b.* Hinges
   ___ *c.* Closers
   ___ *d.* Panic hardware
   ___ *e.* Push/pull bars and plates

___ *f.* Kick plates
___ *g.* Stops and holders
___ *h.* Thresholds
___ *i.* Weatherstripping
___ *j.* Door tracks and hangers

___ 6. <u>Materials</u>
___ *a.* Aluminum
___ *b.* Brass
___ *c.* Bronze
___ *d.* Iron
___ *e.* Steel
___ *f.* Stainless steel

___ 7. <u>Finishes</u>

BHMA #	US #	Finish	Base material
___ 600	US P	Primed for painting	Steel
___ 601	US 1B	Bright japanned	Steel
___ 602	US 2C	Cadmium plated	Steel
___ 603	US 2G	Zinc plated	Steel
___ 605	US 3	Bright brass, clear coated	Brass*
___ 606	US 4	Satin brass, clear coated	Brass*
___ 611	US 9	Bright bronze, clr. coat	Bronze*
___ 612	US 10	Satin bronze, clear coated	Bronze*
___ 613	US 10B	Oxidized satin bronze, oil rubbed	Bronze*
___ 618	US 14	Bright nickel plated, clear coated	Brass, Bronze*
___ 619	US 15	Satin nickel plated, clear coated	Brass, Bronze*
___ 622	US 19	Flat black coated	Brass, Bronze*
___ 624	US 20A	Dark oxidized statuary bronze, clr. coat	Bronze*
___ 625	US 26	Bright chromium plated	Brass, Bronze*
___ 626	US 26D	Satin chromium plated	Brass, Bronze*
___ 627	US 27	Satin aluminum, clr. coat	Aluminum
___ 628	US 28	Satin aluminum, clear anodized	
___ 629	US 32	Bright stainless steel	
___ 630	US 32D	Satin stainless steel	
___ 684	—	Black chrome, bright	Brass, Bronze*
___ 685	—	Black chrome, satin	Brass, Bronze*

*Also sometimes applicable to other base materials.

___ 8. <u>ADA-Accessible Hardware</u>

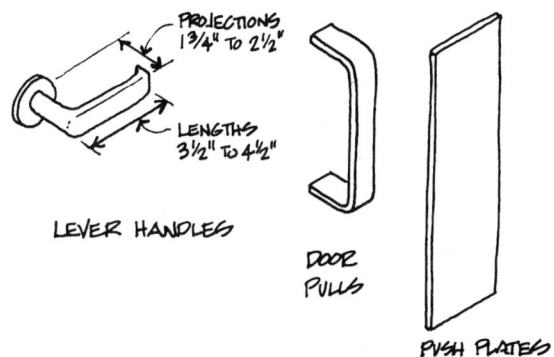

LEVER HANDLES

DOOR
PULLS

PUSH PLATES

___ 9. Costs:

Residential: **$115/door (80% M and 20% L)**
                          Variation −30%, +120%

Commercial:
Office:
  Interior: **$235/door (75% M and 25% L)**
  Exterior: **$450/door (add ≈ $425 for exit devices)**
  Note: **Special doors, such as for hospitals, can cost up to $685/door**

NOTES

___ D. GLASS (26) (27) (45A) (49)

___ 1. General

Glass is one of the great modern building materials because it allows the inside of buildings to have a *visual relationship* with the outside. However, there are a number of *problems* associated with its use.

___ 2. Legal Requirements

The IBC requires *safety glazing* at locations hazardous to human impact. Safety glazing is *tempered glass, wired glass,* and *laminated glass.* Hazardous locations are:

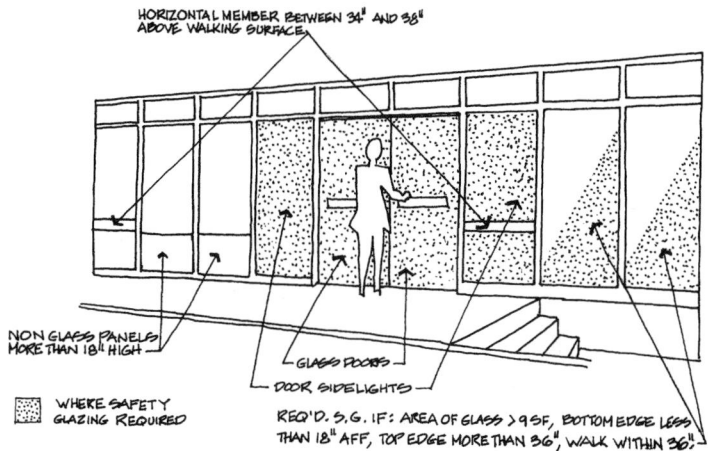

HORIZONTAL MEMBER BETWEEN 34" AND 38" ABOVE WALKING SURFACE

NON GLASS PANELS MORE THAN 18" HIGH

GLASS DOORS

DOOR SIDELIGHTS

▨ WHERE SAFETY GLAZING REQUIRED

REQ'D. S.G. IF: AREA OF GLASS > 9 SF, BOTTOM EDGE LESS THAN 18" AFF, TOP EDGE MORE THAN 36", WALK WITHIN 36"

SAFETY GLAZING

___ 3. Costs:

¼″ clear float glass: $8.50 to $10.80/SF (45% M and 55% L)

Modifiers:

Thickness:

⅛″ glass	−30%
⅜″ glass	+40%
½″ glass	+110%

Structural:

Tempered	+20%
Laminated	+100%

Thermal:

Tinted or reflective	+20%
Double-glazed and/or low E	+100%

NOTES

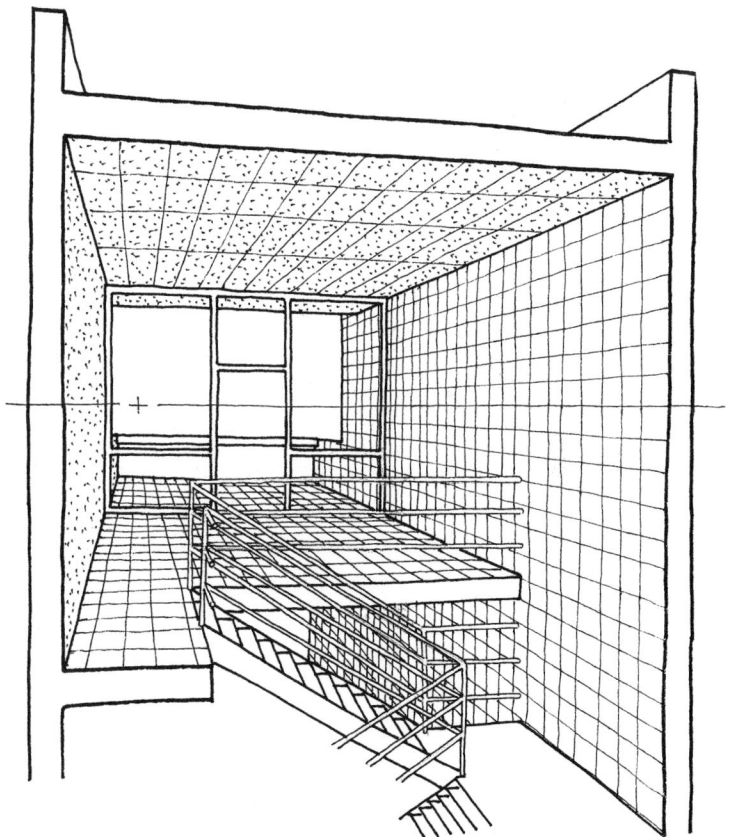

9 FINISHES

NOTES

___ A. PLASTER

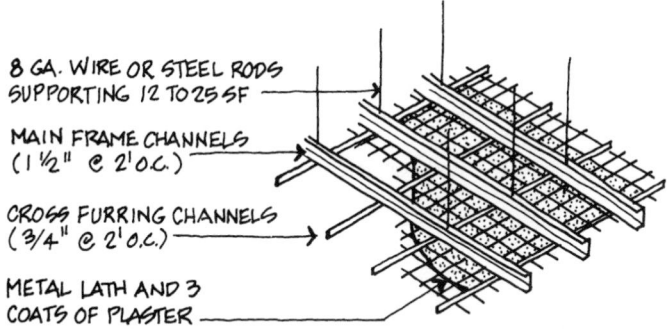

8 GA. WIRE OR STEEL RODS
SUPPORTING 12 TO 25 SF

MAIN FRAME CHANNELS
(1½" @ 2' O.C.)

CROSS FURRING CHANNELS
(¾" @ 2' O.C.)

METAL LATH AND 3
COATS OF PLASTER

**TYPICAL
CEILING**

___ 1. Exterior (stucco) of cement plaster.
___ 2. Interior of gypsum plaster.
___ 3. Wall supports usually studs at between 12″ and 24″ oc. If wood, use 16″ oc min.
___ 4. Full plaster—3 coats (scratch, brown, and finish), but masonry walls can have 1 or 2 coats.
___ 5. Joints: Interior ceilings: 30 oc max.
Exterior walls/soffits: 10 to 20 oc.
___ 6. Provide vents at dead air spaces (½″/SF).
___ 7. Curing: 48 hours moist curing, 7 days between coats.

Costs:

Ceilings with paint, plaster, and lath	**$2.40 to $7.80/SF (25% M and 75% L), can vary up to +60% for plaster**
Walls of stucco with paper-backed wire lath	**$2.65/SF for stucco + $1.20/SF for lath (50% M and 50% L)**

__ B. GYPSUM WALLBOARD (DRYWALL)

___1. **Usually in** 4' × 8' (or 12')
sheets from ¼" to 1" thick
in about ⅛" increments.

___2. **Attach** (nail or screw)
against wood or metal
framing—usually at 16"
(fire rating) to 24" oc.

___3. **Type "X"**, ⅝" will give 1 hr.
fire rating. Roughly each
additional ½" layer will
give 1 hr. rating up to
4 hours, depending on
backing and application.

___4. **Water-resistant** (green)
available for wet areas or
exterior.

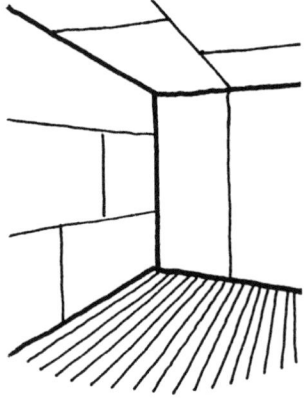

Costs:

½" gyp. bd.   **$.85/SF ceilings**
on wood   **$1.20/SF columns and beams**
frame   **$.85/SF walls**
(Approx. 50% M and 50% L)

**Increase 5% for metal frame. Varies about 15% in cost for ⅛' ea.
thickness. Add $.10/SF for fire resistance. Add $.17/SF for water
resistance. Add $.50/SF for joint work and finish.**

EXAMPLE:

FIND THE COST OF ⅝" GYPB'D. WALL ON FRAME, READY
FOR PAINT.
       ½" = $0.85/SF (WALL) + 12¢ (15% FOR EXTRA
       ⅛" THICKNESS) + $0.50 FOR FINISH.
   ∴   ⅝" = $1.47 /SF, SAY $1.50/SF

___ C. TILE ④ ⑫

___ 1. **Settings**
  ___ *a.* Thick set (¾″ to 1¼″ mortar bed) for slopes.
  ___ *b.* Thin set (⅛″ mortar or adhesive) for faster and less expensive applications.

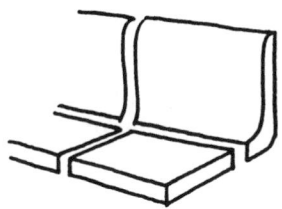

___ 2. **Joints:** ⅛″ to ¼″ (can be epoxy grouted for quarry tile floors).

___ 3. **Types**
  ___ *a.* Ceramic glazed and unglazed for walls and floors of about ¼″ thick and 4–6″ SQ. Many trim shapes available.
  ___ *b.* Ceramic mosaic for walls and floors of about ¼″ thick and 1″ to 2″ SQ.
  ___ *c.* Quarry tile of earth tones for strong and resistant flooring. Usually ½″ to ¾″ thick by 4″ to 9″ SQ.

Typical Costs:

Note: Costs can vary greatly with special imports of great expense.

Glazed wall tile: $7.10/SF (50% M and 50% L), variation of −25%, +100%

Unglazed floor mosaic: $10.65/SF (65% M and 35% L), variation of +35%, −10%

Unglazed wall tile: $7.90/SF (40% M and 60% L), variation of +35%, −15%

Quarry tile: $11.65/SF (same as above), variation of ±10%

Bases: $11.65/LF (same as above), variation of ±10%

Additions: color variations: +10 to 20%
abrasive surface: +25 to 50%

__ D. TERRAZZO

___ 1. A poured material
(usually ½″ thick) of
stone chips in a cement
matrix, usually with a
polished surface.
___ 2. Base of sand and concrete.
___ 3. To prevent cracking,
exposed metal dividers
are set approx. 3′ to 6′
oc each way.

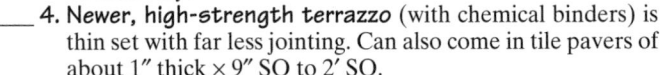

___ 4. Newer, high-strength terrazzo (with chemical binders) is
thin set with far less jointing. Can also come in tile pavers of
about 1″ thick × 9″ SQ to 2′ SQ.

Costs: $10.80/SF to $18.00/SF (45% M and 55% L)
Tiles: $20.40 to $33.50/SF

__ E. ACOUSTICAL TREATMENT

___ 1. Acoustical Ceilings: Can consist of small (¾″ thick × 1′ SQ)
mineral fiber tiles
attached to wallboard
or concrete (usually
glued). Also, acousti-
cal mineral fibers with
a binder can be shot
on gypsum board or
concrete.

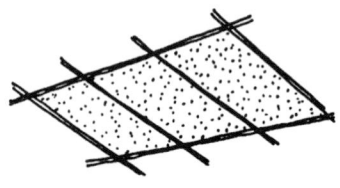

Costs: Small tiles $1.30 to $1.80/SF (40% M and 60% L)

___ 2. Suspended Acoustical Tile Ceilings: Can be used to create a plenum space to conceal mechanical and electrical
functions. Typical applications are 2′ SQ or 2′ × 4′ tiles in
exposed or concealed metal grids that are wire-suspended
as in plaster ceilings. The finishes can vary widely.

Costs: Acoustical panels $1.20 to $2.40/SF (70% M and 30% L)
Suspension system $1.20 to $1.50/SF (80% M and 20% L)
When walls do _not_ penetrate ceilings, can save $0.10 to
$0.25/SF.

___ 3. <u>Other</u>

    ___ a. Other types of suspended ceilings can range from exposed metal to wood, to luminous baffles.

Costs: Special suspended ceilings and suspension system:
    Metal w/ acoustic batts above: \$9.95/SF (50% M and 50% L)
    Luminous, plastic: \$4.90/SF (75% M and 25% L)

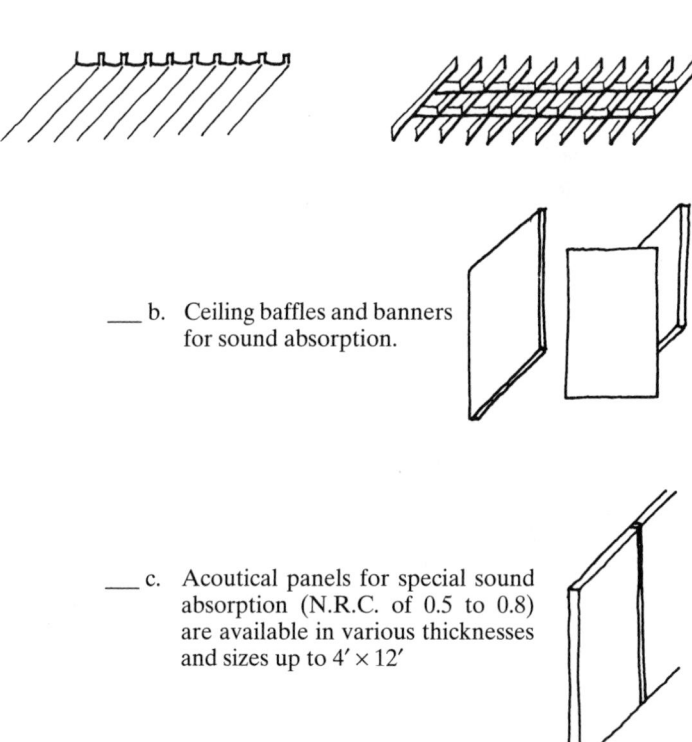

    ___ b. Ceiling baffles and banners for sound absorption.

    ___ c. Acoutical panels for special sound absorption (N.R.C. of 0.5 to 0.8) are available in various thicknesses and sizes up to 4′ × 12′

Costs: ¾″ fabric-covered panels, wall-mounted:
    \$9.70/SF (75% M and 25% L)

___ F. WOOD FLOORING ④ ⑫

___ 1. Finished flooring can be of hardwoods or softwoods, of which oak, southern pine, and Douglas fir are the most commonly used.

___ 2. All-heartwood grade of redwood is best for porch and exterior flooring.

___ 3. If substrate is concrete, often flooring is placed on small wood strips (sleepers); otherwise flooring is often nailed to wood substrates (plywood or wood decking).

___ 4. Because wood is very susceptible to moisture, allowance must be made for movement and ventilation. Allow expansion at perimeters. Vapor barriers below concrete slabs are important.

___ 5. Use treated material in hot, humid climates.

___ 6. Three types of wood flooring:
    ___ *a.* Strip
    ___ *b.* Plank
    ___ *c.* Block (such as parque)

UP TO 3¼" WIDE

OVER 3¼" WIDE

Typical Costs:
    Wood strip fir $6.00/SF (70% M and 30% L)
    Oak +90% +10% finish
    Maple +100% clean and wax = $0.45/SF

___ G. MASONRY FLOORING ④ ⑫

See Part 4 on materials.

Typical Costs:
    1¼" × 4" × 8" brick: $11.00/SF
    (65% M and 35% L)
    Add 15% for special patterns.

___ 1. **Resilient Flooring Consists of:**

   ___ *a.* *Sheet vinyl:* Most common of sheet flooring. Use sheet vinyl where it is desirable to have the fewest joints such as in high-maintenance wet areas. Can be located below, on, or above grade. Not the most resilient. Durability is good to moderate. Poor in quietness.

   ___ *b.* *Vinyl tiles:* Vinyl tiles are the most commonly used of resilient floorings. Can be of homogeneous or composition materials. Can be located below, on, or above grade. Moderately good to poor resilience. Best in durability. Poor in quietness. Also used for wall bases.

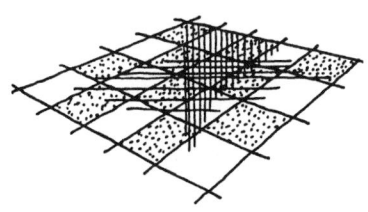

   ___ *c.* *Rubber tiles:* Can come in special studded designs for greater slip resistance. Can be located below, on, or below grade. Have good resilience, durability, and quietness. Also used for static resistance and wall base as well as stair treads.

   ___ *d.* *Cork tiles:* Most comfortable to walk on because of their high resilience. Also the quietest, but the worst for durability. Can be located on grade.

   ___ *e.* *Linoleum sheet or tiles:* Has mostly passed out of use, along with asphalt tiles. Can be used on slab on grade only. Poor for resilience and quietness. Has moderate durability.

___ 2. **Is approx. ¹⁄₁₆″ to ⅛″ thick with tiles being 9 to 12″ square.**

___ 3. **Applied to substrate with mastic.** Substrate may be plywood flooring, plywood or particleboard over wood deck, or concrete slabs.

___ 4. **Vapor barriers are often required under slabs.**

___ 5. **Vinyl or rubber base is often applied at walls for this and other floor systems.**

___ 6. **Protect resilient flooring from furniture weights by using cups or casters.**

A wide range of colors and patterns is available for flooring.

Typical Costs:

Solid vinyl tile ⅛″ × 12″ × 12″	**$3.60/SF (75% M and 25% L), can go up 20% for various patterns and colors; double for "conductive" type.**
Sheet vinyl	**$3.30/SF (90% M and 10% L), variation of −70% and +100% for various patterns and colors.**
Vinyl wall base	**$2.10/LF (40% M and 60% L). Can vary +15%.**
Stair treads	**$9.60/LF (60% M and 40% L). Can vary from −10% to +40%.**

___ 1. **Most wall-to-wall carpeting** is produced by looping yarns through a coarse-fiber backing, binding the backs of the loops with latex, then applying a second backing for strength and dimensional stability. Finally the loops may be left uncut for a rough, nubby surface or cut for a soft, plush surface.

___ 2. **The quality of carpeting** is often determined by its *face weight* (ounces of yarn or pile per square yard), not its total weight. Weights run:

    ___ *a.* Low traffic: 20–24 oz/SY
    ___ *b.* Medium traffic: 24–32 oz/SY
    ___ *c.* High-end carpet: 26–70 oz/SY

___ 3. **A better measure of comparison:**

$$weight\ density\ factor = \frac{face\ weight \times 36}{pile\ height} = oz/CY$$

Ideally, this should be as follows:

    ___ *a.* Residential: 3000 to 3600 oz/CY
    ___ *b.* Commercial: 4200 to 7000 oz/CY

___ 4. **Flame spread:** see p. 172.

___ 5. **There are two basic carpet installation methods:**

    ___ *a.* *Padded and stitched* carpeting: Stretched over a separate pad and mechanically fastened at joints and the perimeter. Soft foam pads are inexpensive and give the carpet a soft, luxurious feel. The more expensive jute and felt pads give better support and dimensional stability. Padding adds to foot comfort, helps dampen noise, and some say, adds to the life of the carpet.

    ___ *b.* *Glued-down* carpets: Usually used in commercial areas subject to heavily loaded wheel traffic. They are usually glued down with carpet adhesive with a pad. This minimizes destructive flexing of the backing and prevents rippling.

___ 6. **Maintenance Factors**

    ___ *a.* Color: Carpets in the midvalue range show less soiling than very dark or very light colors. Consider the typical regional soil color. Specify patterned or multicolored carpets for heavy traffic areas in hotels, hospitals, theaters, and restaurants.

    ___ *b.* Traffic: The heavier the traffic, the heavier the density of carpet construction. If rolling traffic is a factor, carpet may be of maximum density for minimum resistance to rollers. Select only level-loop or dense, low-cut pile.

___ 7. Carpet Materials:

Fiber	Advantages	Disadvantages
Acrylic (rarely used)	Resembles wool	Not very tough; attracts oily dirt
Nylon (most used)	Very tough; resists dirt, resembles wool; low-static buildup	None
Polyester deep pilings	Soft and luxurious	Less resilient; attracts oily dirt
Polypropylene indoor-outdoor	Waterproof; resists fading and stains; easy to clean	Crushes easily
Wool	Durable; easy to clean; feels good; easily dyed	Most expensive

___ 8. **Costs:** **(90% M and 10% L) (Variation ±100%) See p. 477 for interiors wholesale/retail advice. Figure 10% waste.**

Repair/level floors: $2.15 to $8.00/SY (45% M and 55% L)

Padding
Sponge: $7.40/SY (70% M and 30% L) Variation ±10%
Jute: –10%
Urethane: –25%
Carpet
Acrylic, 24 oz, med. wear: $27.10/SY
28 oz, med./heavy: $33.60/SY
Residential
Nylon, 15 oz, light traffic: $19.50/SY
28 oz, med. traffic: $23.75/SY
Commercial
Nylon, 28 oz, med. traffic: $25.00/SY
35 oz, heavy: $29.50/SY
Wool, 30 oz, med. traffic: $39.60/SY
42 oz, heavy: $55.00/SY
Carpet tile: $3.30 to $6.60/SY

CARPET TYPES

| TYPE OF WEAVE | CHARACTERISTICS AND BEST USES |

 **LEVEL LOOP :** EVEN HEIGHT, TIGHTLY SPACED UN-CUT LOOPS. TEXTURE IS HARD AND PEBBLY. HARD WEARING AND EASY TO CLEAN. IDEAL FOR OFFICES AND HIGH TRAFFIC AREAS.

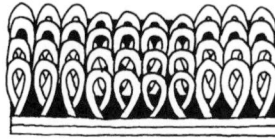 **MULTI-LEVEL LOOP :** UNEVEN HEIGHT IN PATTERNS, TIGHTLY SPACED UNCUT LOOPS. TEXTURE IS HARD & PEBBLY. HARD-WEARING & EASY TO CLEAN. IDEAL FOR OFFICES AND HIGH TRAFFIC AREAS.

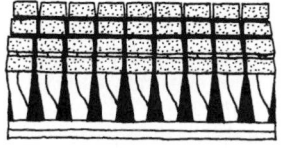 **PLUSH 'CUT' PILE :** EVENLY CUT YARNS WITH MINIMAL TWIST. EXTREMELY SOFT, VELVETY TEXTURE. VACUUMING AND FOOTPRINTS APPEAR AS DIFFERENT COLORS, DEPENDING ON LIGHT CONDITIONS. IDEAL FOR FORMAL ROOMS W/ LIGHT TRAFFIC.

 **FRIEZE 'CUT' PILE :** EVENLY CUT YARNS WITH TIGHT TWIST. EXTREMELY SOFT, VELVETY TEXTURE. VACUUMING AND FOOTPRINTS AP-PEAR AS DIFFERENT COLORS, DEPENDING ON LIGHT CONDITIONS. IDEAL FOR FORMAL RM'S WITH LIGHT TRAFFIC.

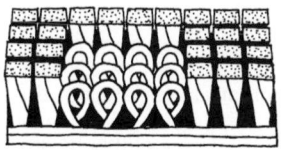 **CUT AND LOOP :** COMBINATION OF BOTH PLUSH AND LEVEL-LOOP. HIDES DIRT FAIRLY WELL. IDEAL FOR RESIDENTIAL APPLICATIONS.

 **INDOOR-OUTDOOR :** CUT, TIGHTLY TWISTED YARNS THAT TWIST UPON THEMSELVES. TEXT-URE IS ROUGH. HIDES DIRT EXTREMELY WELL AND IS NEARLY AS TOUGH AS LEVEL-LOOP. IDEAL FOR RESIDENTIAL APPLICATIONS.

___ J. PAINT AND COATINGS (12) (49)

___ 1. **General**

    ___ *a.* Paints and coatings: are liquids (the "vehicle") with pigments in suspension, to protect and decorate building surfaces.

    ___ *b.* Applications: brushed, rolled, sprayed

    ___ *c.* Failures: 90% are due to either moisture problems or inadequate preparation of surface.

    ___ *d.* Surface Preparation:

       ___ (1) Wood: Sand if required; paint immediately.

       ___ (2) Drywall: Let dry (0 to 7 days). If textured surface is required, prime prior to texturing.

       ___ (3) Masonry and stucco: Wait for cure (28 days).

    ___ *e.* Qualities:

       ___ (1) Thickness

          ___ (*a*) Primers (and "undercoats"): ½ to 1 dry mills/coat.

          ___ (*b*) Finish coats: 1 to 1½ dry mills/coat.

       ___ (2) Breathability: Allowing vapor passage to avoid deterioration of substrate and coating. Required at (see p. 306):

          ___ (*a*) Masonry and stucco: 25 perms

          ___ (*b*) Wood: 15 perms

          ___ (*c*) Metals: 0 perms

    ___ *f.* Paint Surfaces:

       ___ (1) Flat: Softens and distributes illumination evenly. Reduces visibility of substrate defects. Not easily cleaned. Usually used on ceilings.

       ___ (2) Eggshell: Provides most of the advantages of gloss without glare.

       ___ (3) Semigloss

       ___ (4) Gloss: Reflects and can cause glare, but also provides smooth, easily cleanable, nonabsorbtive surface. Increases visibility of substrate defects.

    ___ *g.* Legal Restrictions:

       ___ (1) Check state regulations on paints for use of volatile organic compounds (VOC), use of solvents, and hazardous waste problems.

       ___ (2) Check fire department restrictions on spraying interiors after occupancy or during remodelling.

___ 2. <u>Material Types</u>
    ___ *a.* Water-repellent preservatives: For wood.
    ___ *b.* Stains: Solid (opaque), semitransparent, or clear.
    ___ *c.* Wood coatings: Varnish, shellac, lacquers.
    ___ *d.* Wood primer-sealer: Designed to prevent bleeding through of wood resin contained in knots and pitch pockets, and to seal surface for other coatings. Usually apply 2 coats to knots. Since primer-sealer is white, cannot be used on clear finishes.
    ___ *e.* Latex primer: Best first coat over wallboard, plaster, and concrete. Adheres well to any surface except untreated wood.
    ___ *f.* Alkyd primer: Used on raw wood. Latex "undercoats" can also be used.
    ___ *g.* CMU filler: A special latex primer for reducing voids and to smooth surface on masonry. Does not waterproof.
    ___ *h.* Latex paint: A synthetic, water-based coating, this is the most popular paint because it complies with most environmental requirements, is breathable, and cleans up with water. Use for almost all surfaces including primed (or undercoated) wood. Adheres to latex and flat oils. Avoid gloss oils and alkyds other than primers. Subdivided, as follows:
        ___ (1) Polyvinyl acetate (PVA): Most commonly used. Provides 25+ perms.
        ___ (2) Acrylic: Smoother, more elastic, more durable, often used as a primer. Provides less than 5 perms.
    ___ *i.* Alkyd paint: A synthetic semisolvent-based coating, replacing the old oils. This seems to be going out of use due to environmental laws. Used for exterior metal surfaces. Not breathable.
___ 3. <u>Paint Systems and Costs</u> **(30% M and 70% L) The following was developed for the SW U.S. Add 10% for the rest of the country.**

Comparison of Paint Finish Systems: ICI Paints

Product Type	Applied Cost per sq. ft.	Areas of Use	Benefits	Liabilities
Interior: (premium finishes)				
Latex Flat Wall Paint (GWB = 2 coats)	30¢ - 35¢	Any interior surface where a flat appearance is desired.	Almost unlimited color selection, washable, low odor, dries fast, excellent touch-up and coverage.	Flats are considered washable but not scrubbable(like enamels).
Latex Eggshell Enamel (1 coat primer, 2 coats finish)	40¢ - 50¢	Where a slightly higher sheen is desired instead of a flat.	Same as flat, but more durable and scrubbable than a flat with greater moisture resistance; best quality eggshell enamels have good "block resistance".	More scrubbable than a flat, but less durable than a higher sheen.
Latex Semi-Gloss Enamel (1 coat primer, 2 coats finish)	40¢ - 50¢	Kitchens, bathrooms, doors and trim, cabinets, etc., wherever a medium sheen is needed.	Same as flat, but another step up in durability, scrubbability and moisture resistance compared to a lower sheen product; best quality semi-gloss enamels have good "block resistance".	Increased scrubbability over any lower sheen product, but not as durable as higher sheens. (Note: As sheens rise, hiding is reduced.)
Latex Gloss Enamel (1 coat primer, 2 coats finish)	40¢ - 50¢	Kitchens, bathrooms, doors and trim, cabinets, etc., wherever a high sheen is required.	Same as flat, but a high gloss product with maximum durability, scrubbability and moisture resistance when compared to a lower sheen product.	Highest degree of scrubbability, but lowest hiding.
Waterborne Semi-Gloss or Gloss Epoxy (2-component) (1 coat primer, 2 coats finish)	70¢ - 85¢	Any interior wall surface, metal, concrete block and wood. Ideal for hard usage areas in schools, hospitals, restaurants, public buildings and factories.	Same as latex products, but provides maximum durability and highest performance in a hard, tough and stain-resistant finish.	Dries for recoat overnight, but does not fully cure for 7 days.
Exterior: (premium finishes)				
Latex Heavy-Bodied Stain (Wood = 2 coats)	35¢ - 40¢	Exterior wood surfaces, especially fascias and soffits, where grain of the wood's natural texture is to be highlighted.	Extensive color selection, water-repellent, mildew-resistant, low odor and guards against wood rot.	Not recommended for wood decks, floors, outdoor wood furniture, or brushed or abraded plywood surfaces.

Product	Cost	Use	Description	Comments
Latex Flat Finish (1 coat primer, 2 coats finish)	40¢ - 50¢	Any exterior surface where a flat appearance is desired.	Almost unlimited color selection, washable, low odor, dries fast, excellent touch-up and coverage, can be applied over a variety of properly primed surfaces; fade, chalk and mildew resistant.	Exterior flats do not tend to be as 'self-cleaning' as enamels.
Latex Satin Enamel (1 coat primer, 2 coats finish)	40¢ - 50¢	Where a slightly higher sheen is desired rather than a flat.	Same as flat, but more durable than a flat with greater 'self-cleaning' attributes. Moisture, fade, chalk and mildew resistant.	Easier to clean than a flat, but less durable than a higher sheen.
Latex Semi-Gloss Enamel (1 coat primer, 2 coats finish)	40¢ - 50¢	Use wherever a medium sheen finish is desired.	Same as flat, but another step up in moisture resistance and durability compared to a lower sheen product. Excellent fade, chalk and mildew resistance.	Increased cleanability over any lower sheen product, but not as durable as higher sheens.
Latex Gloss Enamel (1 coat primer, 2 coats finish)	40¢ - 50¢	Use wherever a high gloss finish is desired.	Same as flat, but a high gloss product with excellent durability, moisture resistance, fade, chalk and mildew resistance, when compared to a lower sheen enamel.	Highest degree of cleanability and durability, but least amount of hide.
Aliphatic Urethane Gloss Enamel (2-component) (1 coat primer, 2 coats finish)	70¢ - 85¢	A high-performance, chemically-cured urethane enamel.	Exceptional gloss and color retention, excellent abrasion and chemical resistance, wide color selection (including safety colors), excellent resistance to marring, chipping and scratching.	Product cost is highest of any exterior enamel, but length of service is longest with highest UV protection. Has also been used as anti-graffiti coating. Best applied by a painting contractor with product experience.
Power-washing	8¢ - 12¢	Exterior concrete and CMU surfaces.	Best method for cleaning exterior surfaces prior to repainting.	None.

NOTES

___ K. WALL COVERINGS

___ 1. <u>Wallpaper</u>
    ___ a. *General*
        Comes in many different patterns and colors at usu-
        ally moderate cost, but is often subject to soiling,
        abrasion, and fading.
    ___ b. *Types:*
        ___ (1) Prepasted or unpasted
        ___ (2) Trimmed or untrimmed
        ___ (3) Washable and scrubbable
        ___ (4) Lining (used for foils)
        ___ (5) Strippable (paste stays on wall when paper
            is stripped off wall).
    ___ c. *Patterns*
        ___ (1) Repeat (one pattern per roll)
        ___ (2) Random patterns
        ___ (3) Straight match (i.e., plaids)
        ___ (4) Drop-match (repeat every other roll)
    ___ d. Allow a 20% margin for waste.
    ___ e. Other finishes include vinyl, foils, fabrics, felts, and
        wood veneers.

___ 2. **Costs: Regular wallpaper: $0.95 to $1.25/SF (50% M and 50% L)**
Grass cloths: $1.75 to $3.65/SF
Flexible wood veneer: $6.20 to $7.45/SF
Vinyl: $1.25 to $2.60/SF
Aluminum Foil: $2.60/SF

NOTES

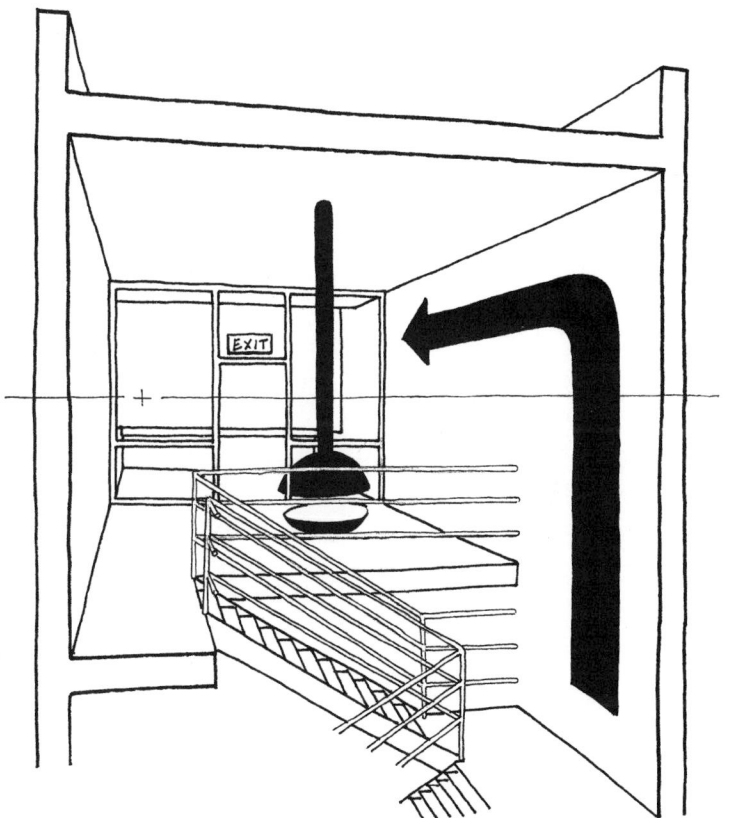

10 SPECIALTIES

NOTES

__ A. VISUAL DISPLAY BOARDS (4) (40)

__ 1. Chalkboards
    __ *a.* *Type*
        Porcelain enamel standard, painted on composition or natural slate of different core construction (except slate).
    __ *b.* *Sizes*
        Thickness: ⅟₃₂ to ⅜″. 4′ × 12′ typical dimensions.
        Costs: Wall hung, $6.00 to $15.35/SF (95% M and 5% L).

__ 2. Bulletin or Tackboards
    __ *a.* *Type*
        Cork or fiberboard (vinyl- or burlap-covered)
    __ *b.* *Sizes*
        Thickness: ⅛ to ½″. Height: 4′. Lengths: 8′, 12′, 14′, 16′+
        Costs: Cork, unframed, ½″ thick: $7.20/SF (65% M and 35% L).

__ B. TOILET PARTITIONS (4)

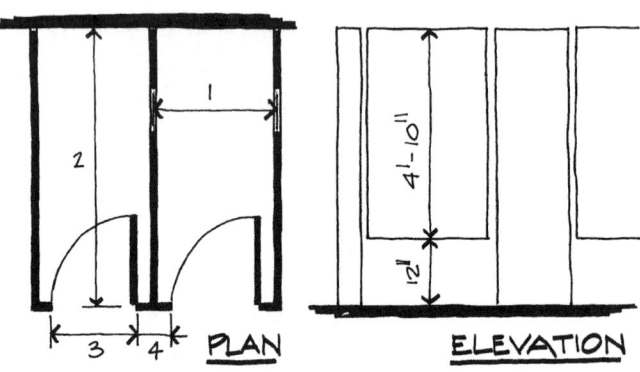

PLAN   ELEVATION

___ 1. **Types:**
    ___ *a.* Floor mounted (with pilasters, as shown).
    ___ *b.* Wall-hung (must provide supports in wall).
    ___ *c.* Ceiling-hung (must provide supports above ceiling).
___ 2. **Finishes:**
    Baked enamel, porcelain enamel, plastic laminate, stainless steel, marble (not wall-hung).
___ 3. **Typical Widths:** 2′6″, 2′8″, 2′10″ (most-used), and 3′0″
___ 4. **Typical Depths:**
    ___ *a.* Open front: 2′6″ to 4′0″
    ___ *b.* Closed front (door): 4′6″ to 4′9″
___ 5. **Typical Doors:** 1′8″, 1′10″, 2′0″, 2′4″, and 2′6″
___ 6. **Typical Pilasters:** 3, 4, 5, 6, 8, or 10 inches
___ 7. For **HC-accessible,** see pp. 96–97.

Costs: Painted metal	(75 to 85% M)
	(plastic lam.: +25 to 35%)
	(stainless steel: +100 to 160%)
	(marble: +130 to 190%)
Floor-mounted:	$625/stall
Wall-hung:	$755/stall
Ceiling-hung:	$755/stall
For HC stall w/ grab bars:	add +$350/stall
For urinal screen:	$360/screen

___ 1. Range from 3″ or 4″ angles cast into concrete or masonry (usually for exteriors) to ½″ clear plastic on interior wood finishes. A typical interior application in a utilitarian space with high traffic (such as a commercial kitchen) might be a surface applied 3″ × 3″ × 3′ high stainless steel angle.

___ 2. **Costs:**
Stainless steel, 16 ga., 3½″ angle: $30/LF
Clear plastic, 2½″ angle: $12.95/LF

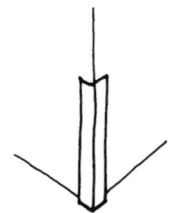

__ D. ACCESS FLOORING (4)

___ 1. **Used in** offices, hospitals, laboratories, open area schools, computer rooms, telecommunications centers, and so forth. They provide mechanical and electrical accessibility and flexibility in placing desks, computers, telephone services, machines, and general office equipment.

___ 2. **Types are** steel, aluminum, steel or aluminum encasing wood or cementous fill or lightweight concrete. Top surface is usually carpet or vinyl.

___ 3. **Sizes are usually** 2′ × 2′ with 1½″ thickness. Clearance below is usually 4 to 12″.

___ 4. **Be sure to check for** added ceiling height required and floor-mounted equipment weight.

___ 5. **Costs: 24″ square steel panels w/carpet: $17.50 (80% M and 20% L)**

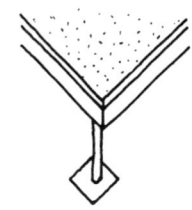

___ 1. **Typical Opening Sizes** (see drawings below):

W	H	D	S
2'	1.5' to 1.75'	1.33' to 1.5'	
3'	2'	1.67'	6½"
4'	2.12'	1.75'	6"
5'	2.5' to 2.75'	2' to 2.17'	9"
6'	2.75' to 3'	2.17' to 2.33'	9"

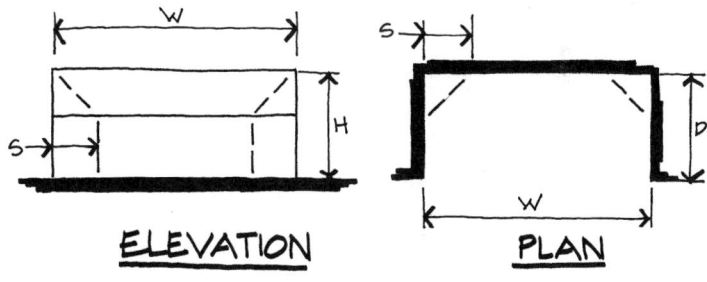

ELEVATION   PLAN

___ 2. **For energy conservation, provide:**
    ___ *a.* Outside combustion air ducted to firebox
    ___ *b.* Glass doors
    ___ *c.* Blower

___ 3. **Per IBC:**
    ___ *a.* Hearth extension to front must be 16" (or 20" if opening greater than 6 SF).
    ___ *b.* Hearth extension to side must be 8" (or 12" if opening greater than 6 SF).
    ___ *c.* Thickness of wall of firebox must be 10" brick (or 8" firebrick).
    ___ *d.* Top of chimney must be 2' above any roof element within 10'.

**Costs: Fabricated metal: $600 to $1800 (75% M and 25% L)
Masonry: $6000 to $18,000**

__ **1. General:** Visual identification and direction by signage is very important for "wayfinding" to, between, around, in, and through buildings. Signage is enhanced by:

- __ *a.* Size
- __ *b.* Contrast
- __ *c.* Design of letter character and graphics.

__ **2. Road Signage:** Can be roughly estimated as follows:

SPEED MPH	VIEWING DISTANCE	ANGLE	SIGN SIZE SF	COPY SIZE INCH HT.
15	220'		8	
30	310'		40	5
40	450'	35°		7
45	660'		90	
50	545'	30°		8½
60	610 - 880'	20°	150	9½

__ **3. Building Signage**
- __ *a.* Site directional/warning signs should be:
    - __ (1) 6' from curb
    - __ (2) 7' from grade to bottom
    - __ (3) 100'–200' from intersections
    - __ (4) 1 to 2.5 FT SQ
- __ *b.* Effective pedestrian viewing distance 20' to 155'
- __ *c.* Effective sign size: ≈10'/inch height (10' max. viewing distance per inch of height of sign).
- __ *d.* Effective letter size: ≈50'/inch height.
- __ *e.* As a rule, letters should constitute about 40% of sign and should not exceed 30 letters in width.
- __ *f.* Materials
    - __ (1) Exterior
        - __ (*a*) Building: fabricated aluminum, illuminated plastic face, back-lighted, cast aluminum, applied letter, die-raised, engraved, and hot-stamped.
        - __ (*b*) Plaque and sign: cast bronze or aluminum, plastic/acrylic, stone, masonry, and wood.

_____ (2) Interior
       _____ (*a*) Permanent mounting: vinyl tape/adhesive backing, silastic adhesive, or mechanical attachment.
       _____ (*b*) Semipermanent: vinyl tape square on inserts.
       _____ (*c*) Changeable: dual-lock mating fasteners, magnets, magnetic tape or tracks.

_____ *g.* Mounting heights

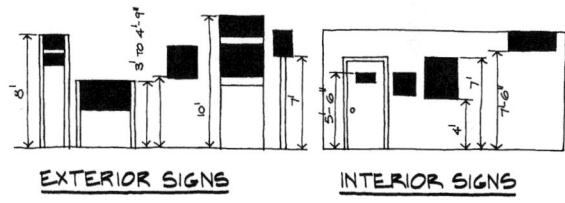

EXTERIOR SIGNS   INTERIOR SIGNS

_____ *h.* Accessibility signage per ADA required at:
    _____ (1) Building entries (when accessible, not required when all are).
    _____ (2) Accessible facilities, such as at rest rooms (when accessible, not required when all are).
    _____ (3) ADA (ANSI) now requires both tactile and visual (with contrast) graphics. Graphics may be mounted on the push side of doors, on side (pull side) of doors (18″), or on nearest adjacent wall when no space is available by the door. Visual graphics (except for elevators) are to be mounted 3′–4″ to 5′–10″ above floor (with ⅝″- to 1¾″-high characters) when viewed from up to 15′; 5′–10″ to 10′ AF (with 2″- to 2¾″-high characters) when viewed from 15′ to 21′; and 10′ AF (with 3″-high characters + ⅛″/ft. beyond 21′) when viewed from greater than 21′. Tactile and braille graphics are to be mounted between 4′ and 5′ AF (except for elevators). Tactile characters are to be ⅝″ to 2″ high and braille ½″ to ¾″. Pictograms (of high contrast) are to have 6″-high backgrounds.

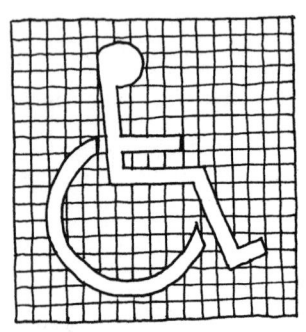

___ *i*   Other common signs and symbols:

BIOLOGICAL HAZARD

RADIATION HAZARD

HIGH VOLTAGE HAZARD

STOP

FIRE EXTINGUISHER

PARKING

PARKING

NO PARKING

FIRST AID

EMERGENCY

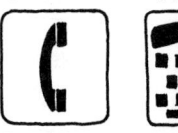
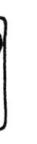

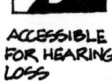

TELEPHONE

TEXT TELEPHONE

ACCESSIBLE FOR HEARING LOSS

VOLUME CONTROL TELEPHONE

INFORMATION

SMOKING PERMITTED

NO SMOKING

RESTROOMS

WOMEN'S RESTROOM

MEN'S RESTROOM

STAIRS

ELEVATOR

EXIT STAIRS

TAXI STAND

ESCALATOR

CAR RENTAL

BUS STOP

AIRPORT

TRAIN STATION

LUGGAGE

TICKET INFORMATION

CURRENCY EXCHANGE

LOCKERS

WAITING ROOM

BANK/CASH MACHINE

LOST AND FOUND

ACCOMMODATION INFORMATION

LOUNGE

DINING

CAFE

DRINKING FOUNTAIN

LITTER RECEPTACLE

COAT ROOM

BARBER

CHANGING TABLE

Costs:

___ **Road/site directional**	**$25 to $50/SF (40% M and 60% L)**
___ **Pylon/monument**	**$12,000 to $48,000 (40% M and 60% L)**
___ **Exterior building, I.D., backlighted, with ind. letters**	**$7200 to $14,400 (same)**
___ **Plaques, cast alum. or bronze**	**$600 to $1080 (85% M and 15% L)**
___ **Plastic, Bakelite**	**$60 to $180/SF (40% M and 60% L)**
___ **Neon, small size**	**$2400 to $4800 (same as above)**
___ **Exit, electrical**	**$360 (45% M and 55% L)**
___ **Metal letters**	**$50 to $95/ea. (60% M and 40% L)**
___ **Plexiglass**	**$100 to $120/SF (95% M and 5% L)**
___ **Vinyl**	**$25 to $35/SF (75% M and 25% L)**

___ G. LOCKERS ④

___ 1. <u>Types:</u> Steel, plastic laminate, wood.
___ 2. <u>Sizes:</u>
    ___ a. Heights: 60 to 72″
    ___ b. Depths: 12″, 15″, 18″, and 21″
    ___ c. Widths: 9″, 12″, 15″, 18″ and 24″
___ 3. **Costs: Steel box locker, 12″ × 15″ × 72″: $150/ea. (80% M and 20% L)**

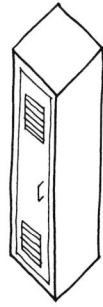

__ H. FIRE EXTINGUISHERS ④ ㊺

__ 1. <u>Class of Fire</u>
  __ a. Class A: Fires of wood, paper, textile, or rubbish. Locate extinguisher within 75′ travel distance.
  __ b. Class B: Fires of gasoline, oil, grease, or fat. Locate extinguisher within 30 to 50′ travel distance.
  __ c. Class C: Fires of an electrical nature.
__ 2. <u>Occupancy Class Requirements</u>
  __ a. *Light hazard:*
    __ (1) Occupancy: Schools, offices, and public buildings
    __ (2) Number: One class A extinguisher per 3000 SF
  __ b. *Ordinary hazard:*
    __ (1) Occupancy: Dry goods shop, warehouse.
    __ (2) Number: One class A extinguisher per 1500 SF.
  __ c. *Extra hazard:*
    __ (1) Occupancy: Paint shops, etc.
    __ (2) Number: One class A extinguisher per 1000 SF.
__ 3. See local fire dept. for exact requirements.
__ 4. Also see p. 440 for cabinets.
  Costs: ABC all-purpose portable:
  2½ lbs: $40/ea
  20 lbs: $120/ea
  Cabinets: $190 to $260/ea.

FIRE CLASSIFICATIONS FOR SELECTING FIRE EXTINGUISHERS

LETTER SYMBOL AND COLOR	PICTURE SYMBOL	DESCRIPTION
GREEN		CLASS A: FIRES INVOLVING ORDINARY COMBUSTIBLE MATERIALS (SUCH AS WOOD, CLOTH, PAPER, RUBBER, AND MANY PLASTICS) THAT REQUIRE THE HEAT ABSORBING (COOLING) EFFECTS OF WATER OR WATER SOLUTIONS, OR THE COATING EFFECTS OF CERTAIN DRY CHEMICALS THAT RETAIRD COMBUSTION.
RED		CLASS B: FIRES INVOLVING FLAMMABLE OR COMBUSTIBLE LIQUIDS FLAMMABLE GASES, GREASES AND SIMILAR MATERIALS THAT ARE BEST EXTINGUISHED BY EXCLUDING AIR (OXYGEN), INHIBITING THE RELEASE OF COMBUSTIBLE VAPORS OR INTERRUPTING THE COMBUSTION CHAIN REACTION.
BLUE		CLASS C: FIRES INVOLVING ENERGIZED ELECTRICAL EQUIPMENT WHERE SAFETY TO THE OPERATOR REQUIRES THE USE OF ELECTRICALLY NONCONDUCTIVE EXTINGUISHING AGENTS.
YELLOW		CLASS D: FIRES INVOLVING COMBUSTIBLE METALS (SUCH AS MAGNESIUM, TITANIUM, ZIRCONIUM, SODIUM, LITHIUM, AND POTASSIUM).

___ I. OPERABLE PARTITIONS

___ 1. Types

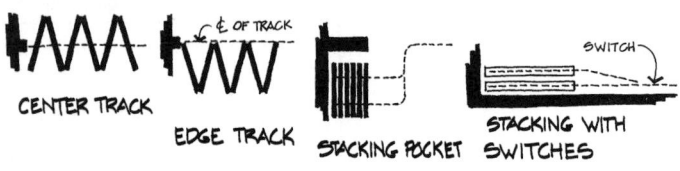

CENTER TRACK

EDGE TRACK

STACKING POCKET

STACKING WITH SWITCHES

___ 2. Data

___ (1) Stack widths:
    ___ (*a*) Accordion: 5″ to 12″
    ___ (*b*) Panels: 15″ to 17″
___ (2) Stack depths: Usually ⅙ to ⅛ of opened width.
___ (3) Panels usually 48″ wide.
___ (4) Acoustic: STC 43 to 54 available.
___ (5) Flame spread: Class I available.

Costs:

___ **Folding, acoustical, vinyl, wood-framed: $75 to $100/SF (70% M and 30% L) Variation: −35% to +50%.**

___ **Accordion, vinyl-faced: $18.50 to $45.00/SF, Variation: ±20%**

___ J. BATHROOM ACCESSORIES

Costs given are for average quality. For better finishes (i.e., brass), add 75% to 100%:

___ **Mirrors**	**$35/SF (90% M and 10% L) variation of ±25%**
___ **Misc. small items (holders, hooks, etc.)**	**$20 to $35/ea. (double, if recessed)**
___ **Bars**	
Grab **$40 to $55/ea.**	
Towel **$20 to $35/ea.**	
___ **Medicine cabinets**	**$95–$120/ea.**
___ **Tissue dispensers**	**$35 to $70/ea.**
___ **Towel dispensers**	**$155 to $480/ea. (increase by 2½ times if waste receptacle included)**

___ K. STORAGE SHELVING

___ 1. Solid metal (industrial), typical.
  ___ a. Widths: 24″ to 48″ in 6″ increments.
  ___ b. Depths: 9″ to 18″ in 3″ increments. 24″ to 36″ in 6″ increments.
  ___ c. Heights: 3′–3″, 6′–3″, 7′–3″, 8′–3″, and 10′–3″.
  ___ d. Shelves can be adjusted in 1″ increments.
___ 2. Shelving comes in other sizes than those above. Other types include wire, wood, etc. (prefabricated or built in).

Costs: Metal industrial shelving 18 GA, 6 shelves:
  ___ 12" deep   **$40/LF**
  ___ 18" deep   **$50/LF**
  ___ 24" deep   **$60/LF**

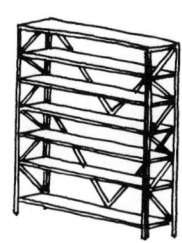

NOTES

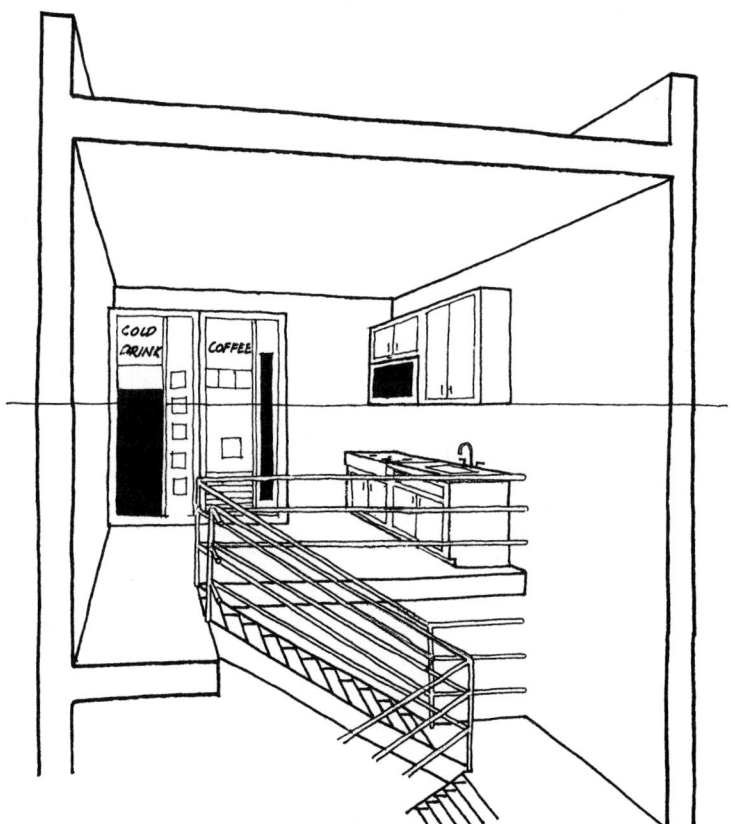

11 EQUIPMENT

NOTES

__ A. CENTRAL VACUUM CLEANING

Costs: $1200 for first 1080 SF installed and $0.20 for each SF added.

__ B. SAFES

Costs: Office, 4-hour, 1.5′ × 1.5′ × 1.5′ = $3600
Jeweler's, 63″ × 25″ × 18″ = $19,000

__ C. CHURCH (TYPICAL SIZES)

__ 1. <u>Pews</u>
Allow minimum of 18″ per person, with each pew no longer than 21′.
Costs: $95 to $180/LF

__ 2. <u>Pulpits</u>
2′ square, movable
5′ square, fixed
Costs: $1620 to $9500

__ 3. <u>Altar</u>
Costs: $2500 to $12,800

__ 4. <u>Communion Table</u>
3′ × 7′

__ 5. <u>Baptismal Font</u>
2′ × 2′
Costs: $850

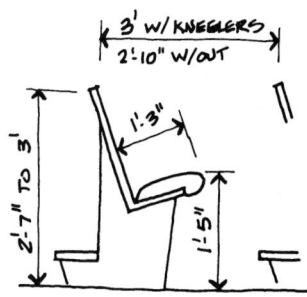

__ D. LIBRARY

__ 1. Stacks

For general sizing of stack area, plan on 16 books per SF.
Allow 3′ to 4.5′ aisles between stacks. Shelves are 12 to 14″
vertically. Shelving units can be steel or wood. Typical
depths are 8″, 10″, and 12″. Double for 2 sides. Typical
heights of units are 42″, 60″, 82″, and 90″. Typical widths are
3′. Track shelving can save up to 45% space.

Costs: Metal, double-face, 10″ shelf × 90″ high: $160/LF

__ 2. Study Carrels

Plan on 2′ D × 3′ W × 4′ H for each carrel seating space.

Costs: Carrel, hardwood: $870 to $1130
Add for wood chair: $120

__ E. THEATER STAGE EQUIPMENT

Costs: Total stage equipment: $115 to $600/SF of stage
Audience Seating: $150 to $340/seat

__ F. BARBER EQUIPMENT

Costs: $3000 to $6420/chair

__ G. CASH REGISTER/CHECKOUT STATIONS

Costs: Supermarket, single conveyor: $2670
Restaurant or store register: $680 to $3000

___ H. LAUNDRY ROOM EQUIPMENT

___ 1. <u>Residential</u>

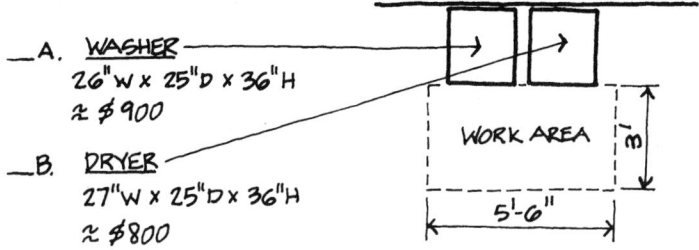

___ A. <u>WASHER</u>
26"W x 25"D x 36"H
≈ $900

___ B. <u>DRYER</u>
27"W x 25"D x 36"H
≈ $800

___ 2. <u>Commercial</u>

   ___ a. For general planning a coin-operated laundry of about 700 SF will have 14 washers and 9 tumblers (dryers) with folding tables, seating, and vending machines.

   ___ b. Coin-operated washers' sizes vary from 20 lb = 28″ W and 30″ D to 50–60 lb = 40″ W and 40″ D. Typical height 44″. Allow 18″ clearance at sides, 24″ behind, and 48″ in front.

   ___ c. Coin-operated dryers' sizes vary from 30 lb = 31″ W and 44″ D to 200 lb = 72″ W and 54″ D. Allow 24″ behind & 48″ in front.

   ___ d. Both washers and dryers need 2′ behind for drainage trough and venting.

   ___ e. Commercial laundries must be accessible per ADA by using a front-loaded machine.
Costs: Commercial dryer, coin-operated dryer: $3300/ea.
Washer: $1400/ea.

___ I. PROJECTION SCREENS

Costs:
Wall or ceiling hung: $7.55 to $11.60/SF

___ J. FOOD SERVICE EQUIPMENT

Costs:
Restaurants: $107 to $175/SF kitchen area
Office buildings: $85 to $140/SF kitchen area
Hospitals: $115 to $185/SF kitchen area

___ K. RESIDENTIAL KITCHEN EQUIPMENT

Costs:
Refrigerator, 33″ W × 32″ D × 66″ H: $840 to $2600
Dishwasher, 23″ W × 24″ D × 33″ H: $600 to $990
Sink: $480 to $790
Range/Oven, 36″ W × 27″ D × 36″ H: $1200 to $3300
Wall oven (microwave), 25″ W × 22″ D × 18″ H: $380 to $815

___ 1. <u>Typical Types and Sizes</u> (in inches)

Type	W	D	H
Money	12	18	38
Candy	34	28	72
Cold Drinks	37	26	72
Hot Drinks	38	33	72
Refer Food	41	36	72
Microwave	22	13	12

___ 2. **A.D.A. prefers** (does not require) the reach heights shown above for accessibility.

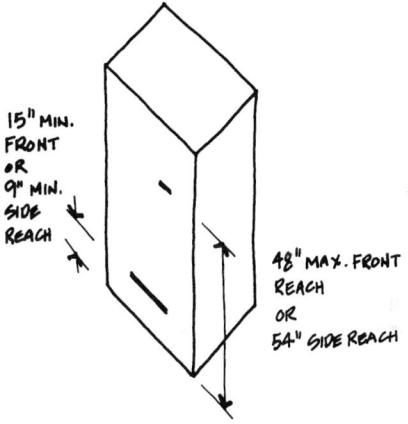

15" MIN. FRONT oR 9" MIN. SIDE REACH

48" MAX. FRONT REACH OR 54" SIDE REACH

NOTES

12

INTERIOR FURNISHINGS

NOTES

__ A. GENERAL COSTS

Costs for furniture and interior objects will vary more than any other item for buildings. These can vary as much as −75% to +500% (or more). Costs given in this part are a reasonable middle value and are "list" prices. See p. 5 for discounting.

__ B. FABRICS

__1. Association of Contract Textiles (ACT) Recommendations
Check for following:
___ *a.* Fire Tests Standards
The following standards and tests may have to be followed, depending on jurisdiction. Check applicable building or fire code, or check applicable building or fire departments for requirements. In some cases it may be state law that rules. Check product literature for compliance.

Common Test Name	Description	Standard Name/Number	Type of Rating
___ Steiner Tunnel Test	Flame-spread and smoke density test for wall and ceiling finishes. Also see item 3 p. 172.	ASTM E84 NFPA 255 UL 723 Chamber Test or UL 992	Class rating
___ Radiant Panel Test	For interior floors. Also see item (d) on p. 174.	ASTM E648 NFPA 253 NBS IR75-950	Class rating
___ PILL Test	Carpet.	DOC FF1-70 DOC FF2-70	Pass or fail
___ Vertical Flame Test	For vertical carpets, draperies, window shades, wall hangings, and plastic films.	NFPA 701 ASTM D6413	Pass or fail

Common Test Name	Description	Standard Name/Number	Type of Rating
___ Corner Test (applied finish)	For napped, tufted, or looped textiles, or carpets on walls or ceilings.	NFPA 265 UBC 42-2 NBC 42-2 NBC Rm. corner Fire Test SBC Std. Test Method for evaluating room fire growth contribution of textile wall coverings:	Pass or fail
(upholstered materials)		NFPA 286	Ranked
___ Smolder Resistance Test (applied finish)	Also known as Cigarette Ignition Tests. Determines the flammability of upholstered furniture.	NFPA 260 CAL 116 ASTM 1353	Pass or fail
(mock-up)		NFPA 261 CAL 117 ASTM 1352	Class rating
___ Smoke Density Test	Measures how much and how dense a material will release when burned.	ASTM E662 NFPA 258	Class rating
___ Toxicity Test	Measures the amount of toxicity emitted when a material burns.	LC 50 Pitts Test	Ranked
___ Upholstered Seating Tests (Full scale)	Measures how much furniture burns.	NFPA 266 CAL 133 ASTM E1537 UL 1056	Pass or fail
(Small scale)		NFPA 272 ASTM 1474	Ranked
___ Mattresses		NFPA 267 ASTM E1590 CAL 129	Ranked
		FF 4-72	Pass or fail

___ *b.* Abrasion resistance

a	A	Test
15,000 double rubs	30,000 double rubs 40,000	Wyzenbeek Martindale

___ *c.* Colorfastness to light
Must pass Class 4 (40 to 60 hours exposure for UV).

___ *d.* Colorfastness to wet and dry crocking (Pigment colorfastness in fabric).

___ *e.* Miscellaneous other physical properties
    ___ (1) Brush pill test: measures tendency for ends of a fiber to mat into fuzz balls.
    ___ (2) Yard/seam slippage test: establishes fabric's likeliness to pull apart at seams. Must pass 25 lbs for upholstery and 15 lbs on drapery.
    ___ (3) Breaking/tensile strength test: evaluates fabric's breaking or tearing. Must pass:
        Upholstery   50 lbs
        Panel fabrics   35 lbs
        Drapery over 6 oz 25 lbs
               under 6 oz 15 lbs

___ 2. Costs:
Upholstery Fabrics

Blends:	**$20 to $95/SY**
Nylons:	**$25 to $60/SY**
Polyester:	**$40 to $55/SY**
Silk:	**$55 to $90/SY**
Wool:	**$50 to $95/SY**
Vinyl:	**$25 to $50/SY**

Add $5/SY for flameproofing.

__ C. ARTWORK AND ACCESSORIES

__ 1. <u>Artwork</u> (photos, reproductions, etc.)
    Costs: $35 to $720/ea.
__ 2. <u>Ash Urns and Trash Receptacles</u>
    Costs: $90 to $360/ea.

__ D. MANUFACTURED CASE WORK

See p. 297.

__ E. WINDOW TREATMENTS

__ 1. <u>Draperies</u>
    __ *a.* Any fabric hung at windows in straight, loose folds. Usually of heavy fabric. Drawn from one or two sides.
    __ *b.* Should be hung 2″ to 4″ from glass and slightly off floor.
    __ *c.* Fullness is the amount of material for folds:
        100% = 2 times window width
        200% = 3 times window width
    __ *d.* Linings are used behind, for black out and to lessen heat flow at window.
    __ *e.* **Costs: $25 to $130/SY**

__ 2. <u>Shades</u>
    __ a. Operate from top down in rolling mechanism. Shade material may be translucent or opaque. Shades reduce light while providing privacy. May be of fabric, vinyl, bamboo, or woven wood.
    __ b. **Costs: $2 to $25/SF**

___ 3. <u>Blinds</u>

    ___ *a.* Are of slats that are stacked when gathered. Materials can be metal, vinyl, or fabric on vinyl.

    ___ *b.* Horizontal are usually 1″ to 2″ wide and curved for better light reflectance.

    ___ *c.* Vertical are usually 3″ up to 7″ wide and are straight.

    ___ *d.* **Costs: $5.50 to $8.75/SF**

___ 4. <u>Curtains</u>

    ___ *a.* Are usually of a sheer fabric to let light in but keep privacy.

___ 5. <u>Shutters</u>

    ___ *a.* Are usually thin louvered wood doors with adjustable slats.

__ F. FURNITURE

See p. 383 for advice on furnishings costs. Costs given in this section are list price (retail), rounded, and approximate range depending on fabric options.

See p. 51 for space planning.

This section advises on different types of furnishings and their costs. This includes both residential and commercial furnishings. As shown in the Table of Contents, this section has 10 categories of furniture examples.

___ 1. <u>Site Furnishings</u>
See p. 251.
___ 2. <u>Lounge/Lobby/Living Room</u>

___ a.	**General Average Costs**	**Residential**	**Commercial**
	Lounge chair, 36″ × 36″:	**$1100**	**$1260**
	Sofa, 36″ × 84″:	**$2760–$4800**	**$4800**
	Coffee table, 30″ × 60″:	**$600–$960**	**$625–$840**

UPHOLSTERED SOFA
COST: $6720 – $9100

UPHOLSTERED CHAIR
COST: $3800 – $5160

UPHOLSTERED CHAIR
COST: $3480– $4800

UPHOLSTERED CHAIR
COST: $3240 - $3600

388

LOUNGE CHAIR
COST: $1525

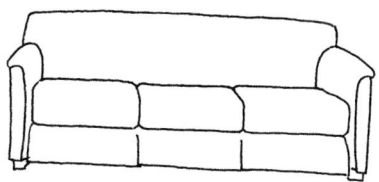

THREE SEAT SOFA
COST: $2740

CHAIR
COST: $2000 - $3500

CHAIR
COST: $1100 - $1850

TWO SEAT
COST: $1430 - $3264

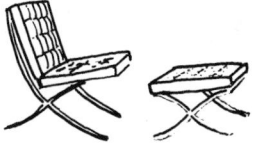

BARCELONA CHAIR & STOOL
COST: CHAIR $9670
        STOOL $4900

STACK CHAIR
COST: $390 - $480

STACK CHAIR W/ ARMS
COST: $400 - $500

ARMCHAIR
COST: $740 - $960

LOVESEAT
COST: $985 - $1280

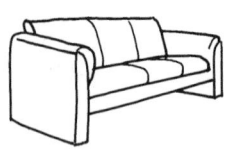

COUCH
COST: $1300 - $1700

LOUNGE CHAIR
COST: $1193 - $4750

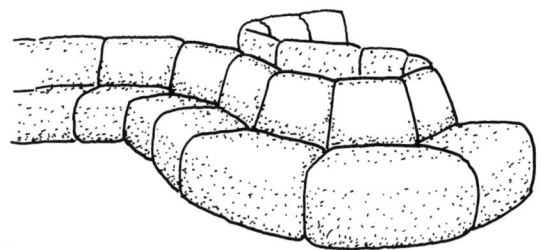

UPHOLSTERED CURVED RECEPTION SEATING
COST:
INSIDE SEGMENT   STRAIGHT SEGMENT
         $1275 - $1800 $1200 -
                                                   $1645

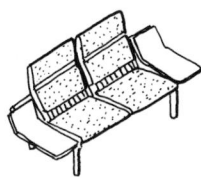

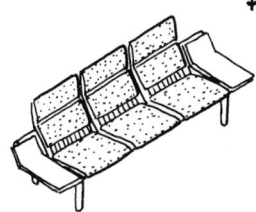

FREE-STANDING UPHOLSTERED SEATING
    2-SEAT UNIT   3-SEAT UNIT
        $4920-$6130 $6330-$8070

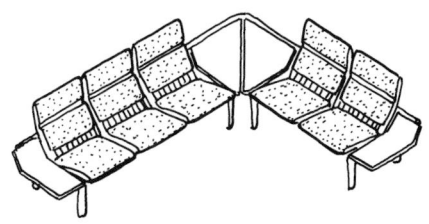

FREE-STANDING UPHOLSTERED SEATING
3-SEAT UNIT AND 2-SEAT UNIT
COST: $13,200 TO $16,800

___ 3. _General Use_

PLASTIC STACK CHAIR
COST: $70

METAL FOLDING CHAIR
COST: $30

ADJUSTABLE HEIGHT FOLDING
TABLE W/ P.B. TOP, 30" x 72"
COST: $320

PERFORATED PLASTIC SIDE
CHAIR
COST: $280 - $320

LIGHT WEIGHT BANQUET
TABLE, 60" DIA., METAL

UPHOLSTERED CHAIR W/ARMS
COST: $440 - $660

___ 4. <u>Benches, Side Tables, Low Tables, Stools, Bookcases</u>

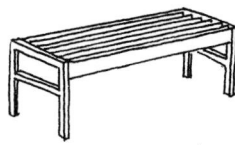

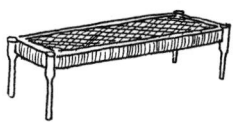

SLATTED WOOD BENCH
COST: $480

TEAK / ROPE BENCH
COST: $480

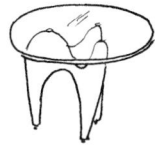

GLASS, METAL, WOOD
END TABLE
COST: $840

SIDE TABLE OF GLASS &
METAL
COST: $755

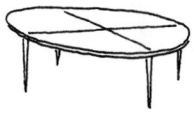

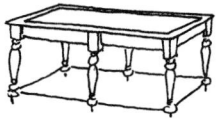

COCKTAIL TABLE
48" W   30" D   17" H
COST: $2160

COCKTAIL TABLE
52" W   26" D   19" H
COST: $290

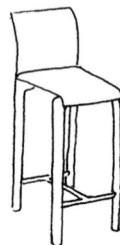

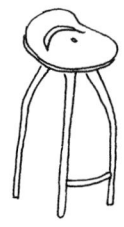

BARSTOOL OF LEATHER
& CHROME
COST: $170

BARSTOOL
COST: $310

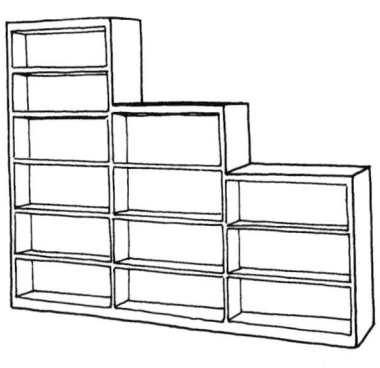

WOOD BOOKSHELVES
DOUBLE 48" x 84"
COST: $155

WOOD BOOKSHELVES
COST: 72"H x 30"W: $230
48"H x 30"W: $170
36"H x 30"W: $130

___ 5. <u>Office</u>

TASK CHAIR
COST: $840 - $1100

LOW BACK
COST: $1100- $1320

EXECUTIVE HIGH-BACK
UPHOLSTERED CHAIR
COST: $865- $1120

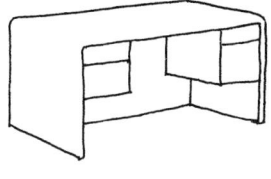

DESK, CONTEMPORARY WOOD
COST: $1980

DESK, LAMINATE WOOD,
COST: $1595
SIDE TABLE: 48"⌀: $1150

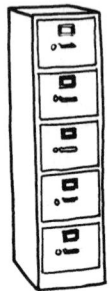

METAL VERTICAL FILE
5 DRAWER
COST: LETTER W/LOCK 15" x 28" = $740
       LEGAL   18" x 28" = $825

METAL LATERAL FILE,
5 DRAWER 67" H
COST: 30" W = $1330
      42" W = $1740

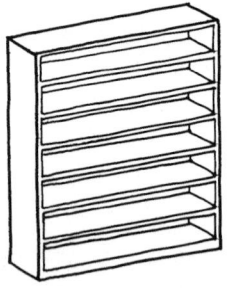

LAMINATE BOOKCASE, 6 SHELF
36" W x 11½" D x 84" H
COST: $360

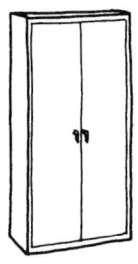

METAL STORAGE CABINET,
5 ADJUSTABLE SHELVES
36" W x 72" H
COST: 24" D $625
      18" D $535

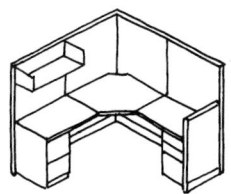

WORK STATION FOR
SECRETARY/WORD PROCESS-
ING, 6'×6'×64" HIGH
COST: $4500

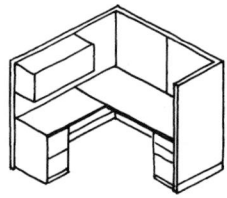

WORK STATION FOR SECRE-
TARY/CLERK
6'×6'×64" HIGH
COST: $4080

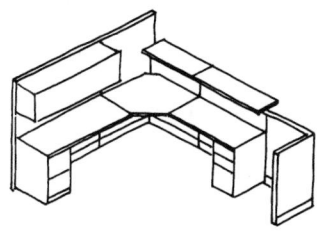

WORK STATION FOR
EXECUTIVE SECRETARY
8'× 9.5'× 64" HIGH
COST: $5830

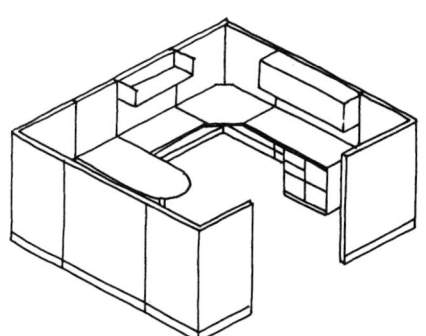

WORK STATION FOR
MANAGER
11.5'× 12.5'× 64" HIGH
COST: $9720

___ 6. <u>Dining/Conference</u>

ROUND, ANELLO TABLE
COST: 42" DIA. = $6840 - $8600
       60" DIA. = $8740 - $13130

SQUARE ANELLO TABLE
COST: 42" DIA = $7720-$9470
       60" DIA= $8965-$13360

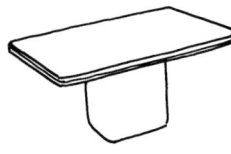

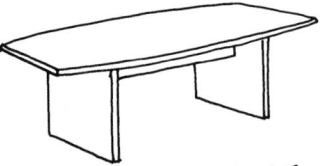

RECTANGULAR ANELLO TABLE
COST: 36" X 72" = $7440-$10600

BOAT-SHAPED LAMINATE
CONFERENCE TABLE
144" L X 48" W
COST: $680 - $790

BANQUET TABLE,
45"W X 120"L
COST: x $5095

GAMING TABLE,
48"DIA.
COST: x $2500

___ 7. <u>School Desks and Chairs</u>

FOLDING TABLET ARM
CHAIR

COST: $95

FOLDING STOOL TABLES
W/P.L. TOPS,
12 STOOLS, 27" OR 29"W X 12'L
COST: $3090

ARM   ARMLESS   TABLET

THERMOPLASTIC AND UPHOLSTERED STEEL,
STACK CHAIRS,
COST: ARM = $260- $625/EA.
         ARMLESS = $200 – $565/EA.
         TABLET = $440 – $800/EA.

___ 8. <u>Bedroom</u>

	Residential	Commercial
___ **General Average Costs**		
Bed:	$720 to $2040	
Side table, 24″ × 30″:	$240–$480	$175
Writing table, 24″ × 48″:	$900	$570
Chest of drawers, 24″ × 72″:	$2040	$1320

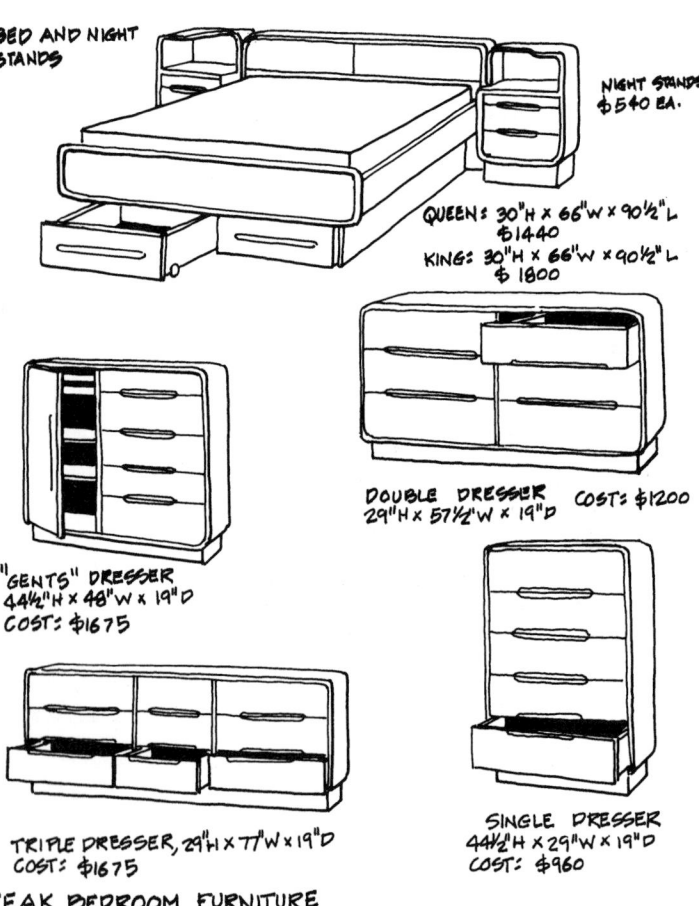

BED AND NIGHT STANDS

NIGHT STANDS
$540 EA.

QUEEN: 30"H × 66"W × 90½"L
$1440
KING: 30"H × 66"W × 90½"L
$1800

DOUBLE DRESSER   COST: $1200
29"H × 57½"W × 19"D

"GENTS" DRESSER
44½"H × 48"W × 19"D
COST: $1675

TRIPLE DRESSER, 29"H × 77"W × 19"D
COST: $1675

SINGLE DRESSER
44½"H × 29"W × 19"D
COST: $960

TEAK BEDROOM FURNITURE

___ 9. <u>Lamps</u>

See p. 471 for Lighting.
Typical costs: Table lamp $300
Floor lamp $360

"CIAO" TABLE LAMP
COST: $330 - $360

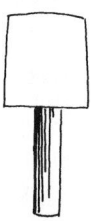

CYLINDER TABLE LAMP,
BRASS OR CHROME
COST: $380 - $400

"DYNASTY" TORCH
COST: $470 - $590

"PHARMACY" FLOOR LAMP
COST: $350 - $360

TABLE LAMP
COST: $335

TEAK TABLE LAMP
COST: $200

___ **10.** <u>*Other Average Costs*</u>
   ___ a. Theater seating
        $210 to $300/seat
   ___ b. Church pews
        $100 to $190/LF
   ___ c. Dormitory furnishings
        $240 to $4620/student
   ___ d. Hospital beds
        $1115 to $3120/bed
   ___ e. Hotel furnishings
        $2640 to $10,410/room
   ___ f. Multiple seating
      ___ (1) Classroom
           $85 to $155/seat
      ___ (2) Lecture hall
           $190 to $530/seat
      ___ (3) Auditorium
           $210 to $335/seat
   ___ g. Restaurants
      ___ (1) Tabletop and base (4 top)
           $180
      ___ (2) Tabletop and base (2 top)
           −25%
      ___ (3) Tabletop and base (6 top)
           +25%
      ___ (4) Chairs
           $180
      ___ (5) Booth
           $40 to $80/LF depth (table and base excluded)
      ___ (6) Banquette
           $15/LF length (table and chairs excluded)

__ G. RUGS AND MATS

Also see p. 345 for carpet.

__ 1. **Oriental Rugs**

    __ a. *Age*

        __ (1) Antique: over 75 years old

        __ (2) Semiantique: less than 75 years old

        __ (3) New: 10 to 15 years old

    __ b. *Construction*

        __ (1) Kelims: Smooth on both sides

        __ (2) Sumak: Woven into herringbone effect.

        __ (3) Knotted: Pile on top and smooth on back.

    __ c. *Type*

        __ (1) Persian

            __ (a) Motifs: curvilinear floral designs, vases, birds, etc. Medallion in center.

            __ (b) Sizes: $10' \times 15'$, $13' \times 24'$

        __ (2) Turkish

            __ (a) Motifs: Prayer rugs, altar designs, tulips. Reds and greens.

            __ (b) Sizes: $3' \times 5'$, $5' \times 7'$, $6' \times 10'$

        __ (3) Caucasian

            __ (a) Motifs: geometric figures of men, birds, and animals. Dark reds.

            __ (b) Sizes: $5' \times 7'$, $6' \times 10'$

        __ (4) Turkoman

            __ (a) Motifs: Rows of rectilinear octagons. Dark reddish browns.

            __ (b) Sizes: $6' \times 10'$, $15' \times 24'$

        __ (5) Chinese

            __ (a) Motifs: Chinese symbols, cloud bands, flowers, birds.

            __ (b) Sizes: $5' \times 8'$, $9' \times 12'$

        __ (6) Other: Moroccan, Afghan, Pakistani, etc.

    __ d. **Costs: New: $1200+**
        Antique: $100,000+

___ 2. <u>Entry Mats</u>
   Costs: Recessed, rubber, ⅜″ thick: $30.00/SF

___ H. INTERIOR PLANTS

___ 1. <u>Profiles of Common Interior Plants</u>

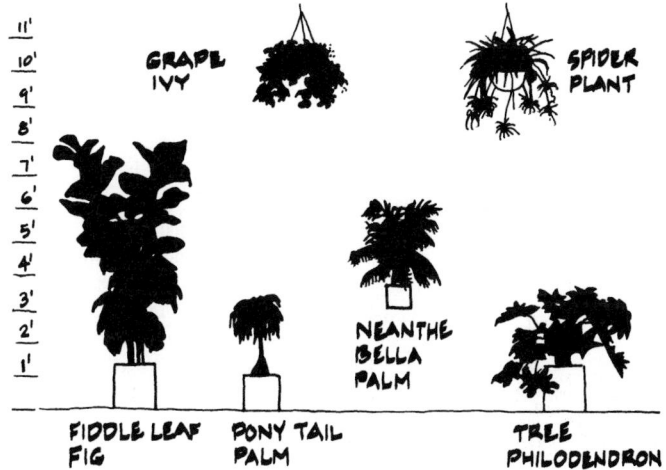

____ 2. Some Plants Requiring Low Light Levels:

Common Name	Botanical Name
Cleveland Peace Lily	*Spathiphyllumx "Clevelandii"*
Corn Plant Dracaena	*Dracaena Fragrans "Massageana"*
Franscher Evergreen	*Aglaomema X Fransher*
Golden Evergreen	*Aglaomema "Pseudobracteatum"*
Janet Craig Dracaena	*D. Deremensis "Janet Craig"*
Malay Beauty Aglaomena	*A. "Malay Beauty"*
Mauna Loa Peace Lily	*Spathiphyllum X "Mauna Loa"*
Neanthe Bella Palm	*Chamaedorea Elegans "Bella"*
Parrot Jungle Evergreen	*Aglaonema "Parrot Jungle"*
Pewter Aglaonema	*A. "Malay Beauty"*
Silver King Evergreen	*Aglaonema X "Silver King"*
Silver Queen Evergreen	*A. X "Silver Queen"*
Snow Queen Evergreen	*A. X "Snow Queen"*
Striped Dracaena	*D. Deremensis "Warneckei"*
Variegated Chinese Evergreen	*Aglaonema Commutatum*
Warneck Dracaena	*D. Deremensis "Warneckei"*
White Rajah	*Aglaonema "Pseudobracteatum"*

___ 3. <u>Interior Plants Information</u>
    ___ a. <u>*Types:*</u>
        ___ (1) Multistem tree
        ___ (2) Standard tree
        ___ (3) Plant clump
        ___ (4) Hanging plants
    ___ b. <u>*Needs:*</u>
        ___ (1) Light:
            ___ (a) Need 10–12 hours on a regular daily basis, but some plants can survive on 2 hours per day.
            ___ (b) Should have 50 fc min. on ground plane or 75 fc min. on desk height.
            ___ (c) Light needs to be full-spectrum. Specify "U.V. full-spectrum" lighting.
        ___ (2) Water:
            ___ (a) Provide for regular watering by hand, hose, or drip. Provide water w/low concentrations of floride and chlorine, and PH value of 5 to 6.
            ___ (b) Provide proper drainage, even if it is necessary to connect into building's plumbing system with sump pumps or syphon tubes.
        ___ (3) Soil:
            Needs to be porous with 50% solids, 25% moisture, and 25% air.

___ 4. <u>Costs</u>:
    Live:
    5′ tree without pot = $55/ea.
    Bush: $25 to $55/ea.
    Table Top Plants: $10/ea.
    Hanging Plants: $10 to $25/ea.
    Artificial Silk: Double above costs.
    Pots: $5 to $30/ea.

NOTES

NOTES

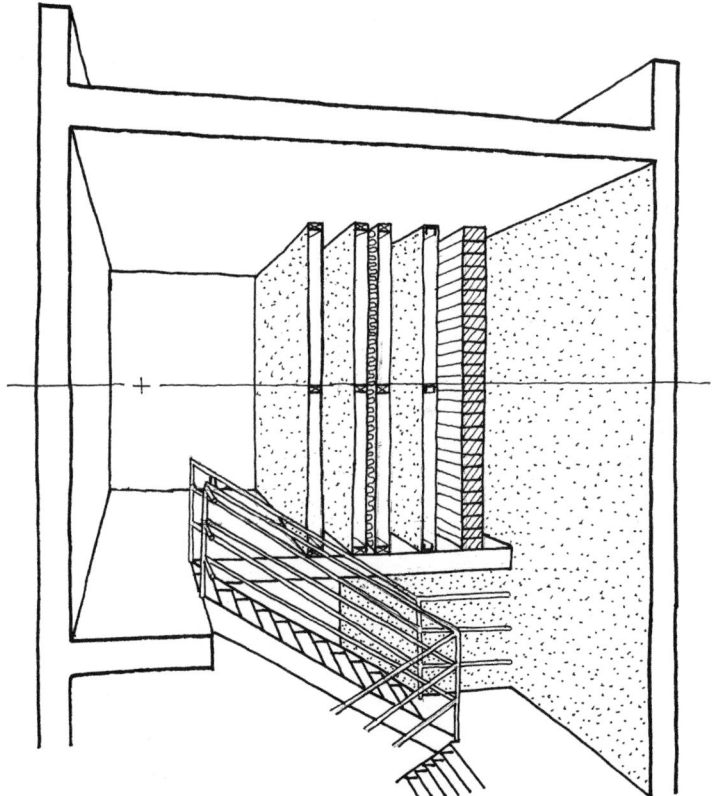

13 ASSEMBLIES

NOTES

__ A. INTERIOR WALL ASSEMBLIES

The following wall assemblies give:
__ Width
__ Sound transmission rating (STC). Also see p. 234.
__ Fire rating in hours resistance.
__ Costs
__ Finishes are *not* included

__ 1. **Drywall Partitions**
    __ a. <u>Wood frame</u>:

½″ gypsum board, each side
2 × 4 studs at 24″ OC

STC = 30 to 34
Fire rating = 0 hour
Cost: $2.80/SF

⅝″ gypsum board, type X, each side
2 × 4 studs at 16″ OC

STC = 30 to 34
Fire rating = 1 hour
Cost: $3.70/SF

⅝″ gypsum board, type X, each side
½″ wood fiber board, each side
2 × 4 studs at 16″ OC

STC = 45 to 49
Fire rating = 1 hour
Cost: $5.70/SF

⅝″ gypsum board, type X, each side
Resilient channel at 24″ OC, each side
2 × 4 studs at 16″ OC

STC = 45 to 49
Fire rating = 1 hour
Cost: $3.35/SF

⅝″ gypsum board, both sides
Resilient channel
1½″ sound insulation
2 × 4 studs at 16″ OC

STC = 50 to 54
Fire rating = 1 hour
Cost = $5.15/SF

2 layers, ⅝″ type X gypsum board
Resilient channel
2 × 4 studs at 16″ OC
⅝″ gypsum board
½″ gypsum board
⅜″ gypsum board

STC = 60 to 64
Fire rating = 1 hour
Cost = $7.80/SF

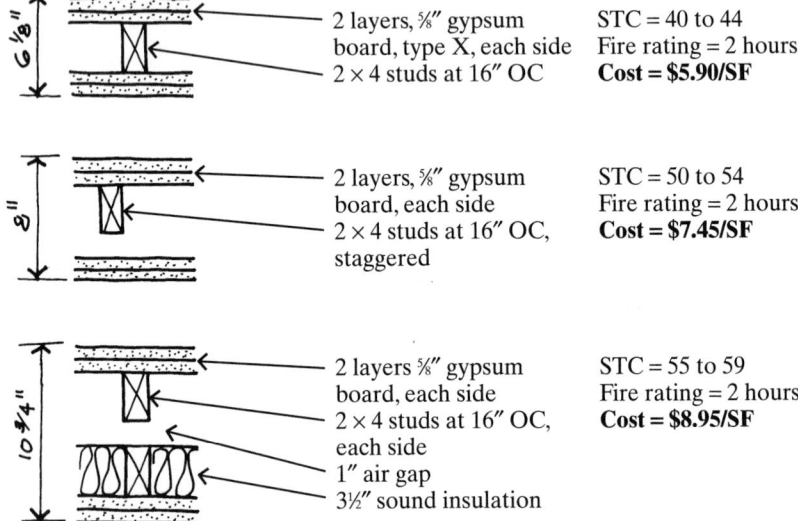

2 layers, ⅝″ gypsum board, type X, each side
2 × 4 studs at 16″ OC

STC = 40 to 44
Fire rating = 2 hours
Cost = $5.90/SF

2 layers, ⅝″ gypsum board, each side
2 × 4 studs at 16″ OC, staggered

STC = 50 to 54
Fire rating = 2 hours
Cost = $7.45/SF

2 layers ⅝″ gypsum board, each side
2 × 4 studs at 16″ OC, each side
1″ air gap
3½″ sound insulation

STC = 55 to 59
Fire rating = 2 hours
Cost = $8.95/SF

___ b. Metal frame:

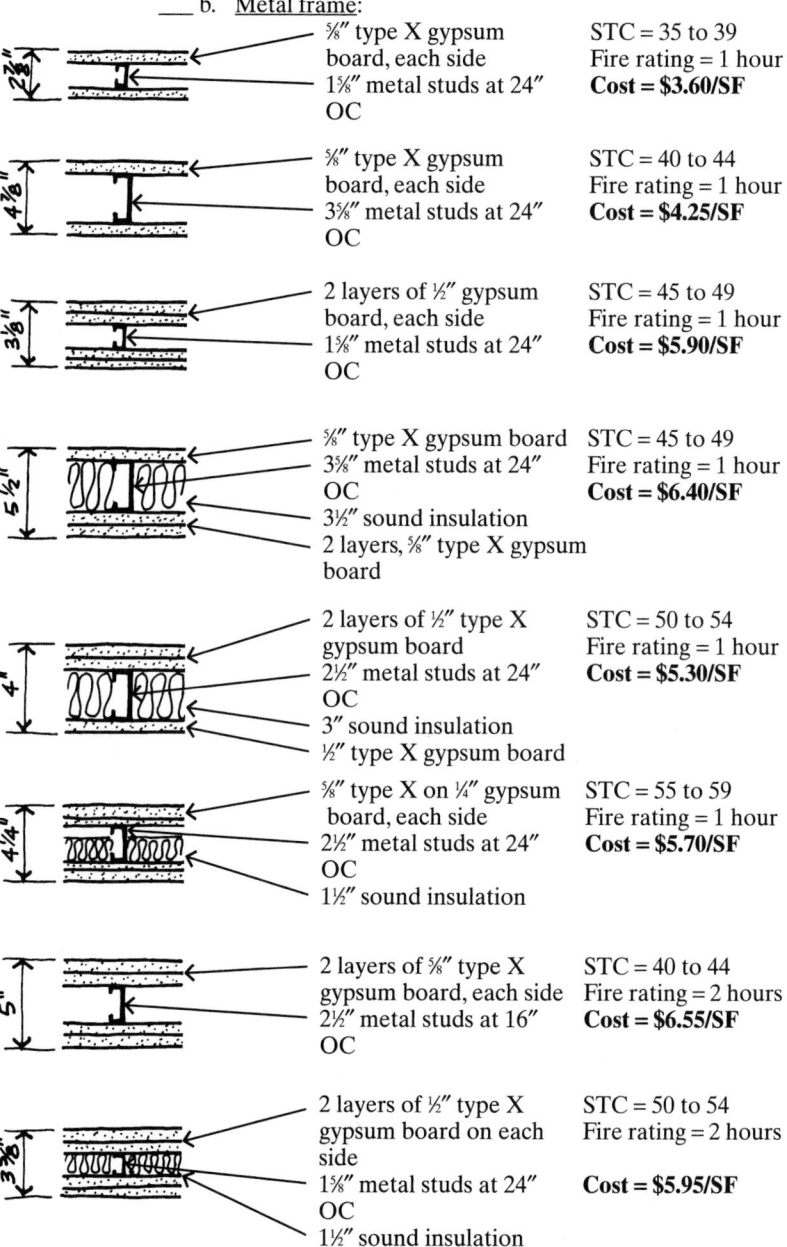

⅝″ type X gypsum board, each side
1⅝″ metal studs at 24″ OC

STC = 35 to 39
Fire rating = 1 hour
Cost = $3.60/SF

⅝″ type X gypsum board, each side
3⅝″ metal studs at 24″ OC

STC = 40 to 44
Fire rating = 1 hour
Cost = $4.25/SF

2 layers of ½″ gypsum board, each side
1⅝″ metal studs at 24″ OC

STC = 45 to 49
Fire rating = 1 hour
Cost = $5.90/SF

⅝″ type X gypsum board
3⅝″ metal studs at 24″ OC
3½″ sound insulation
2 layers, ⅝″ type X gypsum board

STC = 45 to 49
Fire rating = 1 hour
Cost = $6.40/SF

2 layers of ½″ type X gypsum board
2½″ metal studs at 24″ OC
3″ sound insulation
½″ type X gypsum board

STC = 50 to 54
Fire rating = 1 hour
Cost = $5.30/SF

⅝″ type X on ¼″ gypsum board, each side
2½″ metal studs at 24″ OC
1½″ sound insulation

STC = 55 to 59
Fire rating = 1 hour
Cost = $5.70/SF

2 layers of ⅝″ type X gypsum board, each side
2½″ metal studs at 16″ OC

STC = 40 to 44
Fire rating = 2 hours
Cost = $6.55/SF

2 layers of ½″ type X gypsum board on each side
1⅝″ metal studs at 24″ OC
1½″ sound insulation

STC = 50 to 54
Fire rating = 2 hours

Cost = $5.95/SF

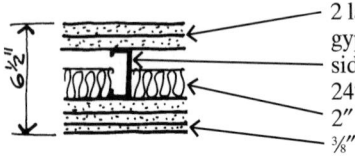

2 layers, ⅝″ type X gypsum board on each side 3⅝″ metal studs at 24″ OC

2″ sound insulation

⅜″ gypsum board

STC = 55 to 59
Fire rating = 2 hours
Cost = $7.70/SF

___ **2. Masonry Walls**

   ___ *a.* Hollow concrete block (unfinished):

Width	STC	Fire rating	Costs
4″	37	1 hour	**$7.20/SF**
6″	42	1½ hour	**$7.80/SF**
8″	47	2 hours*	**$9.00/SF**

*3 to 4 hours if grouted solid

   ___ b. Common brick

Wythe	Width	STC	Fire rating	Costs
1	4″	41	1 hour	**$13.20/SF**
2	8″	49	3-4 hours	**$27.60/SF**

___ **3. Glazed Wall** (¼″ tempered glass)

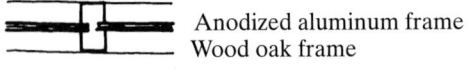

	Costs
Anodized aluminum frame	**$40.00–$45.60/SF**
Wood oak frame	**$26-$32/SF**

CEILING SELECTION TABLE

ASSEMBLY	COST COMPARISON	WEIGHT (LBS/SF)	SOUND ABSORB.	SOUND TRANSM.	LATERAL LOAD	IMPACT	UPLIFT	DEFLECTION	HUMIDITY	FINISH PRE-	FINISH IN PLACE TEXT.	FINISH IN PLACE PAINT	NOTES
SUSP. ACOUSTIC TILE W/ EXPOSED GRID	1 TO 2	2 TO 3	G	F TO P	F TO P	P TO F	P TO G	G	P TO F	●			
SUSP. ACOUSTIC TILE W/ CONCEALED GRID	4 TO 5	→	→	→	F	P	G	→	P	●			
SUSP. PLASTER	3 TO 5	4 TO 6	P →	G →	P	→	→	F TO P	F TO P		●	●	
SUSP. GYPBOARD	2 TO 3	3 TO 4	→	→	G	→	→	→	P TO F		●	●	
ACOUSTIC TILE ATTACHED	4 TO 5	2 TO 3	G →	G →	N/A	→	N/A	N/A →	P TO F			●*	* "NON-BRIDGING" PAINT
SPRAY ON	1 TO 2	.2 TO .3	→	→	→	→	→	→	P		●		

● DENOTES COMMON USAGE
P = POOR
F = FAIR
G = GOOD

FLOOR TO CEILING SPACE

FLOOR
14 TO 20" TYP.
8-10" TYP.
6-8" TYP.
SEE STRUCT.
STRUCT.
HVAC DUCTS
LIGHTS
CEILING - F (± 1" thick)

FLOORING SELECTION TABLE

TYPE	COST COMPARISON	WEIGHT (PSF)	COMFORT	MOISTURE			TRAFFIC						CLEANING		LOCATION				SUBSTRATE		SLIP RESIST. *	CONDUCTIVE	OTHER
				DRY	OCC. WET	FREQ. WET	FOOT LOW	FOOT MOD	FOOT HIGH	WHEEL RUB.	WHEEL STEEL	IMPACT	MILD	HEAVY	OUTSIDE	BELOW GR.	ON GRADE	ABOVE GR.	WOOD	CONC.			
STONE	.9 TO 3	15 TO 40	P	●	●	●	○	●	●	○		●	●	○	●	●	●	●		●	○		
BRICK	4-10.9	20 TO 40	P	●	●	●	●	●	●	●		●	●	●	●	●	●	●	●	●	○		
CONCRETE	1.2 - 2.5	10 TO 15	P	●	●		●	●	●	●	○	●	●	●	●	●	●	●	●	●	●	○	
C.T.	2 - 4	4 - 6	P	●	●	●	●	●	●	○			●	●	●	●	●	●	●	●	●	●	
Q.T.	4 - 5.5	4 - 6	P	●	●	●	●	●	●	○			●	●	●	●	●	●	●	●	●		
RESILIENT	.6 - 2	1 - 2	G	●	●		●	●	●	●		●	●	●		●	●	●	●	●	●	●	
WOOD	3 - 5	1 - 10	F	●	●		●	●	●	●		○	●			●	●	●	●	●	●		
CARPET	1.5 - 5	.5 - 1	G	●			●	●	●			●	●			●	●	●	●	●	●	○	
EPOXY	3 - 5	3 - 7	F	●	●	●	●	●	●	●	●	●	●	●	●	●	●	●	●	●	●	●	

● DENOTES COMMON USAGE OR SUITABILITY

○ DENOTES POSSIBLE OR LIMITED USAGE OR SUITABILITY

* SLIP RESISTANCE

RECOMMENDATIONS FOR STATIC COEFFICIENT OF FRICTION:
NORMAL = 0.5 MIN, H.C. (ADA) = 0.6 MIN, RAMP = 0.8 MIN.

0.2 OR LESS IS VERY SLICK. 0.3 TO 0.4 IS SMOOTH. BROOM FINISH CONCRETE IS USUALLY 0.5 TO 0.7. GRIT STRIPES FOR STAIRS OR RAMPS ARE 0.8 OR ABOVE.

THE COEFFICIENT OF FRICTION IS THE RATIO OF HORIZONTAL FORCE TO VERTICAL FORCE. VALUES SHOULD MEET ASTM D-2047.

NOTES

NOTES

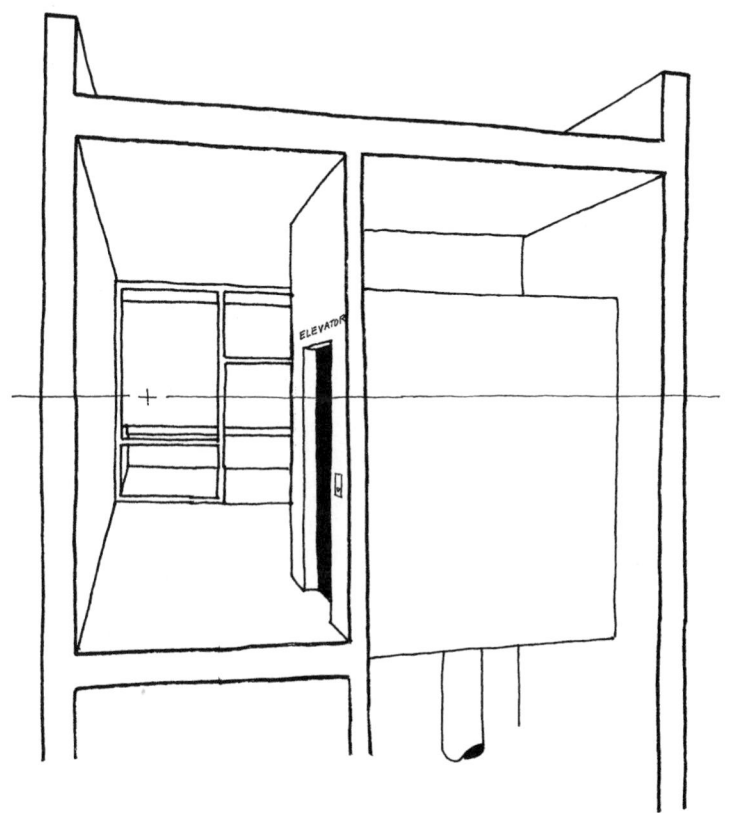

14

CONVEYING SYSTEMS

Per A.D.A. one elevator is required in any building more than 3 stories high or with more than 3000 SF of area on each floor.

___ 1. <u>Hydraulic:</u> The least expensive and slower type. They are moved up and down by a piston. This type is generally used in low-rise buildings (2 to 4 stories) in which it is not necessary to move large numbers of people quickly.

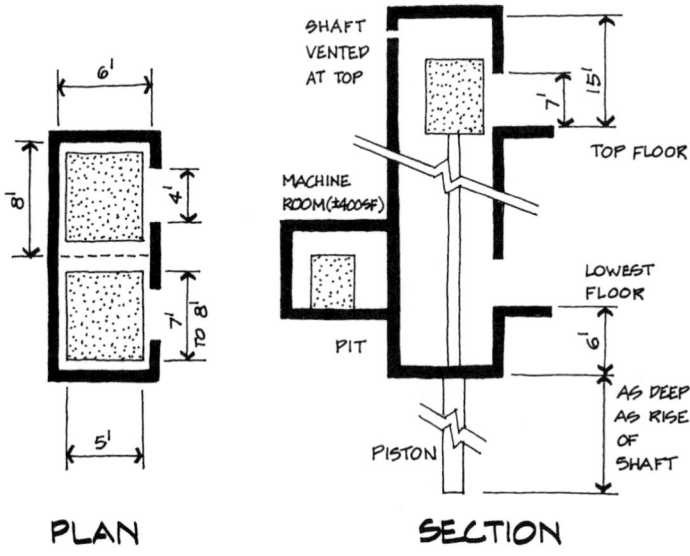

PLAN   SECTION

Costs:

Passenger elevators	**$61,200 (50 fpm, 2000 lbs) to $76,800 (150 fpm, 3000 lbs) per shaft. 3 stops, 3 openings. Add: 50 fpm/stop = +$4200; 500 lb/stop = +$4200; stop = +$6300; custom interior = +$6000.**
Hydraulic freight elevators	**$68,100 (50 fpm, 3000 lbs) to $81,700 (150 fpm, 6000 lbs).**

___ 2. <u>**Traction Elevators:**</u> Traction elevators hang on a counter-weighted cable and are driven by a traction machine that pulls the cable up and down. They operate smoothly at fast speeds and have no limits. Typically, penthouse floor area equals twice the shaft area. A machine room is located either next to the penthouse or on any floor next to the shaft. The shafts, penthouse, pit, and landings are all major special components in the building and often comprise more than 10% of its costs.

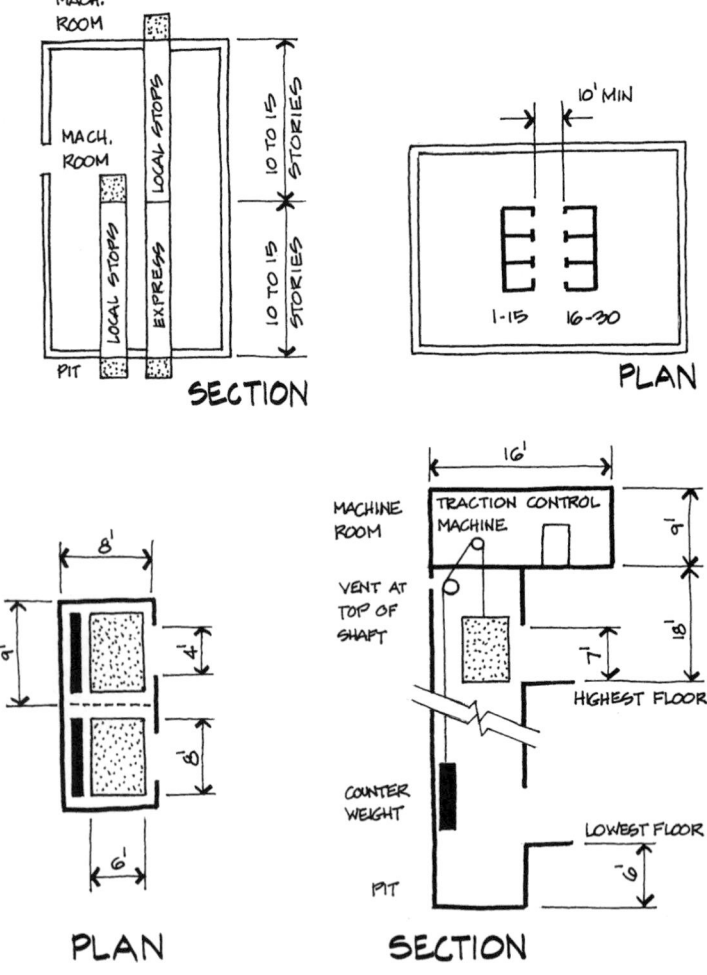

Costs:

Passenger elevators/shaft	**$79,200 (50 fpm, 2000 lbs) to $152,000 (300 fpm, 4000 lbs) for 6 stops, 6 openings. Add: stop = +$4800; 50 fpm/stop = +$2400; 500 lb/stop = +$2400; opening per stop = +$5400; custom interior = +$5760.**
Freight elevators (2 stops)/shaft	**$120,000 (50 fpm, 3500 lbs) to $141,500 (200 fpm, 5000 lbs).**

___ 3. <u>Elevator Rules of Thumb</u>

___ *a.* <u>Commercial</u>

___ (1) One passenger elevator for each 30,000 SF of net floor area.

___ (2) One service elevator for each 300,000 SF of net floor area.

___ (3) Lobby width of 10′ minimum.

___ (4) Banks of elevators should consist of 4 or fewer cars so that people can respond easily to the arrival of an elevator.

___ (5) In high buildings, the elevator system is broken down into zones serving groups of floors, typically 10 to 15 floors. Elevators that serve the upper zones express from the lobby to the beginning of the upper zone. The elevators that serve the lower zones terminate with a machine room above the highest floor served.

___ (6) Very tall buildings have sky lobbies served by express elevators. People arriving in the lobby take an express elevator to the appropriate sky lobby where they get off the express elevator and wait for the local elevator system.

___ (7) Lay out so that maximum walk to an elevator does not exceed 200′.

___ (8) Per ADA, accessible elevators *are* required at *shopping centers* and offices of *health care* providers. Elevators are *not* required in facilities that are less than 3 stories or less than 3000 SF per floor. But, if elevators are provided, at least one will be accessible (see p. 424).

   ___ *b.* Residential

      ___ (1) In hotels and large apartment buildings, plan on one elevator for every 70 to 100 units.

      ___ (2) In a 3- to 4-story building, it is possible to walk up if the elevator is broken, so one hydraulic elevator may be acceptable.

      ___ (3) In the 5- to 6-story range, two elevators are necessary. These will be either hydraulic (slow) or traction (better).

      ___ (4) In the 7- to 12-story range, two traction elevators are needed.

      ___ (5) Above 12 stories, two to three traction elevators are needed.

      ___ (6) Very tall buildings will require commercial-type applications.

      ___ (7) Plan adequate space and seating at lobby and hallways.

  ___ *c.* Where elevators are provided in buildings of 4 stories or more, at least one must accommodate an ambulance stretcher.

  ___ *d.* ADA-accessible elevators (see item 8, p. 423):

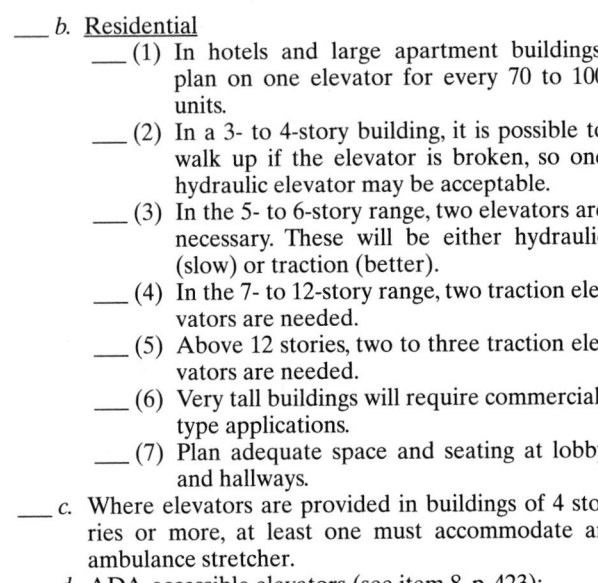

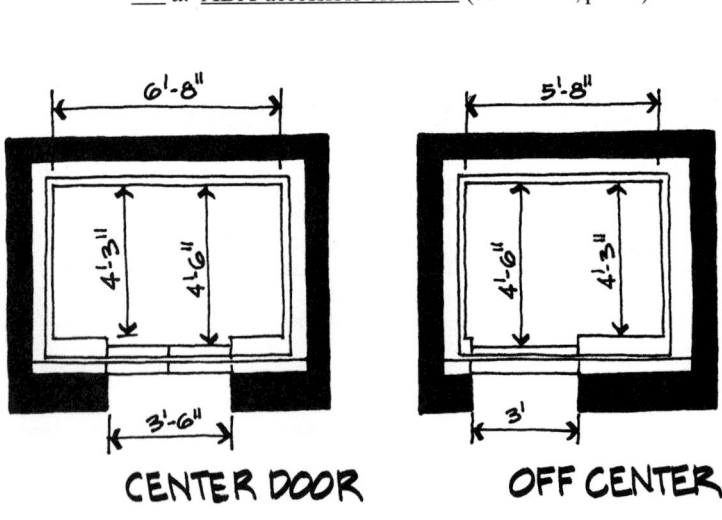

CENTER DOOR   OFF CENTER

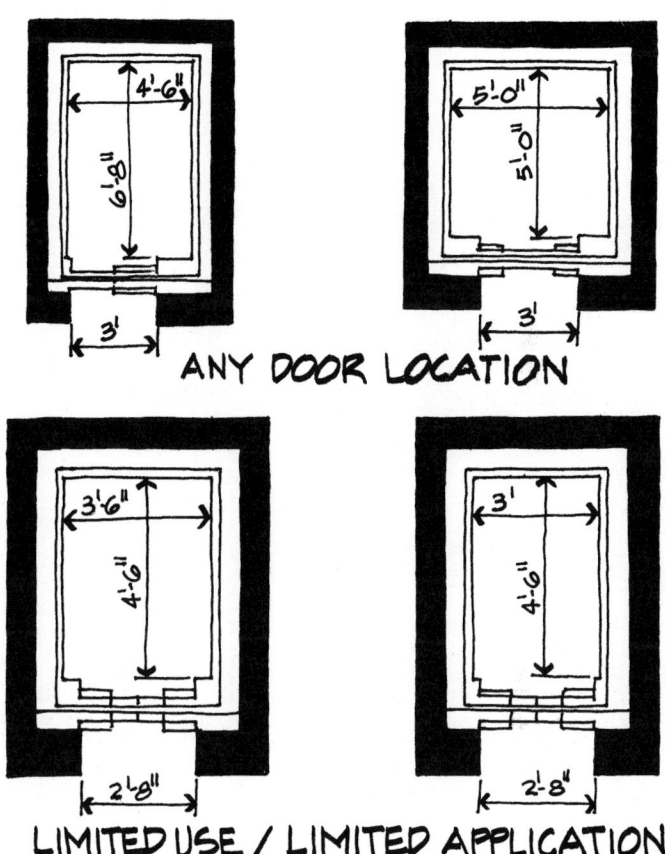

ANY DOOR LOCATION

LIMITED USE / LIMITED APPLICATION

NOTES

__ B. ESCALATORS

Escalators require ⅓ the floor area of elevators to deliver the same passenger loads, need not have pits or penthouses, and can traverse tall floor-to-floor heights. But above 2 levels, riders prefer elevators.

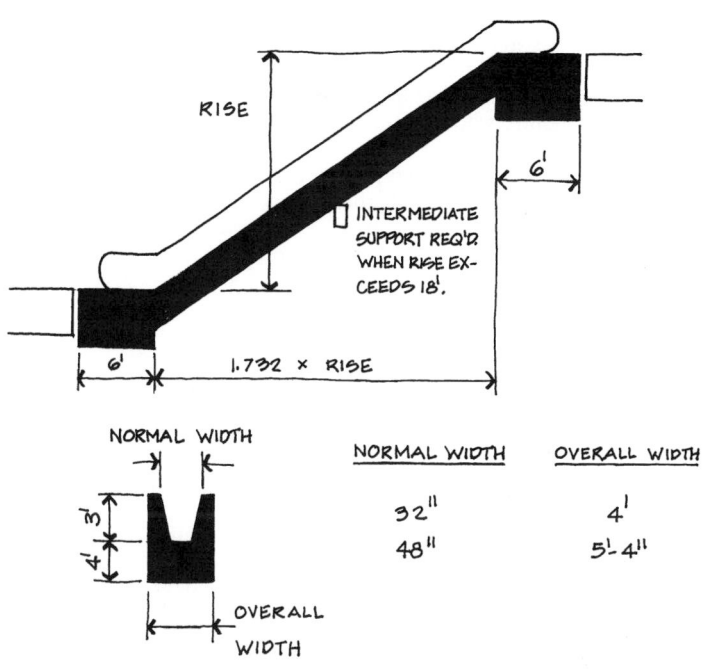

Typically, ceiling-to-floor heights are 4′ due to underside machinery at each end. Risers are 8″ and step slopes are 30°.

Costs:
Escalator costs range from $106,500 for 12′ rise, 32″ width, to $167,500 for 25′ rise, 48″ width. For glass side enclosure add $13,800 to $16,200.

Rules of Thumb

1. All escalators rise at a 30-degree angle.
2. There needs to be a minimum of 10′ clear at top and bottom landings.
3. Provide beams at top and bottom for the escalator's internal truss structure to sit on.
4. The escalator will require lighting that does not produce any distorting shadows that could cause safety problems.
5. Escalators need to be laid out with a crowded flow of people in mind. Crossover points where people will run into each other must be avoided.
6. Current trends in the design of retail space use the escalators as a dramatic and dynamic focal feature of open atrium spaces.
7. Because escalators create open holes through building floor assemblies, special smoke and fire protection provisions are necessary.

NOTES

NOTES

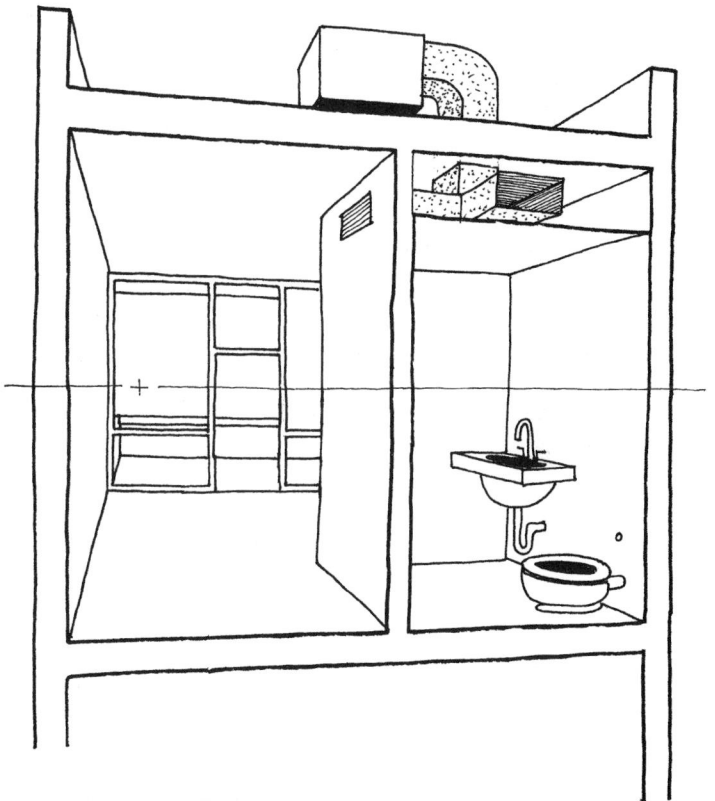

15 MECHANICAL

NOTES

___ A. THE PLUMBING SYSTEM

① ⑤ ⑩ ⑫ ㉕ ㉗A

See pp. 87 thru 99 for toilet rms.
See p. 434 for fixture count 2000 IPC and p. 435 for UPC.

The following systems need to be considered:
___ 1. **Fixture count by code**
___ 2. **Water supply** (p. 435)
___ 3. **Sanitary sewer** (p. 437)
___ 4. **Fire protection** (p. 438)
___ 5. **Gas**
___ 6. **Solar HW systems**

> The interior designer is not likely to be involved in plumbing design, except for fixture selection or maybe toilet room layout. Nevertheless, this section is included for background information.

Costs: **As a rough rule of thumb, estimate $1200 to $1800/fixture (50% M and 50% L) for all plumbing within the building. Assume 30% for fixtures and 70% for lines. Also, of the lines, assume 40% for waste and 60% for supply. For more specifics on fixtures, only:**

Fixture	Residential			Commercial
	Low	**Medium**	**High**	
WC	$200	$660	$1120	$130 to $395
Lavatories	$130	$200	$330	same
Tub/shower	$130	$530	$1060	
Urinals				$330 to $790
Kitchen sinks	$200	$400	$600	

Initial cost is typically only a portion of projected life cycle costs. In commercial buildings a fixture's cost is usually no more than the cost to maintain it for a few months, so any fixtures that reduce maintenance cost usually pay for themselves quickly. When renovating existing buildings, all old piping should be thoroughly cleaned or replaced.

___ 1. *Fixtures Required by Code*

Presently, two plumbing codes rule: the International Plumbing Code (IPC; usually associated with the IBC building code) and the Uniform Plumbing Code (UPC; usually associated with the NFPA building code). You will need to know which code governs to determine your required fixture count.

___ *a.* IPC Requirements:

    ___ (1) See p. 442 for fixture count table.

    ___ (2) Urinals may be substituted for water closets up to 67%.

    ___ (3) Number of users based on occupant load by the IBC. Usually this will be split 50% for each sex.

    ___ (4) Separate toilets for each sex are not required for private facilities or where occupant load is less than 15. Can use a unisex toilet.

    ___ (5) The IBC requires an extra unisex toilet at assembly and mercantile occupancies when the combined fixture count is 6 or more water closets.

    ___ (6) Toilets for *public* must be within 500′ (300′ in malls) and not more than 1 story above or below.

    ___ (7) Toilets for employees must be within 500′ (300′ from tenant space) and not more than 1 story above or below.

    ___ (8) The IBC does not allow toilet rooms to open directly to food service kitchens.

___ *b.* UPC requirements:

    ___ (1) See p. 449 for fixture count table.

    ___ (2) The number of users is based on occupant load determined by the building code and is split 50% for each sex, except in residences and where occupant load is 10 or less (in which a single-user toilet may be used).

    ___ (3) Toilet rooms must be no more than 1 story away.

    ___ (4) Toilet rooms used in private offices do not count toward the number required.

    ___ (5) Business and mercantile occupancies:

        ___ (*a*) A single-user toilet may be used when area is not more than 1500 SF.

___ (*b*) Customer and employee toilets may be combined but must use count that is highest in each type.

___ (*c*) For stores, toilet rooms for customers cannot be more than 500′ away. But at stores of 150 SF or less, employee toilets cannot be more than 300′ away.

___ (6) Food service establishments: When occupant load is 100 or more, must have separate toilet rooms for customers and employees.

___ (7) Water conservation required as follows:

Water closets:	1.6 gal/flush, max.
Urinals:	1.0 gal/flush, max.
Lavatories:	2.2 gal/min, max. (or not more than ¼ gal per use).
Kitchen sinks:	2.2 gal/min, max.
Shower heads:	2.5 gal/min, max.

___ 2. *Plumbing Fixtures*

The men's and women's *restrooms* need to be laid out to determine their size and location in the building. Economical solutions are shared plumbing walls (toilet rooms back to back) and for multistory buildings, stacked layouts.

___ In *cold* climates, chases for plumbing lines should not be on exterior walls, or if so, should be built in from exterior wall insulation.

___ Public buildings should have one janitor sink per 100 occupants, on each floor.

___ 3. *Water Supply*

There are four kinds of water demand: occupancy, special loads, climate control, and fire protection. The water supply is under pressure, so there is flexibility in layout of the water main to the building. In warm climates the *water meter* can be outside, but in cold climates it must be in a heated space. For small buildings allow a space of *20″ W × 12″D × 10″H*. After entering the building the water divides into a hot- and cold-water distribution system at the hot water heater. For small buildings allow for a *gas heater* a space *3′ sq. dia. × 60″H* and for *electric heaters, 24″ dia. × 53″H*. Where bathrooms are spread far apart, consideration should be given to multiple hot water heaters or circulated hot water.

Costs: Residential: $660 to $1980/ea.
Commercial: $1980 to $3900/ea. (80% M and 20% L).

Electric is cheaper for small buildings but high for large buildings.

___ If the water is "hard" (heavy concentration of calcium ions), a *water softener* may be needed. Provide *18″ dia. × 42″H* space.

Costs: ≈ $890

___ If water is obtained from a private *well,* a pump is needed. If the well is *deep,* the pump is usually at the bottom of the well. For this case provide space for a pressure tank that is *20″ dia. × 64″H.* If the well is *shallow* (20′ to 25′ deep) the pump may be provided inside the building. Space for pump and tank should be *36″W × 20″D × 64″H.*

Costs: $185 to $265 per LF of well shaft

___ Water supply *pipes* are usually copper or plastic and range from ½″ to 2″ for small buildings, but 2½″ to 6″ for larger buildings or higher-water-use buildings. Hot and cold pipes are usually laid out parallel. Piping should be kept out of exterior walls in cold climates to prevent winter freeze-ups. The cost of insulation is usually quickly returned by savings in reduced heat loss. As the pipe diameter increases, this becomes more so due to greater volume and surface area.
___ The city water pressure will push water up 2 or 3 stories. Buildings taller than this will need a *surge tank and water pressure pumps.* This equipment takes approximately *100 to 200 SF* of space.

Costs: $6000 to $24,000

___ **4. Sanitary Sewer**
Sewage flow is usually considered to be 95% of supply water flow. Horizontal runs of drainage piping are difficult to achieve inside the building. Pipe pitches should be at least ¼″/ft. Straight runs should not exceed 100′ for metal (or 30′ for plastic) piping. Cleanouts should be located at every direction change exceeding 45°, and every 50′ to 100′. The best arrangement is to bring the plumbing straight down (often along a column) and make connections horizontally under the building. Piping should not pass within 2 vertical feet above any electrical service unless contained in secondary piping.

___ The *sanitary drainage system* collects waste water from the plumbing fixtures, which flows by gravity down through the building and out into the city sewer. Because of the slope requirement, long horizontal runs of drainage pipe will run out of ceiling space to fit in. Ideally, sanitary drainage pipes (called plumbing stacks) should run vertically down through the building collecting short branch lines from stacked bathrooms. A

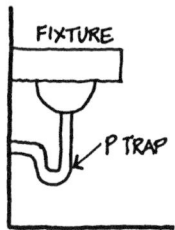

4″ stack can serve approx. *50* WCs and accompanying lavatories. A *6″* stack can serve approximately *150* WCs and lavatories. Pipes are typically of cast iron or plastic (ABS). Each fixture is drained through a "P" trap with a water seal. This, and venting the system to the roof, keeps sewer gases from entering the building.

___ The *building drain* runs horizontally under the building collecting waste water from multiple vertical stacks. A *4″ to 6″* pipe requires a minimum slope of *1%,* and an *8″* pipe requires a minimum slope of $\frac{1}{2}$%. The lowest (or basement) *floor elevation* needs to be set higher than the rim elevation of the next upsteam manhole of the sewer main. If the building drain is below the sewer main, an automatic underground *ejector pump* is needed.

___5. *Fire Protection*

   ___ *a.* A *sprinkler system* is the most effective way to provide fire safety. Research indicates sprinklers will extinguish or contain 95% of fires that start. Water supply and pressure are critical. At a minimum, should have 1-hour water supply.

      ___ (1) The IBC and NFPA standards require sprinklers at certain occupancies (see p. 175). Also see item I in App. A.

      ___ (2) Sprinkler *spacing* (maximum coverage per sprinkler):

         ___ Light hazard (Class I)

            200 SF for smooth ceiling and beam-and-girder construction

            225 SF if hydraulically calculated for smooth ceiling, as above

            130 SF for open wood joists

            168 SF for all other types of construction

At both sides of a fire barrier (horizontal exit adjacent to egress doorways). Also, accessible roofs require one outlet.

___ Ordinary hazard (Class II)

130 SF for all types of construction, except:

100 SF for high-pile storage (12′ or more).

___ Extra hazard (Class III)

90 SF for all types of construction

100 SF if hydraulically calculated.

___ High-piled storage (CHS) warehouses containing combustible items that are stored more than 15′ high.

___ (3) Notes

___ (*a*) Most buildings will be the *225 SF* spacing.

___ (*b*) Maximum spacing for light and ordinary hazard = 15′

High-pile and extra hazard = 12′

___ (*c*) Small rooms of light hazard, not exceeding *800 SF:* locate sprinklers max. of *9′* from walls.

___ (*d*) Maximum distance from walls to last sprinkler is ½ *spacing* (except at small rooms). Minimum is *4″*.

___ (*e*) City ordinances should be checked to verify that local rules are not more stringent than IBC/NFPA requirements.

___ (*f*) The sprinkler riser for small buildings usually takes a space about *2′6″ square.* Pumps and valves for larger buildings take up to about *100 to 500 SF.*

___ (*g*) Types

___ 1. Wet-pipe (water is always in pipe up to sprinkler head)

___ 2. Antifreeze

___ 3. Drypipe (water no further than main; used where freezing is a problem)

___ 4. Preaction (fast response)

___ 5. Deluge

___ 6. Foam water (petroleum fires)

Costs: Wet pipe systems: $1.20 to $4.20/SF
For dry pipe systems, add $.60/SF.

___ *b.* Large buildings often
also require a *stand-*
pipe, which is a large-
diameter water pipe
extending vertically
through the build-
ing with fire-hose
connections at every

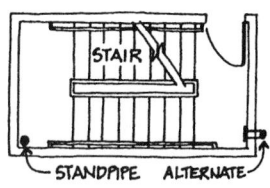

floor. The system is either wet or dry. The IBC de-
fines three classes.

___ (1) *Class I* is dry with 2½″ outlets. There is a
connection point on every landing of every
required stairway above or below grade,
and on both sides of a horizontal exit door.
This type of standpipe is for the fire department to connect their large hoses to.

___ (2) *Class II* is wet with 1½″ outlets and a hose.
This type is located so that every part of the
building is within 30′ of a nozzle attached to
100′ of hose. This type is for use by building
occupants or the fire department.

___ (3) *Class III* is wet with 2½″ outlets and 1½″
hose connections. These are located according to the rules for both Class I and II.

SIAMESE FITTING

Often, two *Siamese* fittings are required in
readily accessible locations on the outside
of the building to allow the fire department
to attach hoses from pumper trucks to the
dry standpipe and to the sprinkler riser.

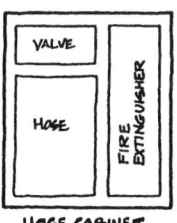

HOSE CABINET

Also, when required, *fire hose cabinets* will
be located in such a way that every point on
a floor lies within reach of a *30′* stream from
the end of a *100′* hose. A typical recessed
wall cabinet for a wet standpipe hose and
fire extinguisher is *2′9″ W × 9″D × 2′9″ H.*
See UBC Table 9-A below for standpipe
requirements.

___ *c.* *Fire alarm and detection systems:* One of the most
effective means of occupant protection in case of a
fire incident is the availability of a fire alarm system.
An alarm system provides early notification to
occupants of the building in the event of a fire,

thereby providing a greater opportunity for everyone in the building to evacuate or relocate to a safe area. Where required (occupancies):

___ (1) Group A, when 300 or more occupants.

___ (2) Group B and M, when over 500 occupants. Where more than 100 persons occupy spaces above or below the lowest level of exit discharge, a manual fire alarm must be installed.

___ (3) Group E, where occupant load is 50 or more.

___ (4) Group F, where multileveled and occupant load is 500 or more is housed above or below the level of exit discharge.

___ (5) Group H, semiconductor fabrication or manufacture of organic coatings.

___ (6) Group I, both manual fire alarm and automatic fire detection system.

___ (7) Residential: Certain residential structures require fire alarm and smoke detectors. This applies to hotels and other R-1 buildings. There is an exception to the required manual alarm system for such occupancies less than 3 stories in height where all guest rooms are completely separated by minimum 1-hour fire partitions and each unit has an exit directly to a yard, egress court, or public way. In R-2 buildings, alarms are required where more than 16 DUs are located in a single structure, or DUs are placed a significant distance vertically from the egress point at ground level.

___ 6. *Gas:* To allow for the gas meter and piping, provide a space $1'6"W \times 1'D \times 2'H$. Where natural gas is not available, propane, butane, and other flammable gases can be used to heat spaces and run stoves and hot-water heaters in homes, low-rise apartments, and small commercial buildings. A typical installation is a large cylinder located just outside the building and should be accessible by truck and well ventilated. The line to the building should be flexible (no iron or steel) and, at best, 10' from windows and stairways.

___7. *Solar Hot Water Systems*

   ___ *a.* In U.S., average person uses *20 gal.* of HW/day.

   ___ *b.* Mount collectors at tilt equal to about the site latitude.

   ___ *c.* Typical collectors are 4′ × 8′ and 4′ × 10′.

   ___ *d.* Typical relationship between collector area and storage volume is *1:3* to *1:7* gal. per SF of collector.

   ___ *e.* Type of systems

      ___ (1) *Open loop, recirculation:* The most widely used system in climates where freezing is of little concern.

      ___ (2) *Open loop, drain down:* Includes valving arrangement from collectors and piping when water temperature approaches freezing.

      ___ (3) *Closed loop, drain back:* Use of separate fluid (such as water) circulated through collectors where it is heated and transferred to HW storage through heat exchanger.

      ___ (4) *Closed loop, antifreeze:* Most widely used with heat exchanger.

   ___ *f.* Auxiliary heat: Typically an electric element in HW tank top.

   ___ *g.* For rough estimates:

      ___ (1) Northeast U.S.: 60 SF collector and 80 gallon tank will provide 50% to 75% need of a family of four.

      ___ (2) Southwest U.S.: 40 SF of collector will do same.

Costs: $3960 to $12,000 per system

TABLE 403.1
MINIMUM NUMBER OF REQUIRED PLUMBING FIXTURES[a]
(See Sections 403.2 and 403.3)

NO.	CLASSIFICATION	OCCUPANCY	DESCRIPTION	WATER CLOSETS (URINALS SEE SECTION 419.2)		LAVATORIES		BATHTUBS/ SHOWERS	DRINKING FOUNTAIN[e, f] (SEE SECTION 410.1)	OTHER
				MALE	FEMALE	MALE	FEMALE			
1	Assembly	A-1[d]	Theaters and other buildings for the performing arts and motion pictures	1 per 125	1 per 65	1 per 200		—	1 per 500	1 service sink
		A-2[d]	Nightclubs, bars, taverns, dance halls and buildings for similar purposes	1 per 40	1 per 40	1 per 75		—	1 per 500	1 service sink
			Restaurants, banquet halls and food courts	1 per 75	1 per 75	1 per 200		—	1 per 500	1 service sink
		A-3[d]	Auditoriums without permanent seating, art galleries, exhibition halls, museums, lecture halls, libraries, arcades and gymnasiums	1 per 125	1 per 65	1 per 200		—	1 per 500	1 service sink
			Passenger terminals and transportation facilities	1 per 500	1 per 500	1 per 750		—	1 per 1,000	1 service sink
			Places of worship and other religious services.	1 per 150	1 per 75	1 per 200		—	1 per 1,000	1 service sink

TABLE 403.1, (continued)
MINIMUM NUMBER OF REQUIRED PLUMBING FIXTURES[a]
(See Sections 403.2 and 403.3)

NO.	CLASSIFICATION	OCCUPANCY	DESCRIPTION	WATER CLOSETS (URINALS SEE SECTION 419.2)		LAVATORIES		BATHTUBS/ SHOWERS	DRINKING FOUNTAIN[e, f] (SEE SECTION 410.1)	OTHER
				MALE	FEMALE	MALE	FEMALE			
1 (cont.)	Assembly (cont.)	A-4	Coliseums, arenas, skating rinks, pools and tennis courts for indoor sporting events and activities	1 per 75 for the first 1,500 and 1 per 120 for the remainder exceeding 1,500	1 per 40 for the first 1,520 and 1 per 60 for the remainder exceeding 1,520	1 per 200	1 per 150	—	1 per 1,000	1 service sink
		A-5	Stadiums, amusement parks, bleachers and grandstands for outdoor sporting events and activities	1 per 75 for the first 1,500 and 1 per 120 for the remainder exceeding 1,500	1 per 40 for the first 1,520 and 1 per 60 for the remainder exceeding 1,520	1 per 200	1 per 150	—	1 per 1,000	1 service sink
2	Business	B	Buildings for the transaction of business, professional services, other services involving merchandise, office buildings, banks, light industrial and similar uses	1 per 25 for the first 50 and 1 per 50 for the remainder exceeding 50		1 per 40 for the first 80 and 1 per 80 for the remainder exceeding 80		—	1 per 100	1 service sink

TABLE 403.1, *(continued)*
MINIMUM NUMBER OF REQUIRED PLUMBING FIXTURES[a]
(See Sections 403.2 and 403.3)

NO.	CLASSIFICATION	OCCUPANCY	DESCRIPTION	WATER CLOSETS (URINALS SEE SECTION 419.2)		LAVATORIES		BATHTUBS/ SHOWERS	DRINKING FOUNTAIN[b, f] (SEE SECTION 410.1)	OTHER
				MALE	FEMALE	MALE	FEMALE			
3	Educational	E	Educational facilities	1 per 50	1 per 50	1 per 50	1 per 50	—	1 per 100	1 service sink
4	Factory and industrial	F-1 and F-2	Structures in which occupants are engaged in work fabricating, assembly or processing of products or materials	1 per 100	1 per 100	1 per 100	1 per 100	(see Section 411)	1 per 400	1 service sink

TABLE 403.1, (continued)
MINIMUM NUMBER OF REQUIRED PLUMBING FIXTURES[a]
(See Sections 403.2 and 403.3)

NO.	CLASSIFICATION	OCCUPANCY	DESCRIPTION	WATER CLOSETS (URINALS SEE SECTION 419.2)		LAVATORIES		BATHTUBS/ SHOWERS	DRINKING FOUNTAIN[b, f] (SEE SECTION 410.1)	OTHER
				MALE	FEMALE	MALE	FEMALE			
5	Institutional	I-1	Residential care	1 per 10	1 per 10	1 per 10	1 per 10	1 per 8	1 per 100	1 service sink
		I-2	Hospitals, ambulatory nursing home patients[b]	1 per room[c]	1 per room[c]	1 per room[c]	1 per room[c]	1 per 15	1 per 100	1 service sink per floor
			Employees, other than residential care[b]	1 per 25	1 per 25	1 per 35	1 per 35	—	1 per 100	—
			Visitors, other than residential care	1 per 75	1 per 75	1 per 100	1 per 100	—	1 per 500	—
		I-3	Prisons[b]	1 per cell	1 per cell	1 per cell	1 per cell	1 per 15	1 per 100	1 service sink
			Reformitories, detention centers, and correctional centers[b]	1 per 15	1 per 15	1 per 15	1 per 15	1 per 15	1 per 100	1 service sink
			Employees[b]	1 per 25	1 per 25	1 per 35	1 per 35	—	1 per 100	—
		I-4	Adult day care and child care	1 per 15	1 per 15	1 per 15	1 per 15	1	1 per 100	1 service sink

445

TABLE 403.1, *(continued)*
MINIMUM NUMBER OF REQUIRED PLUMBING FIXTURES[a]
(See Sections 403.2 and 403.3)

NO.	CLASSIFICATION	OCCUPANCY	DESCRIPTION	WATER CLOSETS (URINALS SEE SECTION 419.2)		LAVATORIES		BATHTUBS/ SHOWERS	DRINKING FOUNTAIN[e, f] (SEE SECTION 410.1)	OTHER
				MALE	FEMALE	MALE	FEMALE			
6	Mercantile	M	Retail stores, service stations, shops, salesrooms, markets and shopping centers	1 per 500		1 per 750		—	1 per 1,000	1 service sink
7	Residential	R-1	Hotels, motels, boarding houses (transient)	1 per sleeping unit		1 per sleeping unit		1 per sleeping unit	—	1 service sink
		R-2	Dormitories, fraternities, sororities and boarding houses (not transient)	1 per 10		1 per 10		1 per 8	1 per 100	1 service sink
		R-2	Apartment house	1 per dwelling unit		1 per dwelling unit		1 per dwelling unit	—	1 kitchen sink per dwelling unit; 1 automatic clothes washer connection per 20 dwelling units

TABLE 403.1, *(continued)*
MINIMUM NUMBER OF REQUIRED PLUMBING FIXTURES[a]
(See Sections 403.2 and 403.3)

NO.	CLASSIFICATION	OCCUPANCY	DESCRIPTION	WATER CLOSETS (URINALS SEE SECTION 419.2)		LAVATORIES		BATHTUBS/ SHOWERS	DRINKING FOUNTAIN[e, f] (SEE SECTION 410.1)	OTHER
				MALE	FEMALE	MALE	FEMALE			
7 (cont.)	Residential (cont.)	R-3	One- and two-family dwellings	1 per dwelling unit		1 per dwelling unit		1 per dwelling unit	—	1 kitchen sink per dwelling unit; 1 automatic clothes washer connection per dwelling unit
		R-3	Congregate living facilities with 16 or fewer persons	1 per 10	1 per 10	1 per 10	1 per 10	1 per 8	1 per 100	1 service sink
		R-4	Residential care/assisted living facilities	1 per 10	1 per 10	1 per 10	1 per 10	1 per 8	1 per 100	1 service sink
8	Storage	S-1 S-2	Structures for the storage of goods, warehouses, storehouse and freight depots. Low and Moderate Hazard.	1 per 100		1 per 100		See Section 411	1 per 1,000	1 service sink

TABLE 403.1, *(continued)*
MINIMUM NUMBER OF REQUIRED PLUMBING FIXTURES[a]
(See Sections 403.2 and 403.3)

a. The fixtures shown are based on one fixture being the minimum required for the number of persons indicated or any fraction of the number of persons indicated. The number of occupants shall be determined by the *International Building Code.*

b. Toilet facilities for employees shall be separate from facilities for inmates or patients.

c. A single-occupant toilet room with one water closet and one lavatory serving not more than two adjacent patient sleeping units shall be permitted where such room is provided with direct access from each patient sleeping unit and with provisions for privacy.

d. The occupant load for seasonal outdoor seating and entertainment areas shall be included when determining the minimum number of facilities required.

e. The minimum number of required drinking fountains shall comply with Table 403.1 and Chapter 11 of the *International Building Code.*

f. Drinking fountains are not required for an occupant load of 15 or fewer.

TABLE 4-1
Minimum Plumbing Facilities[1]

Each building shall be provided with sanitary facilities, including provisions for persons with disabilities as prescribed by the Department Having Jurisdiction. Table 4-1 applies to new buildings, additions to a building, and changes of occupancy or type in an existing building resulting in increased occupant load. Exception: New cafeterias used only by employees.

The total occupant load shall be determined in accordance with the Building Code. The type of building or occupancy shall be determined based on the actual use of the various spaces within the building. Building categories not shown in Table 4-1 shall be considered separately by the Authority Having Jurisdiction. The minimum number of fixtures shall be calculated at 50 percent male and 50 percent female based on the total occupant load.

Once the occupant load and uses are determined, the requirements of Section 412.0 and Table 4-1 shall be applied to determine the minimum number of plumbing fixtures required.

Type of Building[2] or Occupancy	Water Closets[14] (Fixtures per Person)		Urinals[5,10] (Fixtures per Person)	Lavatories (Fixtures per Person)		Bathtubs or Showers (Fixtures per Person)	Drinking[3,13,17] Fountains (Fixtures per Person)
Assembly places – theatres, auditoriums, convention halls, etc.– for permanent employee use	Male 1: 1-15 2: 16-35 3: 36-55 Over 55, add 1 fixture for each additional 40 persons.	Female 1: 1-15 3: 16-35 4: 36-55	Male 0: 1-9 1: 10-50 Add one fixture for each additional 50 males.	Male 1 per 40	Female 1 per 40		
Assembly places – theatres, auditoriums, convention halls, etc.– for public use	Male 1: 1-100 2: 101-200 3: 201-400 Over 400, add one fixture for each additional 500 males and 1 for each additional 125 females.	Female 3: 1-50 4: 51-100 8: 101-200 11: 201-400	Male 1: 1-100 2: 101-200 3: 201-400 4: 401-600 Over 600, add 1 fixture for each additional 300 males.	Male 1: 1-200 2: 201-400 3: 401-750 Over 750, add one fixture for each additional 500 persons.	Female 1: 1-200 2: 201-400 3: 401-750		1: 1-150 2: 151-400 3: 401-750 Over 750, add one fixture for each additional 500 persons.

TABLE 4-1, *(continued)*
Minimum Plumbing Facilities[1]

Type of Building[2] or Occupancy	Water Closets[14] (Fixtures per Person)		Urinals[5,10] (Fixtures per Person)	Lavatories (Fixtures per Person)		Bathtubs or Showers (Fixtures per Person)	Drinking[3,13,17] Fountains (Fixtures per Person)
Dormitories[9]– School or labor[16]	Male 1 per 10 Add 1 fixture for each additional 25 males (over 10) and 1 for each additional 20 females (over 8).	Female 1 per 8	Male 1 per 25 Over 150, add 1 fixture for each additional 50 males.	Male 1 per 12 Over 12, add one fixture for each additional 20 males and 1 for each 15 additional females.	Female 1 per 12	1 per 8 For females, add 1 bathtub per 30. Over 150, add 1 bathtub per 20.	1 per 150[12]
Dormitories – for staff use[16]	Male 1: 1-15 2: 16-35 3: 36-55 Over 55, add 1 fixture for each additional 40 persons.	Female 1: 1-15 3: 16-35 4: 36-55	Male 1 per 50	Male 1 per 40	Female 1 per 40	1 per 8	
Dwellings[4] Single dwelling Multiple dwelling or apartment house[16]	1 per dwelling 1 per dwelling or apartment unit			1 per dwelling 1 per dwelling or apartment unit		1 per dwelling 1 per dwelling or apartment unit	
Hospital waiting rooms	1 per room			1 per room			1 per 150[12]
Hospitals – for employee use	Male 1: 1-15 2: 16-35 3: 36-55 Over 55, add 1 fixture for each additional 40 persons.	Female 1: 1-15 3: 16-35 4: 36-55	Male 0: 1-9 1: 10-50 Add one fixture for each additional 50 males.	Male 1 per 40	Female 1 per 40		

TABLE 4-1, *(continued)*
Minimum Plumbing Facilities[1]

Type of Building[2] or Occupancy	Water Closets[14] (Fixtures per Person)		Urinals[5,10] (Fixtures per Person)	Lavatories (Fixtures per Person)		Bathtubs or Showers (Fixtures per Person)	Drinking[3,13,17] Fountains (Fixtures per Person)
Hospitals Individual room Ward room	1 per room 1 per 8 patients			1 per room 1 per 10 patients		per room 1 per 20 patients	1 per 150[12]
Industrial[6] warehouses, workshops, foundries, and similar establishments – for employee use	Male 1: 1-10 2: 11-25 3: 26-50 4: 51-75 5: 76-100 Over 100, add 1 fixture for each additional 30 persons.	Female 1: 1-10 2: 11-25 3: 26-50 4: 51-75 5: 76-100		Up to 100, 1 per 10 persons Over 100, 1 per 15 persons[7,8]		1 shower for each 15 persons exposed to excessive heat or to skin contamination with poisonous, infectious or irritating material	1 per 150[12]
Institutional – other than hospitals or penal institutions (on each occupied floor)	Male 1 per 25	Female 1 per 20	Male 0: 1-9 1: 10-50 Add one fixture for each additional 50 males.	Male 1 per 10	Female 1 per 10	1 per 8	1 per 150[12]
Institutional – other than hospitals or penal institutions (on each occupied floor) – for employee use	Male 1: 1-15 2: 16-35 3: 36-55 Over 55, add 1 fixture for each additional 40 persons.	Female 1: 1-15 3: 16-35 4: 36-55	Male 0: 1-9 1: 10-50 Add one fixture for each additional 50 males.	Male 1 per 40	Female 1 per 40	1 per 8	1 per 150[12]

TABLE 4-1, (continued)
Minimum Plumbing Facilities[1]

Type of Building[2] or Occupancy	Water Closets[14] (Fixtures per Person)	Urinals[5,10] (Fixtures per Person)	Lavatories (Fixtures per Person)	Bathtubs or Showers (Fixtures per Person)	Drinking[3,13,17] Fountains (Fixtures per Person)
Office or public buildings	Male 1: 1-100 2: 101-200 3: 201-400 Female 3: 1-50 4: 51-100 8: 101-200 11: 201-400 Over 400, add one fixture for each additional 500 males and 1 for each additional 150 females.	Male 1: 1-100 2: 101-200 3: 201-400 4: 401-600 Over 600, add 1 fixture for each additional 300 males.	Male 1: 1-200 2: 201-400 3: 401-750 Female 1: 1-200 2: 201-400 3: 401-750 Over 750, add one fixture for each additional 500 persons.		1 per 150[12]
Office or public buildings – for employee use	Male 1: 1-15 2: 16-35 3: 36-55 Female 1: 1-15 3: 16-35 4: 36-55 Over 55, add 1 fixture for each additional 40 persons.	Male 0: 1-9 1: 10-50 Add one fixture for each additional 50 males.	Female 1 per 40 Male 1 per 40		
Penal institutions – for employee use	Male 1: 1-15 2: 16-35 3: 36-55 Female 1: 1-15 3: 16-35 4: 36-55 Over 55, add 1 fixture for each additional 40 persons.	Male 0: 1-9 1: 10-50 Add one fixture for each additional 50 males.	Male 1 per 40 Female 1 per 40		1 per 150[12]
Penal institutions – for prison use Cell Exercise room	1 per cell 1 per exercise room	Male 1 per exercise room	1 per cell 1 per exercise room		1 per cell block floor 1 per exercise room

TABLE 4-1, (continued)
Minimum Plumbing Facilities[1]

Type of Building[2] or Occupancy	Water Closets[14] (Fixtures per Person)		Urinals[5,10] (Fixtures per Person)	Lavatories (Fixtures per Person)		Bathtubs or Showers (Fixtures per Person)	Drinking[3,13,17] Fountains (Fixtures per Person)
Public or professional offices[15]	Same as Office or Public Buildings for employee use[15]		Same as Office or Public Buildings for employee use[15]	Same as Office or Public Buildings for employee use[15]			Same as Office or Public Buildings for employee use[15]
Restaurants, pubs, and lounges[11,15]	Male 1:1-50 2:51-150 3:151-300 Over 300, add 1 fixture for each additional 200 persons.	Female 1:1-5 2:51-150 4:151-300	Male 1:1-150 Over 150, add 1 fixture for each additional 150 males.	Male 1:1-150 2:151-200 3:201-400 Over 400, add 1 fixture for each additional 400 persons.	Female 1:1-150 2:151-200 3:201-400		
Retail or Whole-sale Stores	Male 1:1-100 2:101-200 3:201-400 Over 400, add one fixture for each additional 500 males and one for each 150 females	Female 1:1-25 2:26-100 4:101-200 6:201-300 8:301-400	Male 0:0-25 1:26-100 2:101-200 3:201-400 4:401-600 Over 600, add one fixture for each additional 300 males	1 per 2 water closets			0: 1-30[17] 1:31-150 One additional drinking fountain for each 150 persons thereafter
Schools – for staff use All schools	Male 1:1-15 2:16-35 3:36-55 Over 55, add 1 fixture for each additional 40 persons.	Female 1:1-15 2:16-35 3:36-55	Male 1 per 50	Male 1 per 40	Female 1 per 40		

453

TABLE 4-1, *(continued)*
Minimum Plumbing Facilities[1]

Type of Building[2] or Occupancy	Water Closets[14] (Fixtures per Person)		Urinals[5,10] (Fixtures per Person)	Lavatories (Fixtures per Person)		Bathtubs or Showers (Fixtures per Person)	Drinking[3,13,17] Fountains (Fixtures per Person)
Schools – for student use Nursery	Male 1: 1-20 2: 21-50 Over 50, add 1 fixture for each additional 50 persons.	Female 1: 1-20 2: 21-50		Male 1: 1-25 2: 26-50 Over 50, add 1 fixture for each additional 50 persons.	Female 1: 1-25 2: 26-50		1 per 150[12]
Elementary	Male 1 per 30	Female 1 per 25	Male 1 per 75	Male 1 per 35	Female 1 per 35		1 per 150[12]
Secondary	Male 1 per 40	Female 1 per 30	Male 1 per 35	Male 1 per 40	Female 1 per 40		1 per 150[12]
Others (colleges, universities, adult centers, etc.)	Male 1 per 40	Female 1 per 30	Male 1 per 35	Male 1 per 40	Female 1 per 40		1 per 150[12]
Worship places educational and activities Unit	Male 1 per 150	Female 1 per 75	Male 1 per 150	1 per 2 water closets			1 per 150[12]
Worship places principal assembly place	Male 1 per 150	Female 1 per 75	Male 1 per 150	1 per 2 water closets			1 per 150[12]

TABLE 4-1, *(continued)*
Minimum Plumbing Facilities¹

1   The figures shown are based upon one (1) fixture being the minimum required for the number of persons indicated or any fraction thereof.

2   Building categories not shown on this table shall be considered separately by the Authority Having Jurisdiction.

3   Drinking fountains shall not be installed in toilet rooms.

4   Laundry trays. One (1) laundry tray or one (1) automatic washer standpipe for each dwelling unit or one (1) laundry tray or one (1) automatic washer standpipe, or combination thereof, for each twelve (12) apartments. Kitchen sinks, one (1) for each dwelling or apartment unit.

5   For each urinal added in excess of the minimum required, one water closet shall be permitted to be deducted. The number of water closets shall not be reduced to less than two-thirds (2/3) of the minimum requirement.

6   As required by PSAI Z4.1, *Sanitation in Places of Employment*.

7   Where there is exposure to skin contamination with poisonous, infectious, or irritating materials, provide one (1) lavatory for each five (5) persons.

8   Twenty-four (24) lineal inches (610 mm) of wash sink or eighteen (18) inches (457 mm) of a circular basin, when provided with water outlets for such space, shall be considered equivalent to one (1) lavatory.

9   Laundry trays, one (1) for each fifty (50) persons. Service sinks, one (1) for each hundred (100) persons.

10   General. In applying this schedule of facilities, consideration shall be given to the accessibility of the fixtures. Conformity purely on a numerical basis may not result in an installation suited to the needs of the individual establishment. For example, schools should be provided with toilet facilities on each floor having classrooms.

     a.   Surrounding materials, wall, and floor space to a point two (2) feet (610 mm) in front of urinal lip and four (4) feet (1,219 mm) above the floor, and not less than two (2) feet (610 mm) to each side of the urinal shall be lined with non-absorbent materials.

     b.   Trough urinals shall be prohibited.

TABLE 4-1, (continued)
Minimum Plumbing Facilities[1]

11. A restaurant is defined as a business that sells food to be consumed on the premises.

   a. The number of occupants for a drive-in restaurant shall be considered as equal to the number of parking stalls.

   b. Hand-washing facilities shall be available in the kitchen for employees.

12. Where food is consumed indoors, water stations shall be permitted to be substituted for drinking fountains. Offices, or public buildings for use by more than six (6) persons shall have one (1) drinking fountain for the first one-hundred fifty (150) persons and one (1) additional fountain for each three-hundred (300) persons thereafter.

13. There shall be at least one (1) drinking fountain per occupied floor in schools, theatres, auditoriums, dormitories, offices, or public buildings.

14. The total number of water closets for females shall be equal to the total number of water closets and urinals required for males. This requirement shall not apply to Retail or Wholesale Stores.

15. For smaller-type Public and Professional Offices such as banks, dental offices, law offices, real estate offices, architectural offices, engineering offices, and similar uses. A public area in these offices shall use the requirements for Retail or Wholesale Stores.

16. Recreation or community room in multiple dwellings or apartment buildings, regardless or their occupant load, shall be permitted to have separate single-accommodation facilities in common-use areas within tracts or multi-family residential occupancies where the use of these areas is limited exclusively to owners, residents, and their guests. Examples are community recreation or multi-purpose areas in apartments, condos, townhouses, or tracts.

17. A drinking fountain shall not be required in occupancies of 30 or less. When a drinking fountain is not required, then footnotes 3, 12, and 13 are not applicable.

EXAMPLE:

PROBLEM: FIGURE THE REQUIRED PLUMBING FIXTURES FOR
A 10000 SF OFFICE SPACE. FIGURE FOR BOTH
THE I.P.C. AND THE U.P.C. CODES.

SOLUTION:

A. <u>BY I.P.C.</u> (SEE TABLE A FOR I.P.C. ON P.191)

  1. BUSINESS: 10000 SF ÷ 100 SF/OCC. = 100 USERS

   ÷2 = 50/SEX

  2. FIXTURES: (SEE I.P.C. TABLE 403.1)

	M (50)	W (50)
WC	2̶ 1	2
LAV	1	1
UR	1	

    ALSO: 1 DF REQ'D.

B. <u>BY U.P.C.</u> (SEE TABLE 4-1 OF U.P.C.)

  1. FIGURE OCC. LOAD FOR EXITING AS ABOVE.

        FROM TABLE 4-1 ASSUME

  PUBLIC, NOT EMPLOYEE, USE.

   10000 SF ÷ 100 SF/OCC = 100 OCC ÷2 = 50/SEX

  2. FIXTURES

	M (50)	W (50)
WC	1	3
LAV	1	1
UR	1	

    ALSO: 1 DF REQ'D.

NOTES

__ B. HEATING, VENTILATION, AND AIR CONDITIONING (HVAC) ① ④ ⑩ ⑷₅ₐ

The interior designer may be involved with HVAC when doing a tenant improvement or remodel. This will probably involve working with an architect or engineer. This section is included to give the interior designer a basic understanding of HVAC functions.

See p. 462 for selection and **cost** table.

Costs: Equipment 20% to 30%; distribution system 80% to 70%.

During programming it is useful to do a functional partitioning of the building into major zones for:

___ 1. Similar schedule of use
___ 2. Similar temperature requirements
___ 3. Similar ventilation and air quality
___ 4. Similar internal heat generation
___ 5. Similar HVAC needs

During design, if possible, locate spaces with similar needs together. See p. 224 for energy conservation and equipment efficiency. As a general rule, provide at least 3′ around all HVAC equipment for maintenance.

___ 1. *General*
   HVAC systems can be divided into four major parts:
   ___ a. *Source:*
      ___ (1) The *boiler and chiller* to create heat and cold for the system to use. (In small package systems this is an internal electric coil, gas furnace, or refrigeration compressor).
      ___ (2) *Cooling tower* (or air-cooled condenser) located outside to exhaust heat.
   ___ b. *Distribution:*
      ___ (1) *Air handlers* to transfer heat and cold to air (or at least fresh air) to be blown into the building zones. In large buildings this is in a fan room. (In small package systems this is an internal fan.)
      ___ (2) The *system* of ducts, control boxes, and diffusers to deliver conditioned air to the spaces.
   ___ c. *Delivery* of diffusers, baseboard radiators, unit heaters, convector cabinets, induction units, etc.

___ **2. *Systems for Small Buildings***

   ___ *a. Roof-mounted "package systems"* are typically used for residential and small to medium commercial buildings. They are AC units that house the first three parts in one piece of equipment that usually ends up on the roof. Used usually in warm or temperate climates. Typical sizes:

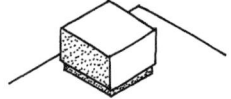 Roof Mounted "Package" Unit

Size	Area served	Dimensions	System
2 to 5 tons	600 to 1500 SF	$6'L \times 4'W \times 4'H$	Single zone constant vol. delivery system; can serve more than one zone with variable air vol. delivery system
5 to 10 tons	1500 to 4500 SF	$10'L \times 7'W \times 5'H$	
15 to 75 tons	4500 to 22,500 SF	$25'L \times 9'W \times 6'H$	

Notes:
  1. Units should have *3' to 4'* of clearance around.
  2. A *ton is 12,000 Btu* of refrigeration.
  3. Each *ton is equal to 400 CFM.*

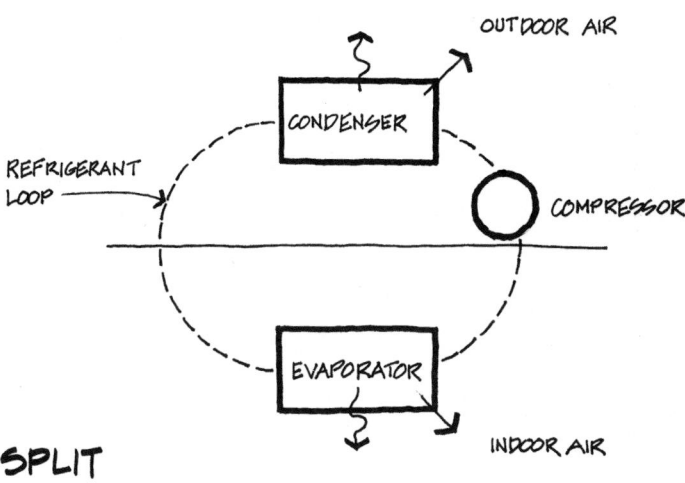

OUTDOOR AIR

CONDENSER

REFRIGERANT LOOP

COMPRESSOR

EVAPORATOR

INDOOR AIR

SPLIT

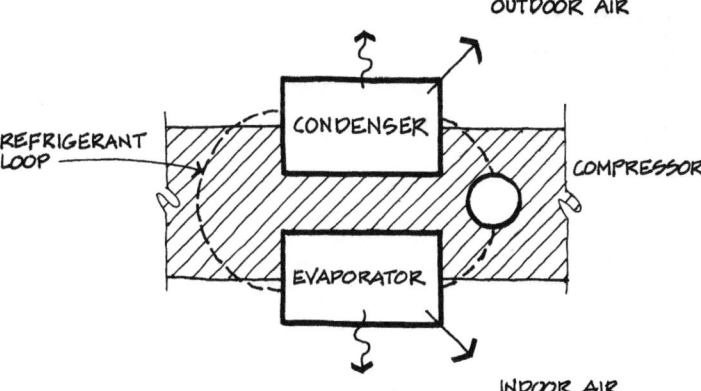

OUTDOOR AIR

CONDENSER

REFRIGERANT LOOP

COMPRESSOR

EVAPORATOR

INDOOR AIR

THROUGH WALL

DIAGRAM OF LOCAL SYSTEMS

#	TYPE	HEATING	COOLING	BUILDING SMALL	BUILDING LARGE	FUEL ELECT.	FUEL GAS	FUEL OTHER	DELIVERY AIR	DELIVERY PIPES	MIN. OPERATING COST IN COLD CLIMATE	MIN. OPERATING COST IN MODERATE CLIMATE	MAX. CONTROL OF AIR VELOCITY & QUALITY	MAX. INDIVIDUAL CONTROL OF TEMP.	MINIMUM NOISE	MINIMUM VISUAL OBTRUSIVENESS	MIN. SPACE FOR EQUIP.	MIN. MAINTENANCE	MIN. FL. TO FL. HT.	MAX. FLEXIBILITY OF RENTAL SPACE	$ PER SF (±10% 50%M & 50%L)	$/TON OF AC
1	ROOF MT'D "PACKAGE" UNITS	●	●	●		●	●		●									●			5.75	1610
2	CENTRAL FORCED AIR (& "SPLIT" SYSTEMS)	●	●	●		●	●		●		●		●	●	●	●		●			7.20	1380
3	FORCED HOT WATER	●		●		●	●			●	●							●			7.20	1840
4	EVAPORATIVE COOLING		●	●		●			●			●		●				●			5.75	2300
5	THROUGH WALL UNITS	●	●	●		●			●					●			●				2.85	690
6	ELECTRIC BASE BOARD	●		●		●						●					●				1.75	690
7	ELECT. FAN UNIT HEATERS	●		●		●						●					●				1.50	
8	RADIANT	●		●		●						●			●						2.90	
9	WALL FURNACE	●		●		●										●	●	●			1.75	
10	PASSIVE SOLAR	●		●				●				●									2.90	
11	ACTIVE SOLAR	●		●				●				●						●				
12	STOVES	●		●				●				●										
13	SINGLE ZONE CONSTANT VOL.	●	●		●	●	●	●	●		●		●			●		●			7.20	2530
14	MULTI ZONE CONSTANT VOL.	●	●		●	●	●	●	●					●							10.00	2875
15	VARIABLE AIR VOLUME	●	●		●	●	●	●	●		●			●		●				●	11.50	3105
16	DOUBLE DUCT	●	●		●	●	●	●	●		●			●						●	14.35	3450
17	INDUCTION	●	●		●	●	●	●	●	●	●			●		●			●	●	11.50	3105
18	FAN COIL WITH AIR	●	●		●	●	●	●	●	●	●			●		●			●	●	11.50	2875
19	FAN COIL UNITS	●	●		●	●	●	●		●	●					●			●	●	10.00	2530
20	HOT WATER BASE BOARDS	●			●	●	●	●		●	●				●			●			5.75	

___ *b.* *Forced-air central heating* is typically used for residential and light commercial buildings. It heats air with gas, oil flame, or elect. resistance at a furnace. A fan blows air through a duct system. The furnace can be upflow (for basements), side flow, or down flow (for attic). The furnace must be vented. Furnace sizes range between $2'W \times 2.5'D \times 7'H$ to $4'W$ to $7'D \times 7'H$. Main ducts are typically $1' \times 2'$ horizontal and $1' \times .33'$ vertical.

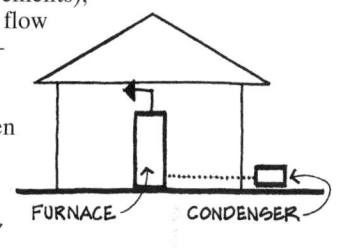

Can add cooling with a "*split system*" by adding evaporator coils in the duct and an exterior condenser. Typical condensers range from $2'W \times 2'D \times 2'H$ to $3.5'W \times 4'D \times 3'H$.

___ *c.* *Forced hot water heating* is typically used for residential buildings and commercial offices. A burner or electric resistance heats water to fin tube convectors (or fan coil unit with blowers). The fueled boiler must be vented and provided with combustible air. Boiler sizes range from $2'W \times 2'D \times 7'H$ to $3'W \times 5'D \times 7'H$. Fin tube convectors are typically $3''D \times 8''H$. Fan coils are $2'W \times 2.5'H$. There is *no* cooling.

___ *d.* *Evaporative cooling* is typically used for residential buildings. It works only in hot, dry climates. A fan draws exterior air across wet pads and into the duct system. There is *no* heating. Cooler size typically is $3'W \times 3'D \times 3'H$. Main duct is typically $1.5'W \times 1.5'D$.

___ *e.* *Through-wall units and package terminal units* are typically used for motels/ hotels as well as small offices. They are self-contained at an exterior wall and are intended for small spaces. These are usually electric (or *heat pump* in mild climates), which are used for *both* heating and cooling. Interior air is recirculated and outside air is added. Typical sizes:

Package Terminal Units	3.5′W × 1.5′D × 1.3′H
Through-Wall Units	2′W × 2′D × 1.5′H

___ *f.* *Electric baseboard convectors* are typically used for residential buildings and commercial offices. They heat by electrical resistance in *3″D × 8″H* baseboards around the perimeter of the room. There is *no cooling*.

___ *g.* *Electric fan-forced unit heaters* are much like item *f* above, but are larger because of internal fans recirculating the air. There is *no cooling*. Typical sizes range from *1.5′ W × 8″D × 8″H* to *2′ W × 1′D × 1.8′H.*

___ *h.* *Radiant heating:* Electrical resistance wires are embedded in floor or ceiling. There is *no cooling*. An alternative is to have recessed radiant panels, typically *2′ × 2′* or *2′ × 4′*. For alternative cooling and heating use water piping. These are typically residential applications.

___ *i.* *Wall furnaces* are small furnaces for small spaces (usually residential). They must be vented. There is *no cooling*. They may be either gas or electric. The typical size is *14″W × 12″D × 84″H.*

___ *j.* Other miscellaneous small systems (typically residential):
___ Passive solar heating (see p. 219)

    ___ Active solar heating
    ___ Heating stoves (must be properly vented!)
___ 3. *Custom Systems for Large Buildings*

These are where the first three parts (see p. 459) must have areas allocated for them in the floor plan. In tall buildings due to distance, mechanical floors are created so that air handlers can move air up and down *10 to 15* floors. Thus mechanical floors are spaced *20 to 30* floors apart.

A decentralized chiller and boiler can be at every other mechanical floor or they can be centralized at the top or base of the building with one or more air handlers at each floor. See p. 467 for equipment rooms. *Note:* An alternative for large buildings (but not high rise) is to go with a number of large "package" units on the roof. Instead of installing one large unit, incorporate several smaller units of the same total capacity, then add one more, so any unit may be serviced without affecting the total operation.

    ___ *a.* Delivery systems
        ___ (1) *Air delivery systems:*

Because of their size, <u>*ducts*</u> are a great concern in the preliminary design of the floor-to-ceiling space. See p. 415. The main supply and return ducts are often run above main hallways because ceilings can be lower and because this provides a natural path of easy access to the majority of spaces served. Ideally, all ducts should be run as straight and clear of obstructions as possible, contain no corners that could collect dirt, and have access portals that allow inspection and cleaning. Horizontal runs should be pitched slightly to prevent moisture collection.

When ducts cross any kind of fire wall or fire barrier, they usually require fire dampers.

Air rates for buildings vary from *1 CFM/SF to 2 CFM/SF* based on usage and climate. Low-velocity ducts require *1 to 2 SF of area per 1000 SF* of building area served. High-velocity ducts require *0.5 to*

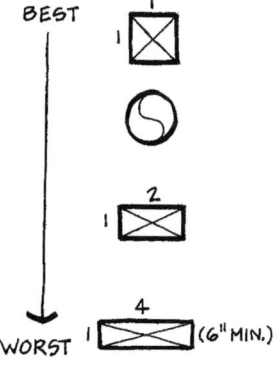

DUCTS

24" × 10" DUCT IN PLAN
WIDTH   HEIGHT

BEST

WORST   (6" MIN.)

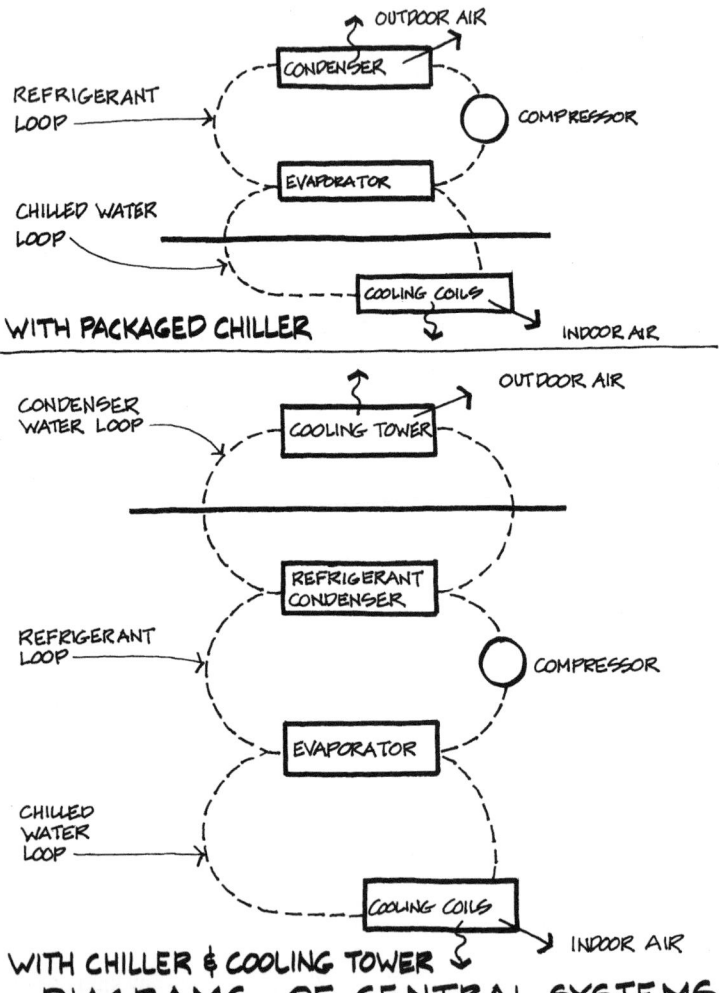

WITH PACKAGED CHILLER

WITH CHILLER & COOLING TOWER

DIAGRAMS OF CENTRAL SYSTEMS

LARGE SYSTEMS

<u>DELIVERY SYSTEM</u> (SEE FOLLOWING PAGES)

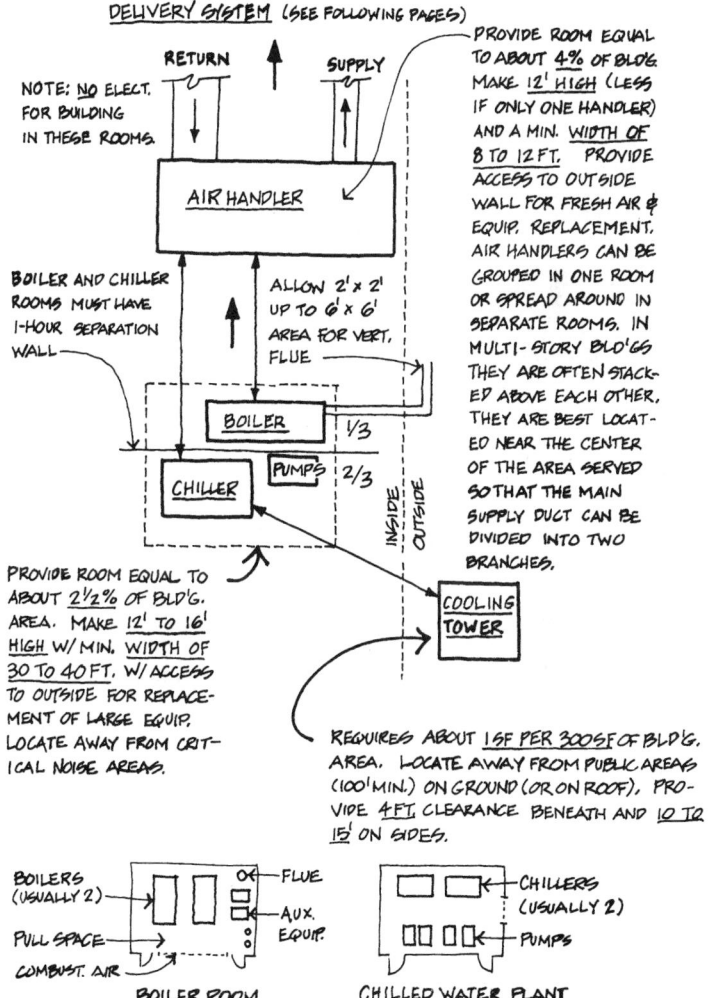

NOTE: <u>NO</u> ELECT. FOR BUILDING IN THESE ROOMS.

RETURN

SUPPLY

AIR HANDLER

PROVIDE ROOM EQUAL TO ABOUT <u>4%</u> OF BLDG. MAKE 12' HIGH (LESS IF ONLY ONE HANDLER) AND A MIN. <u>WIDTH OF 8 TO 12 FT.</u> PROVIDE ACCESS TO OUTSIDE WALL FOR FRESH AIR & EQUIP. REPLACEMENT. AIR HANDLERS CAN BE GROUPED IN ONE ROOM OR SPREAD AROUND IN SEPARATE ROOMS. IN MULTI-STORY BLD'GS THEY ARE OFTEN STACK-ED ABOVE EACH OTHER, THEY ARE BEST LOCAT-ED NEAR THE CENTER OF THE AREA SERVED SO THAT THE MAIN SUPPLY DUCT CAN BE DIVIDED INTO TWO BRANCHES.

BOILER AND CHILLER ROOMS MUST HAVE 1-HOUR SEPARATION WALL

ALLOW 2' x 2' UP TO 6' x 6' AREA FOR VERT. FLUE

BOILER   ⅓

PUMPS   ⅔

CHILLER

INSIDE   OUTSIDE

PROVIDE ROOM EQUAL TO ABOUT <u>2½%</u> OF BLD'G. AREA. MAKE 12' TO 16' HIGH W/ MIN. <u>WIDTH OF 30 TO 40 FT.</u> W/ ACCESS TO OUTSIDE FOR REPLACE-MENT OF LARGE EQUIP. LOCATE AWAY FROM CRIT-ICAL NOISE AREAS.

COOLING TOWER

REQUIRES ABOUT <u>1 SF PER 300 SF</u> OF BLD'G. AREA. LOCATE AWAY FROM PUBLIC AREAS (100' MIN.) ON GROUND (OR ON ROOF), PRO-VIDE <u>4 FT.</u> CLEARANCE BENEATH AND <u>10 TO 15'</u> ON SIDES.

BOILERS (USUALLY 2)

FLUE

AUX. EQUIP.

PULL SPACE

COMBUST. AIR

<u>BOILER ROOM</u>

CHILLERS (USUALLY 2)

PUMPS

<u>CHILLED WATER PLANT</u>

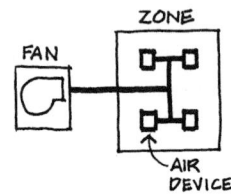

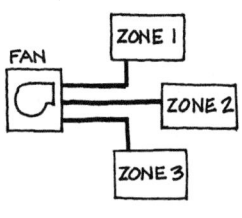

1.0. Air returns are required and are about the same size, or slightly larger, than the main duct supply. (The above-noted dimensions are interior. Typically, ducts are externally lined with 1″ or 2″ of insulation.)

Costs: Ducts typically cost about 40% of total HVAC costs.

___ (*a*) *Single-zone constant-volume systems* serve only one zone and are used for large, open-space rooms without diverse exterior exposure. This is a low-velocity system.

___ (*b*) *Multizone constant-volume systems* can serve up to eight separate zones. They are used in modest-sized buildings where there is a diversity of exterior exposure and/or diversity of interior loads. This is a low-velocity system.

___ (*c*) *Subzone box systems* often modify single-zone systems for appended spaces. They use boxes that branch off the main supply duct to create separate zones. The size of the boxes can be related to the area served:

Box	Area served
4′L × 3′W × 1.5′H	500 to 1500 SF
5′L × 4′W × 1.5′H	1500 to 5000 SF

The main ducts can be high-velocity, but the ducts after the boxes at each zone (as well as the return air) are low-velocity.

___ (*d*) *Variable air-volume* single duct can serve as many subzones as required. It is the dominant choice in many commercial buildings because of its flexibility and energy savings. It is most effectively used for interior zones. At exterior zones hot water or electrical reheat coils are added

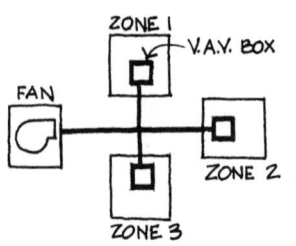

to the boxes. Each zone's temperature is controlled by the volume of air flowing through its box. Typical above ceiling boxes are:

8″ to 11″H for up to 1500 SF served (lengths up to 5′) up to 18″H for up to 7000 SF

___ (*e*) *Double-duct systems* can serve as a good choice where *air quality control* is important. The air handler supplies hot air for one duct and cold air for the other. The mixing box controls the mix of these two air ducts. This system is not commonly used except in retrofits. It is a "caddie" but also a "gas guzzler."

___ (*f*) *Variable air-volume dual-duct systems.* This system is high-end first cost and most likely used in a retrofit. One duct conveys cool air, one other hot air. This system is most common where a dual-duct constant volume system is converted to VAV. The box is generally controlled to provide either heat or cool air as required in varying quantities.

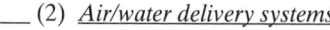

ZONE I

MIXTURE BOX

FAN

ZONE 2

ZONE 3

___ (2) *Air/water delivery systems*

These types of systems *reduce the ductwork* by tempering air near its point of use. Hot and cold water are piped to remote induction or fan coil units. Since the air ducts carry only fresh air, they can be sized at *0.2 to 0.4 SF per 1000 SF* of area served. The main hot and cold water lines will be *2 to 4 inches* diameter, including insulation, for medium size buildings.

___ (*a*) *Induction* is often used for the *perimeter of high-rise office buildings* and is *expensive*. Air from a central air handler is delivered through high-velocity ducts to each induction unit. Hot and cold lines

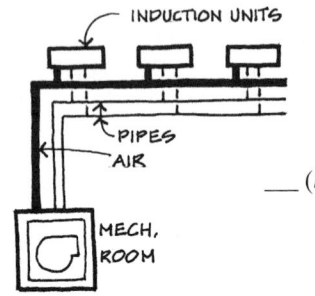

run to each unit. Each unit is located along the outside wall, at the base of the windows. They are *6 to 12 inches deep* and *1 to 3 feet high.*

___ (*b*) *Fan coils with supplementary air* are used where there are *many small rooms needing separate control.* Hot and cold water lines are run through the coils. A fan draws room air through the coil for heating and cooling. A separate duct system supplies fresh air from a remote air handler.

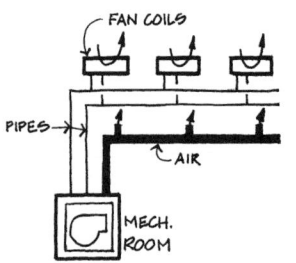

The fan coils are *6 to 12 inches deep and 1 to 3 feet high.* They can also be a vertical shape ($2' \times 2' \times 6'$ *high*) to fit in a closet. They are often stacked vertically in a tall building to reduce piping. Fan coil units can be located in ceiling space.

___ (3) *Water delivery systems* use hot and cold water lines only. No air is delivered to the areas served.

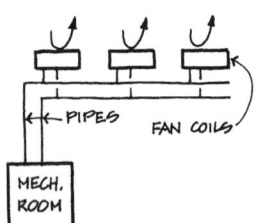

___ (*a*) *Fan coil units* can have hot and/or cold water lines with fresh air from operable windows or an outdoor air intake at the unit.

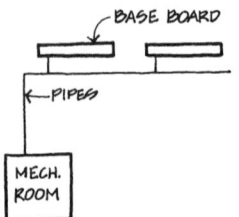

___ (*b*) *Hot water baseboards* supply only heat. Often used in conjunction with a cooling-only VAV system for perimeter zones. Baseboards are *6 inches high by 5 inches deep* and as long as necessary.

___ 4. **Diffusers, Terminal Devices, and Grilles:** Interface the HVAC system with the building interiors for visual impact and thermal comfort. Supply and return grilles or registers should be as far apart as possible in each space (ideally at opposite walls and opposite corners, and one near the ceiling and the other near the floor), and they should be located where occupants or furnishings will not block them. Grilles are side wall devices. Their "throw" should be about ¾ the distance to the other side of the room. Opposite wall should be no greater than about *16′ to 18′* away (can throw up to 30′ in high rooms with special diffusers). Diffusers are down-facing and must be coordinated with the lighting as well as uniformly spaced (at a *distance apart of approximately the floor-to-ceiling height*). Returns should be spaced so as to not interfere with air supply. Assume return air grilles at one per 400 SF to 600 SF.

GRILLE

DIFFUSER

NOTES

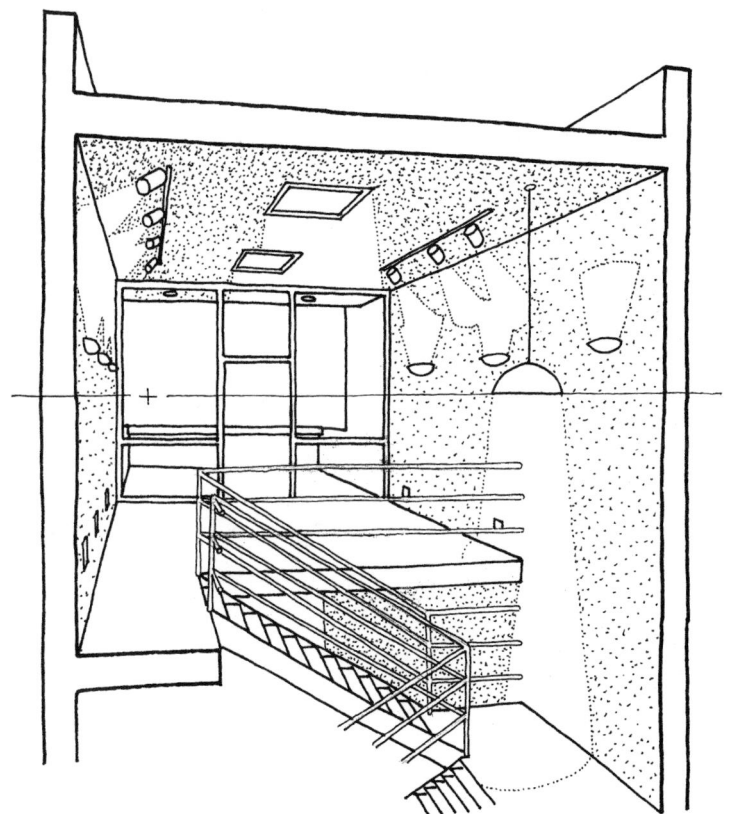

16 LIGHTING AND ELECTRICAL

NOTES

___ 1. *General*

    ___ *a.* Lighting terms and concepts using the analogy of a sprinkler pipe

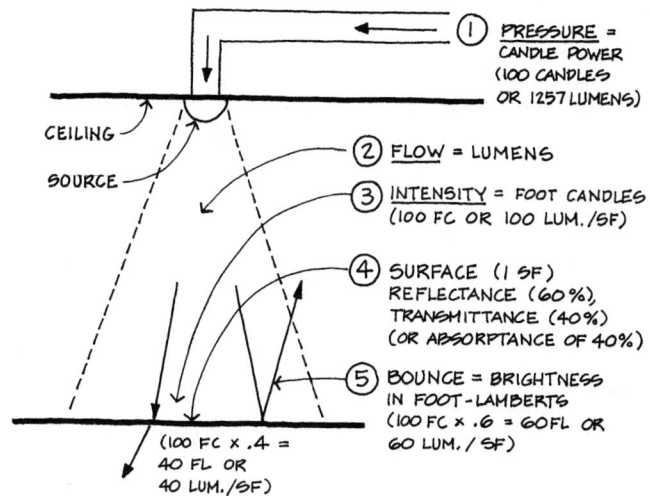

      ___ (1) Visible light is measured in lumens.

      ___ (2) One lumen of light flux spread over one square foot of area illuminates the area to one *footcandle.*

      ___ (3) The ratio of lumens/watts is called *efficacy,* a measure of *energy efficiency.*

      ___ (4) The incident angle of a light beam always equals the reflectance angle on a surface.

      ___ (5) The 1-1-1 Rule: When 1 lumen of light strikes 1SF of perfectly reflective area, 1′ away, at right angles, then 1 lumen of output = 1 fc of incident light = 1 foot-lambert of reflected light.

    ___ *b.* Considerations in seeing

      ___ (1) *Contrast* between the object or area being viewed and its surroundings will help vision. Too little will wash out the object. Too much will create glare. Recommended maximum ratios:

        ___ Task to adjacent area   3 to 1
        ___ Task to remote dark surface   3 to 1

   ___ Task to remote light surface   1 to 1
   ___ Window to adjacent wall   20 to 1
   ___ Task to general visual field   40 to 1
   ___ Focal point: up to   100 to 1

___ (2) _Brightness_ (How much light?). For recommended lighting levels, see d, below, or p. 478.

___ (3) _Size_ of object or material being viewed. As the viewing task becomes smaller, the brightness needs to increase and vice versa.

___ (4) _Time._ As the view time is decreased, the brightness and contrast needs to increase and vice versa.

___ (5) _Glare._ Not only can too much contrast create glare, but light sources at the wrong angle to the eye can create glare. Typically, the nonglare angles are from 30° to 60° from the vertical.

"VEILING REFLECTIONS"

___ (6) _Color._ See p. 245.
___ (7) _Interest._

___ c. Types of overall light sources

   ___ (1) _Task lighting_ is the brightest level needed for the immediate task, such as a desk lamp. Select from table on p. 478.

   ___ (2) _General lighting_ is the less bright level of surroundings for both general seeing and to reduce contrast between the task and surroundings. It is also for less intense tasks, such as general illumination of a lobby. This type of lighting can be natural and/or artificial.

   ___ (3) As a _general rule,_ general lighting should be about _one-third_ that of task lighting down to _20 fc._ Noncritical lighting (halls, etc.) can be reduced to _one-third_ of general lighting down to _10 fc._

   For more details, see p. 478.

___ d. Typical amounts of light

   ___ (1) Residential

      ___ _Casual_ activities: 20 fc
      ___ _Moderate_ activities (grooming, reading, and preparing food):   up to 50 fc

       ___ *Extended* activities
(hobby work, household
accounts, prolonged
reading):   up to 150 fc

       ___ *Difficult* activites (sewing): up to 200 fc

___ (2) Commercial

       ___ *Circulation:*   up to 30 fc

       ___ *Merchandising:*   up to 100 fc

       ___ *Feature* displays: up to 500 fc

       ___ *Specific* activities
(i.e., drafting):   200 fc to 2000 fc

___ (3) For more detailed recommendations, see
p. 478.

___ *e.* For recommended room reflectances, see p. 249.

___ *f.* *Calculation* of a point source of light on an object
can be estimated by:

$$\text{Foot-candles} = \frac{\text{Source}}{\text{distance}^2} \times \text{Cosine of incident angle}$$

SOURCE CAN
BE IN CANDLES,
LUMENS, OR
FOOT-LAMBERTS

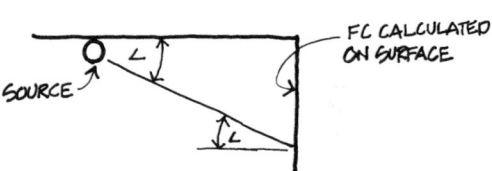

Light hitting a surface at an angle will illuminate the
surface less than light hitting perpendicular to the
surface. The cosine of the incident angle is used to
make the correction. Doubling the distance from
source to surface cuts the illumination of the surface
to one-fourth of its previous intensity. Also, see
p. 491 for other calculations.

DESIGN LIGHTING LEVELS

	TYPE OF ACTIVITY	TYPE OF LIGHTING	FOOTCANDLES X	Y	Z	TYPICAL SPACES
A	PUBLIC SPACES W/DARK SURROUNDINGS	GENERAL AREA LIGHTING THROUGHOUT SPACES	2	3	5	THEATER, STORAGE
B	SIMPLE ORIENTATION FOR SHORT TEMPORARY VISITS		5	7.5	10	DINING, CORRIDORS, CLOSETS, STORAGE
C	WORKING SPACES WHERE VISUAL TASKS ARE ONLY OCCASIONALLY PERFORMED		10	15	20	WAITING, EXHIBITION, LOBBIES, LOCKERS, RESIDENTIAL DINING, STAIRS, TOILETS, ELEVATORS, LOADING DOCKS
D	PERFORMANCE OF VISUAL TASKS OF HIGH CONTRAST OR LARGE SIZE	ILLUMINATION ON TASK	20	30	50	GENERAL OFFICE, EXAM ROOMS, MANUFACTURING, READING ROOMS, DRESSING, DISPLAY
E	PERFORMANCE OF VISUAL TASKS OF MEDIUM CONTRAST OR SMALL SIZE		50	75	100	DRAFTING, LABS, KITCHENS, EXAM ROOM, SEWING, DESKS, FILES, WORK BENCH, READING, MANUFACTURING, CLASSROOMS
F	PERFORMANCE OF VISUAL TASKS OF LOW CONTRAST OR VERY SMALL AREA		100	150	200	ARTWORK AND DRAFTING, DEMONSTRATION, INSPECTION, SURGERY, LABS, FITTING, RECORDS, CRITICAL AT WORK BENCH, DIFFICULT SEWING, MANUFACTURE ASSEMBLY
G	PERFORMANCE OF VISUAL TASKS OF LOW CONTRAST AND VERY SMALL SIZE OVER A PROLONGED PERIOD.	ILLUMINATION ON TASK BY COMBINATION OF GENERAL AND LOCAL LIGHTING	200	300	500	CRITICAL SURGERY, VERY DIFFICULT MANUFACTURING ASSEMBLY, CLOSE INSPECTION
H	PERFORMANCE OF VERY PROLONGED & EXACTING VISUAL TASK		500	750	1000	
I	PERFORMANCE OF VERY SPECIAL TASKS OF EXTREMELY LOW CONTRAST AND SMALL SIZE		1000	1500	2000	

	AGE	% REFL.	SPEED &/OR ACCURACY
X	<40	>70	NOT IMPORT.
Y	40-55	30-70	IMPORT.
Z	>50	<30	CRITICAL

2. _Electric (Artificial) Lighting_

(4) (10) $(11A)$ (43) $(45A)$

For energy conservation, see p. 224.

For energy conservation, see p. 224.

___ a. Lamp types

    ___ (1) _Incandescent_ lamps produce a warm light, are inexpensive and easy to use but have limited _lumination per watt (20 to 40)_ and a short life. _Normal voltage_ lamps produce a point source of light. Most common shapes are A, R, and PAR. _Low voltage_ lamps produce a very small point of intense brightness that can be focused into a precise beam of light (for merchandise or art). These are usually PAR shapes or designed to fit into a parabolic reflector. Sizes are designated in ⅛ inch of the widest part of lamp. Tungsten-Halogen (quartz) and low voltage are a special type of incandescent. Quartz is another type of incandescent that has high-intensity white light with slightly longer life.

    ___ (2) _Gaseous discharge_ lamps produce light by passing electricity through a gas. These lamps require a ballast to get the lamp started and then to control the current.

        ___ (a) _Fluorescent lamps_ produce a wide, linear, diffuse light source that is well-suited to spreading light downward to the working surfaces of desks or displays in a commercial environment with normal ceiling heights (8' to 12'). Lamps are typically 17, 25, or 32 watts. The deluxe lamps have good color-rendering characteristics and can be chosen to favor the _cool_ (_blue_) or the _warm_ (_red_) end of the spectrum. _Dimmers for fluorescents are expensive._ Fluorescent lamps produce more _light per watt of energy (70–85 lumens/watt)_ than incandescent; thus operating costs are low. The purchase price and length of life of fluorescent lamps are greater than for incandescent and less than for HID. Four-feet lamp lengths utilize 40 watts and are most common. *Designations are F followed by wattage, shape, size, color, and a form factor.*

___ (*b*) *High-intensity discharge* (*HID*) lamps can be focused into a fairly good beam of light. These lamps, matched with an appropriate fixture are well-suited to beaming light down to the working place from a high ceiling (*12′ to 20′*). *Dimming HID lamps is difficult. The lamps are expensive but produce a lot of light and last a long time.* If there is a power interruption, HID lamps will go out and cannot come on again for about *10 minutes* while they cool down. Therefore, *in an installation of HID lamps, a few incandescent or fluorescent lamps are needed to provide backup lighting.* Since they operate at high temperatures, they would be a poor choice for low ceilings, wall sconces, or any other close-proximity light source. They would also be a poor choice in assemblies and other occupancies where power outages could cause panic.

___ *Mercury vapor* (MV; the *bluish* street lamps). Because they emit a blue-green light, they are excellent for highlighting foliage, green copper exteriors, and certain signage. Deluxe version is warmer. *35 to 65 lumens/watt.* This is not much used anymore.

___ *Metal halide* (MH) are often *ice blue cool* industrial-looking lamps. Deluxe color rendering bulbs are 50 to 400 watts, and almost as good as deluxe fluorescent for a warmer effect. Efficiency is *80 lumens/watt.*

___ _High-pressure sodium_ (HPS)
produces a _warm golden yellow_
light often used for highways.
Bulbs are 35 to 400 watts. Deluxe color rendering is almost as
cool as deluxe fluorescent for a
cooler effect. Efficiency is _100
lumens/watt._

___ _Low-pressure sodium_ (LPS)
produces a _yellow_ color which
makes all colors appear in
shades of grey. They are excellent for promoting plant
growth indoors. Bulbs are typically 35 to 180 watts. Used for
parking lots and roadways.
Efficiency is _150 lumens/watt._

___ (c) _Cold cathode_ (neon) has a color
dependent on the gas and the color
of the tube. _Can be most any color._
Does not give off enough light for
detailed visual tasks, but does give
off enough light for _attracting atten-
tion,_ indoors or out.

COMMON LAMP SIZES & SHAPES (11A)

LAMP SHAPE TYPE DESIGNATION	LAMP SIZE (NOS = DIAMETERS IN 1/8")						
	I	Q	F	MV	MH	HP	LP
A	15-25	—	—	23	—	—	—
E	—	—	—	23-28	17-37	25	—
ED	—	—	—	17-37	—	17-37	—
G	16-40	—	—	—	—	—	—
T	8-21	3-5	—	—	15	10	17-21
BT	—	—	—	37-56	37-56	—	—
R	14-60	12-30	—	40-60	40-60	—	—
ER	30-40	—	—	—	—	—	—
PAR	16-64	16-64	—	38	38	38	—
	BEAM SPREADS RANGE FROM 5° TO 130°						
MR	—	11-16	—	—	—	—	—
STRAIGHT TUBE	8-10 L=24"	—	5-17 L=4-96"	—	—	—	—
COMPACT	—	—	9-40	—	—	—	—
DOUBLE-ENDED	—	3-6 L=4-10"	—	—	6-8 L=4-14"	—	—

1. LAMP SHAPE DESIGNATIONS: **A** = ARBITRARY OR STANDARD, **E** = ELONGATED, **ED** = ELLIPSOIDAL, **G** = GLOBE, **R** = REFLECTOR, **ER** = ELLIPTICAL REFLECTOR, **PAR** = PARABOLIC ALUMINIZED REFLECTOR, **T** = TUBULAR, **BT** = BLOWN TUBE, **MR** = MULTIFACETED REFLECTOR (SMALL QUARTZ CAPSULE IN A FACETED GLASS REFLECTOR).
2. LAMP TYPE DESIGNATIONS: **I** = INCANDESCENT, **Q** = QUARTZ, **F** = FLUORESCENT, **MV** = MERCURY VAPOR, **MH** = METAL HALIDE, **HP** = HIGH PRESSURE SODIUM, **LP** = LOW PRESSURE SODIUM.

___ *b.* Types of reflectors: (11A)

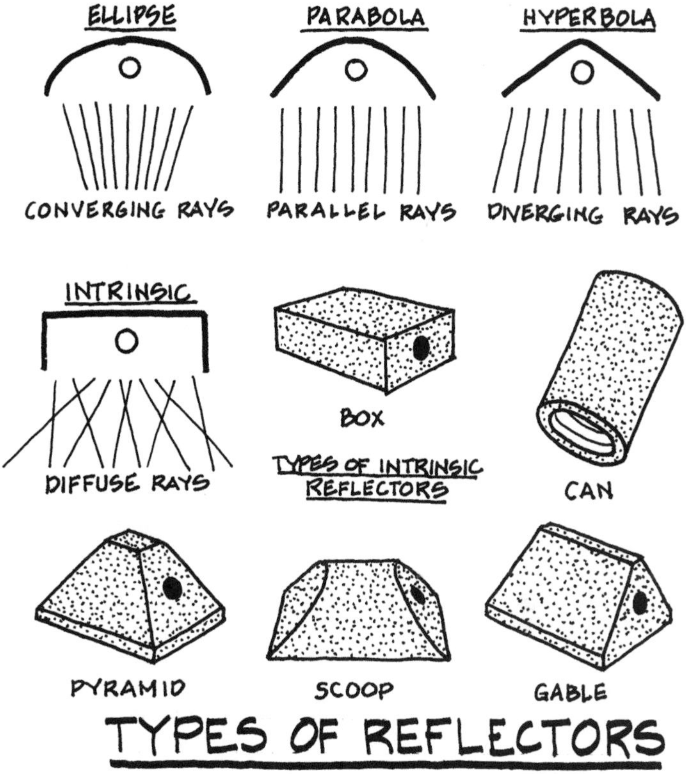

ELLIPSE

PARABOLA

HYPERBOLA

CONVERGING RAYS   PARALLEL RAYS   DIVERGING RAYS

INTRINSIC

DIFFUSE RAYS

BOX

TYPES OF INTRINSIC REFLECTORS

CAN

PYRAMID   SCOOP   GABLE

TYPES OF REFLECTORS

___ *c.* Lighting systems and fixture types

Note: Costs include lamps, fixture, and installation labor, but not general wiring. As a rule of thumb, fixtures are 20% to 30%, and distribution (not included in following costs) is 30% to 70%.

___ (1) *General room lighting*
A large proportion of commercial space requires even illumination on the workplace. This can be done a number of ways.
___ (*a*) *Direct lighting* is the most common form of general room lighting.

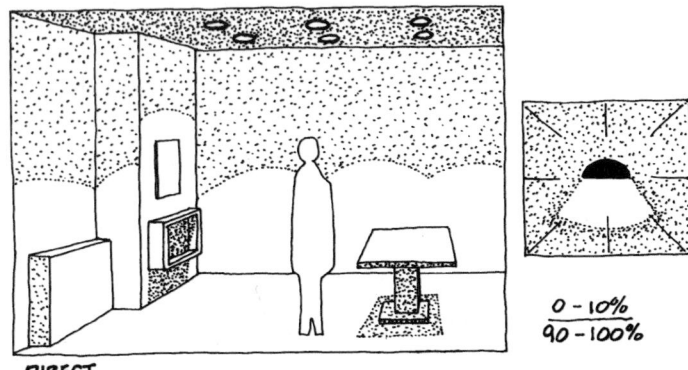

$$\frac{0 - 10\%}{90 - 100\%}$$

DIRECT

All recessed lighting is an example of a direct lighting system, but a pendant fixture could be direct if it emits virtually no light above the horizontal. Unless extensive wall washing, or high light levels (as with fluorescent for general office lighting) are used, the overall impression of a direct lighting system should be one of low general brightness with the possibility of higher intensity accents.

A guide to determine max. spacing is the *spacing-to-mounting-height ratio.* The mounting height is the height from the working place (*usually 2.5′ above floor*) to the level of the height fixtures. Note that the ratio does not apply to the end of oblong fixtures due to the nature of their light distribution.

$$\text{Spacing} = \left(\frac{S}{MH} \right) \times (\text{Mounting Ht.})$$

EXAMPLE:

WHAT IS AN AVERAGE FLUORESCENT FIXTURE SPACING IF THE CEILING IS 9' AND THE S/MH RATIO IS TO BE 1.5?

SPACING = (1.5)(9' - 2.5') = 9.75'
SAY: 10'

Types of direct lighting are:

___ *Wide-beam diffuse lighting* is often fluorescent lights for normal ceiling heights (8' to 12'). The fixtures will produce a repetitive two-dimensional pattern that becomes the most prominent feature of the ceiling plane. Typical S/MH = *1.5*.

Typical recessed fluorescent fixture:

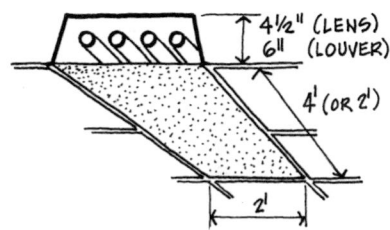

Costs: **2' × 4' = $100 to $170/ea. (85% M and 15% L), variation of −10%, +20%.**

2' × 2' = 10% less

1' × 4' = +10% more

___ *Medium-beam downlighting* is produced with a fixture located in or on the ceiling that creates a beam of light directed downward. In the circulation and lobby areas of a building, *incandescent lamps* are often used. For large areas, *HID lamps* are often selected. In both cases the light is in the form of a *conical beam,* and *scallops* of light will be produced on wall surfaces.

S/MH is usually about *0.7 to 1.3.*

Typical fixture: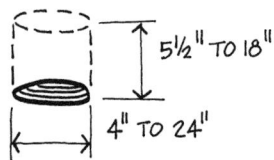

Cost: (per ea. fixture) (Variation of −10 to +35%).

	Res.	Comm.
Low voltage:	**$180**	**$370 (85% M and 15% L)**
Incandescent:	**$80**	**$370 (90% M and 10% L)**
Fluorescent:	**$150**	**$335 (85% M and 15% L)**
HID:	**$180**	**$550 (80% M and 20% L)**

___ *Narrow beam downlights* are often used in the same situation as above, but produce more of a spotlight effect at low mounting heights. This form of lighting is used to achieve even illumination where the ceiling height is relatively high. S/MH is usually *0.3 to 0.9.* Typical fixture same as above.

Cost: Same as medium-beam downlighting above.

___ (*b*) *Semidirect lighting*

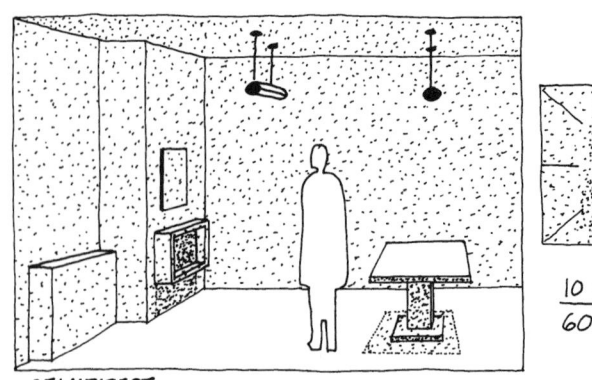

SEMIDIRECT

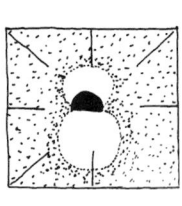

10 − 40%

60 − 90%

All systems other than direct ones necessarily imply that the lighting fixtures are in the space, whether pendant-mounted, surface-mounted, or portable. A semidirect system will provide good illumination on horizontal surfaces, with moderate general brightness.

Typical fixtures:

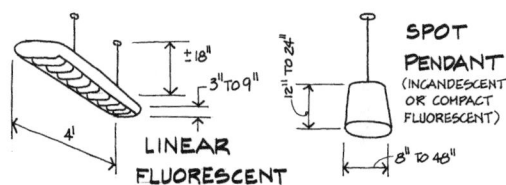

Costs: Fluor.: $395 to $920 (90% M and 10% L)
Pendant: $180 to $550 (90% M and 10% L)

___ (c) _General diffuse lighting_

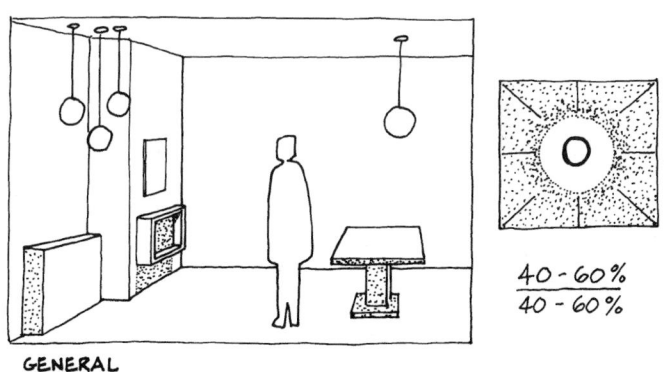

GENERAL
DIFFUSE

A general diffuse system most typically consists of suspended fixtures, with predominantly translucent surfaces on all sides. Can be incandescent, fluorescent, or HID.

Typical fixture: see sketch above
Costs: $90 to $670 (90% M and 10% L)

___ (*d*) <u>*Direct-indirect lighting*</u>

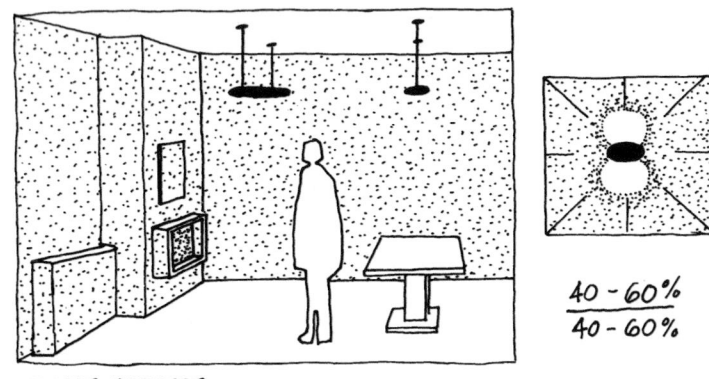

DIRECT - INDIRECT

40 - 60%
——————
40 - 60%

A direct-indirect will tend to equally emphasize the upper and lower horizontal planes in a space (i.e., the ceiling and floor).

Typical fixture: same as semidirect

Costs: Same as Semidirect.

___ (*e*) <u>*Semi-indirect lighting*</u>

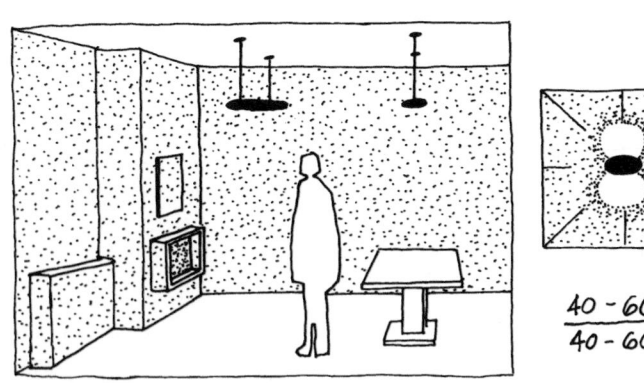

DIRECT - INDIRECT

40 - 60%
——————
40 - 60%

A semi-indirect system will place the emphasis on the ceiling, with some downward or outward-directed light.

Typical fixture:

-8" TO 18"

12" TO 20"

6' TO 8'

WALL SCONCE
(INCAND., FLUOR., HID)

14" TO 48"

-8" TO 18"

PENDANT- UP

Costs: Wall sconce: $215 to $920 (90% M and 10% L)
Pendant: $425 to $2694 (85% M and 15% L)

___ (f) *Indirect lighting*

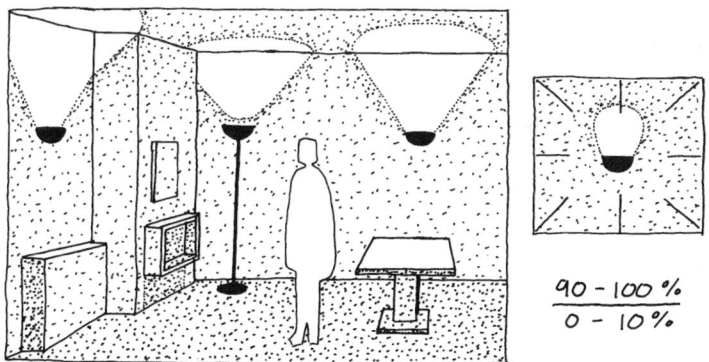

INDIRECT

90 – 100 %
0 – 10 %

A fully indirect system will bounce all the light off the ceiling, resulting in a low-contrast environment with little shadow.

Typical fixture: Same as Direct-Indirect.

Costs: Same as Direct-Indirect.

> *Note:* ADA requires that, along accessible routes, *wall-mounted* fixtures protrude no more than *4″* when mounted lower than *6′8″* AFF.
>
> ___ (g) *Accent or specialty lighting*

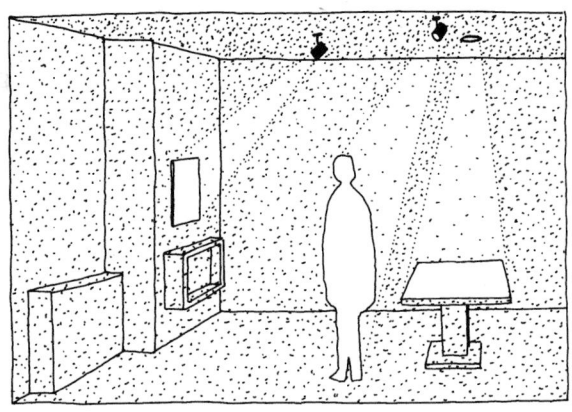

ACCENT

> Used for special effects or spot lighting, such as lighting art objects or products on display.
>
> Typical fixtures:

TRACK   RECESSED ACCENT

Costs: Track: $100 to $550 (90% M and 10%L)
Recessed accent: $180 to $1225 (80% M and 20% L)

> ___ *d.* Simplified calculations
>> ___ (1) For estimating light from one source (such as a painting on a wall lit by a ceiling mounted spot) use the *Cosine Method* shown on p. 477.
>> ___ (2) For general room lighting use the *Zonal Cavity Method*.

ZONAL CAVITY CALCULATIONS METHOD FOR GENERAL LIGHTING

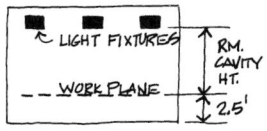

$$\text{ROOM CAVITY RATIO (RCR)} = \frac{(5)(H)(\text{LENGTH} + \text{WIDTH})}{\text{LENGTH} \times \text{WIDTH}}$$

H = HEIGHT FROM THE WORK PLAN (2.5 FT. ABOVE FLOOR) TO BOTTOM OF LIGHT FIXTURES.

LENGTH & WIDTH = ROOM DIMENSIONS

$$\text{NUMBER OF FIXTURES} = \frac{(\text{FOOTCANDLES}) \times (\text{AREA OF ROOM})}{(\text{LUMENS PER FIXTURE}) \times (CU) \times (\text{MAINT. FACTOR})}$$

FOOTCANDLES = THE DESIRED ILLUMINATION ON THE WORK PLANE.
SEE PART 1.

LUMENS PER FIXTURE = (LUMENS PER LAMP) × (NUMBER OF LAMPS
IN THE FIXTURE).

CU = COEFFICIENT OF UTILIZATION

THE COEFFICIENT OF UTILIZATION EXPRESSES THE EFFICIENCY OF THE LIGHT FIXTURE ROOM COMBINATION. IT IS DEPENDENT ON FIXTURE EFFICIENCY, DISTRIBUTION OF LIGHT FROM THE FIXTURE, ROOM SHAPE, AND ROOM SURFACE REFLECTANCES. LIGHT FIXTURE MANUFACTURERS PRINT TABLES LISTING THE CU AS A FUNCTION OF ROOM CAVITY RATIO AND ROOM SURFACE REFLECTANCES FOR EACH INDIVIDUAL LIGHT FIXTURE. SEE NEXT PAGE.

MAINTENANCE FACTOR = VARIES FROM 0.85 TO 0.65. THE MAINT. FACTOR ADJUSTS THE CALCULATION FOR THE FACT THAT LAMPS PRODUCE LESS LIGHT AS THEY GET OLDER AND FIXTURES GET DIRTY AND REFLECT LESS LIGHT OUT OF THE FIXTURE.

TYPICAL COEFFICIENTS OF UTILIZATION

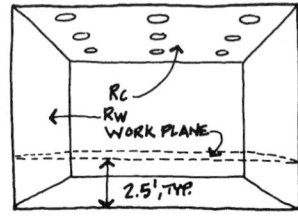

INCANDESCENT PATTERN DOWNLIGHT		
ROOM TYPE	HIGH REFL. FIN.	LOW REFL. FIN.
TYP. SMALLER RMS. (MOD. LOW CL'G.)	0.70 TO 0.80	0.60 TO 0.70
TYP. LARGER RMS.		
RELATIVELY HIGH CL'G.	0.85 TO 0.90	0.80 TO 0.85
RELATIVELY LOW CL'G.	0.90 TO 0.95	0.85 TO 0.90

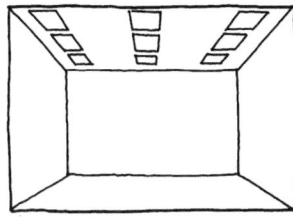

FLUORESCENT, 2×4, (PRISMATIC LENS)		
ROOM TYPE	HIGH REFL. FIN.	LOW REFL. FIN.
TYP. SMALLER RMS. (MOD. LOW CL'G.)	0.35 TO 0.45	0.30 TO 0.40
TYP. LARGER RMS.		
RELATIVELY HIGH CL'G.	0.50 TO 0.60	0.45 TO 0.50
RELATIVELY LOW CL'G.	0.60 TO 0.70	0.55 TO 0.60

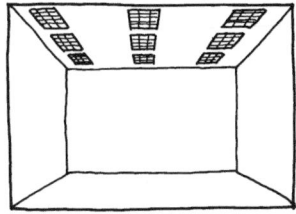

FLUORESCENT, 2×4, (PARABOLIC LOUVER)		
ROOM TYPE	HIGH REFL. FIN.	LOW REFL. FIN.
TYP. SMALLER RMS. (MOD. LOW CL'G.S.)	0.30 TO 0.45	0.25 TO 0.35
TYP. LARGER RMS.		
RELATIVELY HIGH CL'G.	0.55 TO 0.65	0.45 TO 0.55
RELATIVELY LOW CL'G.	0.65 TO 0.75	0.55 TO 0.65

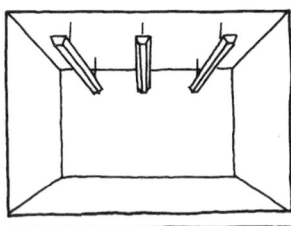

FLUORESCENT PATTERN OF INDIRECT LIGHTING		
ROOM TYPE	HIGH REFL. FIN.	LOW REFL. FIN.
TYP. SMALLER RMS. (MOD. LOW CL'G.S.)	0.35 TO 0.50	0.15 TO 0.20
TYP. LARGER RMS.		
RELATIVELY HIGH CL'G.	0.40 TO 0.65	0.20 TO 0.30
RELATIVELY LOW CL'G.	0.50 TO 0.75	0.30 TO 0.40

H.I.D. PATTERN OF INDIRECT LIGHTING		
ROOM TYPE	HIGH REFL. FIN.	LOW REFL. FIN.
TYP. SMALLER RMS. (MOD. LOW CL'G.S.)	0.28 TO 0.38	0.05 TO 0.15
TYP. LARGER RMS.		
RELATIVELY HIGH CL'G.	0.40 TO 0.55	0.10 TO 0.20
RELATIVELY LOW CL.	0.50 TO 0.65	0.10 TO 0.25

EXAMPLE :

PROBLEM:

DO A PRELM. DESIGN OF A
20' x 30' CLASS ROOM W/
DESK HEIGHT OF 2.5' AND
CEILING HEIGHT OF 9'. USE
2 x 4 LAY IN FLUOR. LIGHTS
WITH 4 - 32 WATT LAMPS.
ASSUME REFLECTANCE OF:
CEILINGS = 80%; WALLS =
50%; AND FLOORS = 40%

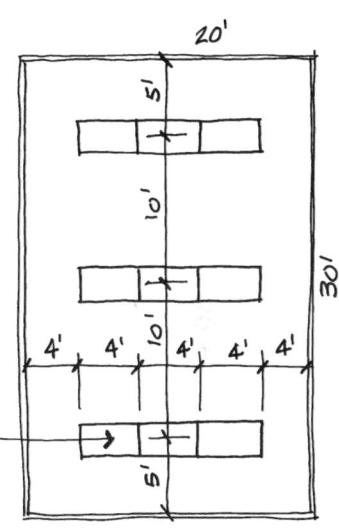

2 x 4 FLUOR. FIXTURE ⟶

SOLUTION:

1. NO. OF FIXTURES = $\dfrac{FC \times A}{LUM/FIX \times CU \times MF}$

   WHERE: FC = DESIRED LIGHT LEVEL, SELECT 75 FC (P. 478)
   A = AREA OF ROOM = 20' x 30' = 600 SF
   LUM./FIX. = ASSUME 80 LUM/WATT (SEE P. 479)
   × 32 WATTS × 4 LAMPS = $\dfrac{10\,240}{LUM./FIX.}$

   CU = COEF. OF UTILIZATION. FROM TYPICAL CU's
   ON P.   , AT FLUOR., 2 x 4, SELECT 0.6
   MF = MAINT. FACTOR, SELECT 0.8

   $= \dfrac{75 \times 600}{10240 \times 0.6 \times 0.8}$ = 9.15

   SAY 9 FIX.

2. SPACING (P. 480) FOR DIRECT FLUOR = S/MH = 1.5
   SPACING = (1.5)(9 - 2.5) = 9.75'   SAY 10'

3. LAYOUT AS SHOWN ABOVE.

NOTES

__ B. POWER AND TELEPHONE

(1) (10) (20) (45A)

For Energy Conservation, see p. 224. The architect needs to contact the utilities early to verify power availability and type.

__ 1. Electrical Power

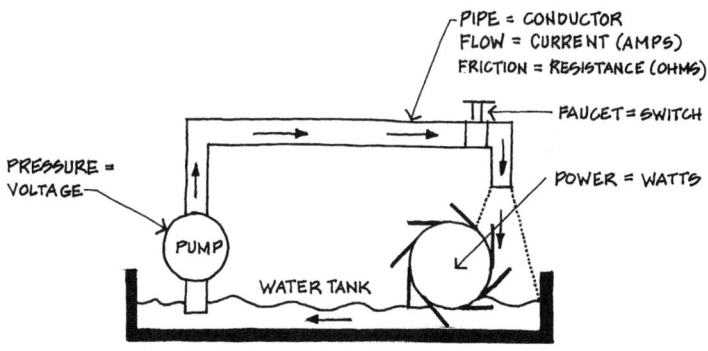

PIPE = CONDUCTOR
FLOW = CURRENT (AMPS)
FRICTION = RESISTANCE (OHMS)

FAUCET = SWITCH

PRESSURE = VOLTAGE

POWER = WATTS

PUMP

WATER TANK

__ *a.* Water analogy (an electrical circuit)
    1 volt = Force needed to drive a current of 1 amp through a resistance of 1 ohm.
    1 watt = Rate at which electrical energy is consumed in a circuit with a force of 1 volt in a current of 1 amp.

__ *b.* Basic formulas
    __ Power formula: Watts = volts × amps
    Used to convert wattage ratings of devices to amps. Wires and circuits are rated by amps.
    __ Ohm's law:

$$\text{Amps} = \frac{\text{volts}}{\text{ohms}}$$

Devices may draw different amperage even though connected to the same voltage.

__ *c.* Modern electronics and computers are increasingly having an impact on building design, requiring more space than ever.

__ *d.* A building's electrical system has three general parts: *service* (where the power enters and is regulated), *distribution* (the network of wires that carry power to all parts of the building), and *circuits* (where the energy

495

is utilized). An electrical system may be classified as small, medium, or large. As a rule, provide 20% to 30% of breaker space for future expansion.

___ *e.* *Building power systems* consist of:

    ___ *Transformer* to reduce voltage from utility power grid. Exterior ones should ideally be 20′ away from the building.

    ___ *Main switchboard* (sometimes called *service entrance section* or *switchgear*) with main disconnect and distribution through circuit breakers or fused switches.

    ___ *Subpanels and branch circuits* to distribute power throughout building.

More detailed description based on building size:

___ (1) *Residential and small commercial buildings* typically use *120/240 volt, single-phase* power, at 60 to 200 amps and one or two panel boxes.

    ___ (*a*) *Transformers* are pole mounted (oil cooled, *18″ dia.* × *3′ H*) or for underground system, oil or dry type pad mounted on ground. Both outside building.

    ___ (*b*) *Main switchboard* usually located at power entry to building and typically sized at *20″ W* × *5″ D* × *30″ H.*

    ___ (*c*) *Branch circuits* should not extend more than *100′* from panel. Panel boards are approx. *20″ W* × *5″ D* × *30″ to 60″ H.* The max. no. of breakers per panel is *42.*

    ___ (*d*) *Clearance* in front of panels and switch boards is usually *3′ to 6′.*

___ (2) *Medium-sized commercial buildings* typically use *120/208 V, 3-phase* power to operate large motors used for HVAC, etc., as well as to provide 120 V for lights and outlets. Service is typically 800 to 1200 amps.

    ___ (*a*) *Transformer* is typically liquid-cooled, pad-mounted outside building and should have *4′* clearance around and be within *30′* of a drive.

The size can be approximated by area served:

Area	No. res. units	Pad size
18,000 SF	50	$4' \times 4'$
60,000 SF	160	$4.5' \times 4.5'$
180,000 SF		$8' \times 8'$

___ (*b*) *Main switchboard* for lower voltage is approx. *6′ W × 2′D × 7′H* (for 2000 amps or less or up to 70,000 SF bld'g.). Provide *3′ to 6′* space in front for access. Higher voltage require access from both sides. *3000 amps* is usually the largest switchboard possible.

___ (*c*) *Branch panels:* For *general* lighting and outlets is same as for residential and small commercial except there are more panels and at least *one per floor.* The panel boards are generally related to the functional groupings of the building.

For *motor* panels, see large buildings.

___ (3) *Large commercial buildings* often use *277/480 V, 3-phase* power. They typically purchase power at higher voltage and step down within the building system. Typically, electrical rooms are required, ideally with two exits (one to the outside). All large electrical components require 3.5′ in front and side, 2.5′ at rear, and 3′ above, for clear access.

___ (*a*) *Transformer* is typically owned by the building and located in a vault inside or outside (underground). Vault should be located adjacent to exterior wall, ventilated, fire-rated, and have two exits. Smaller dry transformers located throughout the building will step the 480 V down to 120 V. See below for size.

___ (*b*) *Main switchboard* is approx. *10 to 15′ W × 5′D × 7′H* with *4′ to 6′* maintenance space on all sides. Typical sizes of transformer vaults and switchgear rooms:

Commercial building	Residential building	Transformer vault	Switchgear room
100,000 SF	200,000 SF	20′ × 20′ × 11′	30′ × 20′ × 11′
150,000 SF	300,000 SF	(30′ × 30′ × 11′ combination)	
300,000 SF	600,000 SF	20′ × 40′ × 11′	30′ × 40′ × 11′
1,000,000 SF	2,000,000 SF	20′ × 80′ × 11′	30′ × 80′ × 11′

Over *3000 amp,* go to multiple services. XFMR vaults need to be separated from rest of the building by at least *2-hr.* walls.

___ (*c*) *Branch panels*

___ Panels for lighting and outlets will be same as for medium-sized buildings except that they are often located in closets with telephone equip. The area needed is approx. *0.005* × the building area served.

___ Motor controller panel boards for HVAC equip., elev's., and other large equipment are often in (or next to) mechanical room, against a wall. A basic panel module is approx. *1′W ×* *1.5′D × 7′H.* One module can accommodate 2- to 4-motor control units stacked on top of one another. Smaller motors in isolated locations require individual motor control units approx. *1′W × 6″D × 1.5′H.*

___ (*d*) *Other:* In many buildings an emergency generator is required. Best location is outside near switchgear room. If inside, plan on a room 12′W × 18′ to 22′L. If emergency power is other than for life safety, size requirements can go up greatly. In any case, the generator needs combustion air and possibly cooling.

___ *f.* Miscellaneous items on electrical
___ (1) Circuit symbols on electrical plans

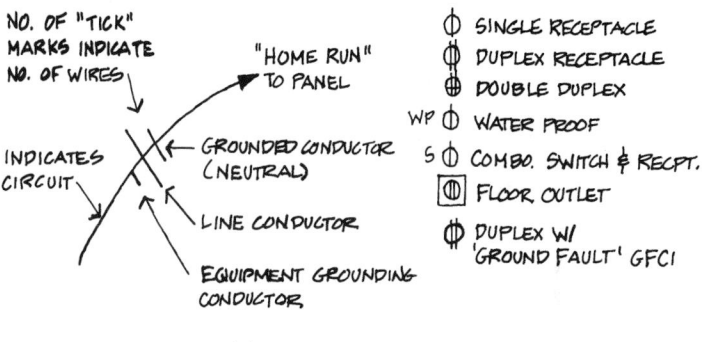

___ (2) Lightning protection: As a rule, a tall building should have at least 2 lightning rods on its roof, with special conductors down to ground terminals.
___ (3) For fire alarms, see p. 439.
___ (4) Residential
___ (*a*) Service drops (overhead lines) must be:
___ *10′* above ground or sidewalk
___ *15′* above driveways
___ *18′* above streets
___ (*b*) A min. of *1 wall switch* controlling lighting outlets required in all rooms (but convenience outlets may apply in main rooms).
___ (*c*) All rooms require a convenience outlet every 12′ along walls, 2′ or longer.
___ (*d*) Provide sufficient 15- and 20-amp circuits for min. of *3 watts of power/SF. One* circuit for every *500 to 600 SF.*
___ (*e*) A min. of *two* #12 wire (copper), *20-amp* small appliance circuits required pantry, dining, family, extended to kitchen.

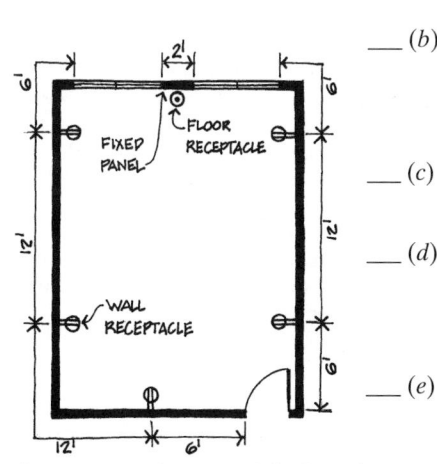

PLAN OF OUTLETS IN A TYP. ROOM

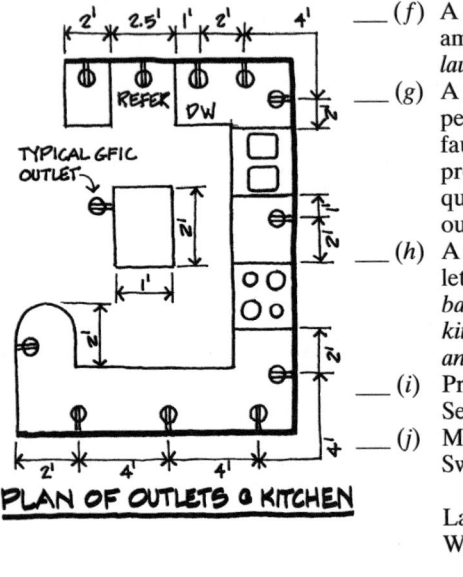

PLAN OF OUTLETS @ KITCHEN

___ (f) A min. of *one* #12 wire, 20-amp circuit required for *laundry* receptacle.

___ (g) A min. of *one* receptacle per *bathroom* with ground fault circuit interrupter protection (GFCI, required within *6'* of water outlet and at exteriors).

___ (h) A min. of *one 20-amp* outlet (GFCI) required in *basement, garage, patios, kitchen counters, wet bars, and crawl spaces.*

___ (i) Provide *smoke detector.* See p. 164.

___ (j) Mounting heights:
Switches, counter receptacles, bath outlets: *4'* AFF
Laundry: *3'6"* AFF
Wall convenience outlets: *12"* AFF

___ (3) For outlets and controls required to be *HC accessible,* per ADA, place between *18" and 4'* AFF.

___ (4) Always check room switches against *door swings.*

___ (5) Check flush-mounted wall panels against *wall depth.*

___ (6) Building must always be *bonded and grounded* by connecting all metal piping to electrical system, and by connecting electrical system into the ground by either a buried rod or plate outside the building or by a wire in the footing (UFER).

___ (7) Consider *lightning protection* by a system of rods or masts on roof connected to a separate ground and into the building elect. ground system.

Costs (20% M and 80% L):
  Outlets including wiring:
    Residential: $50–$70/ea.
    Commercial: $60–$80/ea.
    Hospital: $50–$95/ea.

___ 2. <u>Building Telephone and Signal Systems</u>
   ___ *a.* Small buildings often have a telephone mounting board (TMB) of ¾″ plywood with size up to 4′ × 4′.
   ___ *b.* Medium-size buildings often need a telephone closet of *4′ to 6′*.
   ___ *c.* Large buildings typically have a *400-SF* telephone terminal room. Secondary distribution points typical throughout building (one per area or floor) usually combined with electrical distribution closets (approx. *0.005* × area served).
   ___ *d.* ADA requires that where public phones are provided, at least *one* must be HC-accessible (1 per floor, 1 per bank of phones). See ADA for special requirements.

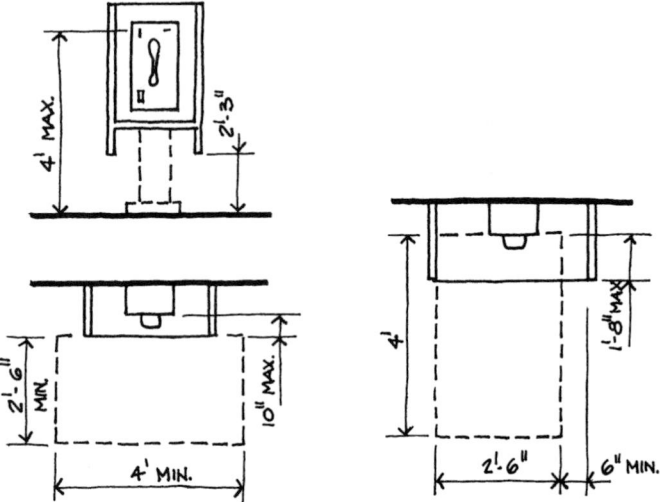

   ___ *e.* ADA/ANSI now has requirements on emergency signals (called "appliances") in buildings. Where required by code, wall-mounted appliances must be either 6'8" to 8' above floor or 4" to 12" below ceiling. For very high rooms, ceiling-mounted appliances must be suspended to be no higher than 30'. Corridors must have appliances every 50' to 100' and 15' from ends. Rooms must have one appliance unless it is not visible everywhere, limited to two, but 80' square rooms or larger may require more.

NOTES

NOTES

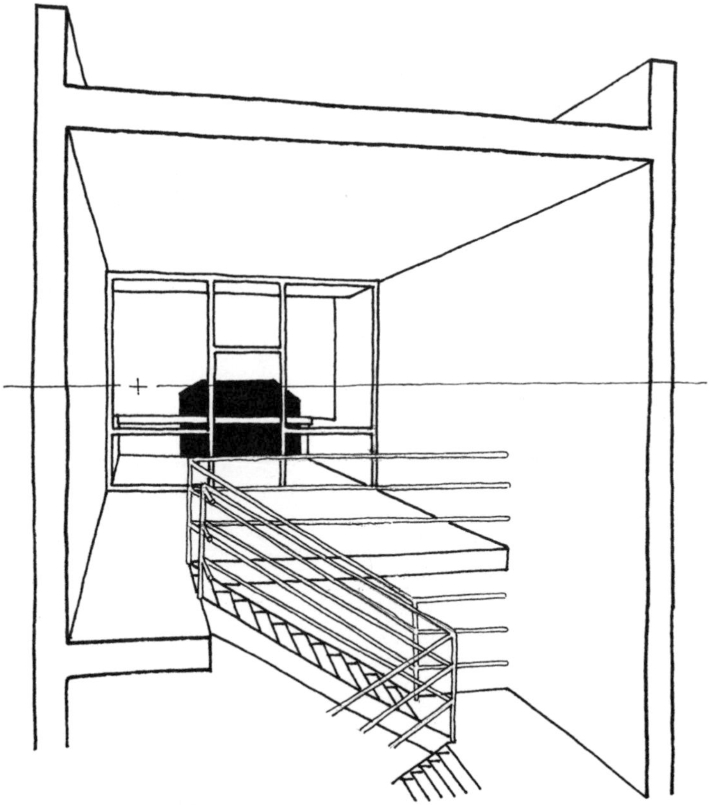

APPENDIX

NOTES

INTERIOR COSTS AND INTERIOR DESIGN FEES BY BUILDING TYPE

Also, see p. 3.

Unfortunately, the interiors industry has not developed much data on nationwide average installation costs and design fees, at least as compared to general construction and architectural fee surveys. Perhaps future data will improve and be available for future editions.

This appendix is the best that can be done with the limited amount of data available.

The following tables are laid out as follows:

___ A. General
>   FF&E installed costs are in $/SF. These costs *may* include some general construction costs (ceilings, walls, etc.) beyond "furniture, fixtures, and equipment." For restaurants, of the FF&E costs, the E (kitchen equipment) is probably excluded. You should note that the source in B, below, gives a breakdown of interior designers' installed costs as about 30% for furniture and fixtures and 70% for built construction. Much of the data is from App. A of *The Architect's Portable Handbook,* 4th ed.
>
>   Interior design fees are a percentage of the above FF&E costs.

___ B. This section is a summary of data available from a survey of the top 100 firms (top gross fees) in the United States. This data is *not* a national average. The data presented is for total work of the firms, not just the building type in question, and therefore is of limited value. The high and low numbers are the highest and lowest of the 20 firms surveyed.
>
>   *Note:* All numbers in Appendix A have been rounded off.

OFFICE

A. OFFICE, LARGE (+50000 SF OR HIGHRISE)

		LOW	MED	HIGH
F.F.&E. INSTALLED COSTS	$/SF	10	25	65
INTERIOR DESIGN FEES	%	3	4	5

OFFICE, MID RISE / SIZE

		LOW	MED	HIGH
F.F.&E. INSTALLED COSTS	$/SF	—	—	—
INTERIOR DESIGN FEES	%	3	4	5

OFFICE, SMALL (UNDER 50000 SF, LOW RISE)

		LOW	MED	HIGH
F.F. & E. INSTALLED COSTS	$/SF	20	40	60
INTERIOR DESIGN FEES	%	3	5	6

4

B. AVERAGE OF 10 FIRMS WITH HIGHEST FEES IN OFFICE TYPE PROJECTS:

AVE. F.F. & E.	$/SF	44.50	89.30	145.70
AVE. FEES	%	2.8	4.1	5.6
AVE. FEES	$/SF	2.40	3.65	8.25

FINANCIAL

		LOW	MED	HIGH
A. BANKS				
F.F. & E. INSTALLED COSTS	$/SF	15	20	25
INTERIOR DESIGN FEES	%			

B. AVERAGE OF 10 FIRMS WITH HIGHEST FEES IN FINANCIAL:

AVE. F.F. & E.	$/SF	35	85	155
AVE. FEES	%	3	5.9	13.18
AVE. FEES	$/SF	2.40	4.30	8.25

HOSPITALITY

		LOW	MED	HIGH
A. RESTAURANTS				
F.F.&E. INSTALLED COSTS	$/SF	35	60	120
INTERIOR DESIGN FEES	%	4	5	5
HOTELS (HIGH RISE)				
F.F.&E. INSTALLED COSTS	$/SF	—	25	—
INTERIOR DESIGN FEES	%	5	7	8
HOTELS (LOW RISE & MOTELS)				
F.F.&E. INSTALLED COSTS	$/SF	—	22	—
INTERIOR DESIGN FEES	%	4	5	6

B. AVERAGE OF 10 FIRMS WITH HIGHEST FEES IN HOSPITALITY:

AVE. F.F.&E.	$/SF	55.40	167.35	450
AVE. FEES	%	1.7	6.5	16.7
AVE. FEES	$/SF	21	10.50	33.75

RETAIL

A. FEES (AS % OF INSTALLED COSTS)

	LOW	MED.	HIGH
SHOPPING MALLS	3	4	5
STRIP SHOPPING CENTERS	3	4	5
RETAIL STORES	3	5	9

B. AVERAGE OF 10 FIRMS WITH HIGHEST FEES IN RETAIL:

		LOW	MED.	HIGH
AVE. F.F. & E.	$/SF	33.50	77.65	120
AVE. FEES	%	2.8	5.5	13.8
AVE. FEES	$/SF	1.85	3.65	4.80

511

MEDICAL / HEALTH CARE / ASSISTED LIVING

A.

		LOW	MED	HIGH
HOSPITALS, NEW				
F.F. & E. INSTALLED COSTS	$/SF	40		65
INTERIOR DESIGN FEES	%	4	5	6
HOSPITALS, REMODEL				
F.F. & E. INSTALLED COSTS	$/SF	3	10	12
INTERIOR DESIGN FEES	%			
CLINICS				
F.F. & E. INSTALLED COSTS	$/SF	4	6	6
INTERIOR DESIGN FEES	%			
MEDICAL OFFICE				
F.F. & E. INSTALLED COSTS	$/SF	15	20	30
INTERIOR DESIGN FEES	%	4	5	7
NURSING HOMES				
F.F. & E. INSTALLED COSTS	$/SF	3	6	7
INTERIOR DESIGN FEES	%			

B. AVERAGE OF 10 FIRMS WITH HIGHEST FEES IN MEDICAL:

AVE. F.F. & E.	$/SF	35	65	90
AVE. FEES	%	3.3	6.5	9
AVE. FEES	$/SF	1.8	3.0	10

EDUCATION

A. *CLASSROOMS* (INTERIOR DESIGN FEES)

		%	$/SF	LOW	MED.	HIGH
				4	5	5
F.F. & E. INSTALLED COSTS						
COLLEGE CLASSROOMS & ADM.				7		20
COLLEGE STUDENT UNION				7		22
COLLEGE LAB				12		30
ELEMENTARY SCHOOL				6		12
JR. HIGH				6		12
SR. HIGH				6		12
VOCATIONAL				6		12

B. AVERAGE OF 10 FIRMS WITH HIGHEST FEES IN EDUCATION:

AVE. F.F. & E.	$/SF	54.70	125	344
AVE. FEES	%	3.3	6.28	12.5
AVE. FEES	$/SF	1.80	7.50	16.30

GOVERNMENT / INSTITUTIONAL

	LOW	MED	HIGH
A. GENERAL			
1. INTERIOR DESIGN FEES (% OF INSTALLED COSTS)			
POSTAL	3	4	5
FEDERAL OFFICE	4	5	7
CORRECTION	3	4	5
STATE & LOCAL GOV. OFFICE BUILDING	4	6	7
2. INSTALLED F.F. & E. COSTS ($/SF)			
COURT HOUSE		35	

B. AVERAGE OF 10 FIRMS WITH HIGHEST FEES IN GOVERNMENT WORK:

		LOW	MED	HIGH
AVE. F.F. & E.	$/SF	31	69.40	100
AVE. FEES	%	3.3	7.5	12.8
AVE. FEES	$/SF	1.80	4.70	10.50

514

RESIDENTIAL

A. GENERAL

		LOW	MED.	HIGH
APTS./CONDO (LARGE, + 100000 SF, HIGH RISE)				
F.F. & E. INSTALLED COSTS	$/SF	12	18	25
INTERIOR DESIGN FEES	%	3	5	8
APTS./CONDOS (SMALL, - 100 000 SF, LOW RISE)				
F.F. & E. INSTALLED COSTS	$/SF	12	18	25
INTERIOR DESIGN FEES	%	4	5	5
SINGLE FAMILY				
F.F. & E. INSTALLED COSTS	$/SF	5	18	60
INTERIOR DESIGN FEES	%	5	8	12
DORMS, HOUSING				
F.F. & E. INSTALLED COSTS	$/SF	4	25	
INTERIOR DESIGN FEES	%		6	7

NOTES

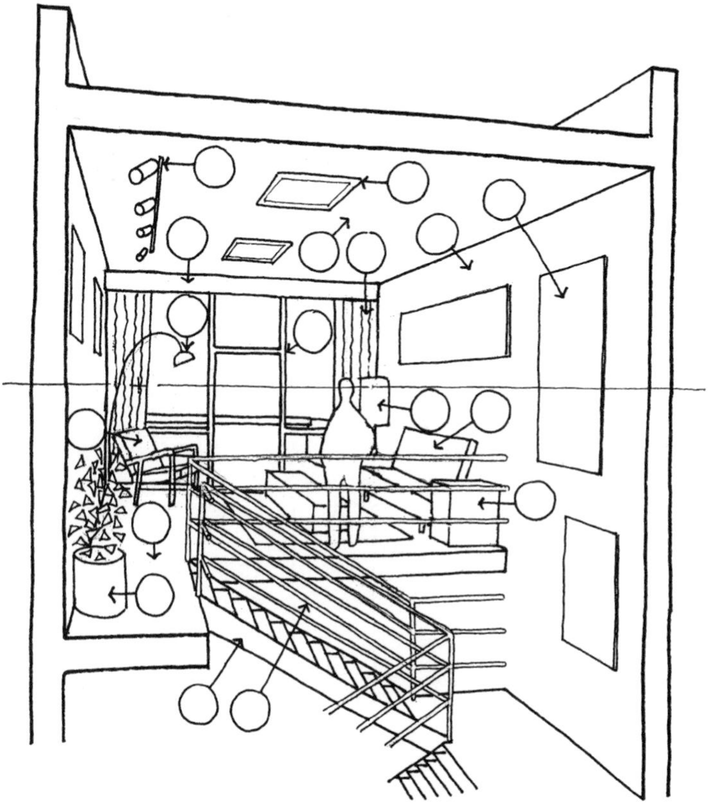

REFERENCES / INDEX

NOTES

__ REFERENCES

This book was put together from a myriad of sources. References shown at the front of a section indicate general background information. A reference shown at a specific item indicates a copy from the reference. The major book references are listed as follows, and many are recommended for architects' libraries:

___ (1) Allen, Edward, and Joseph Iano. *The Architect's Studio Companion, Rules of Thumb for Preliminary Design,* 3rd ed. New York: John Wiley & Sons, Inc., 2002.

___ (2) Ambrose, James. *Simplified Design for Building Sound Control.* New York: John Wiley & Sons, Inc., 1995.

___ (3) American Institute of Architects. *Architectural Graphic Standards,* 6th ed., New York: John Wiley & Sons, Inc., 1970.

___ (4) American Institute of Architects. *Architectural Graphic Standards,* 10th ed., New York: John Wiley & Sons, Inc., 2000.

___ (5) American Institute of Architects. *The Architect's Handbook of Professional Practice.* David Haviland, editor, 1988.

___ (6) Austin, Richard. *Designing the Interior Landscape.* New York: Van Nostrand Reinhold, 1985.

___ (7) Ballast, David Kent. *Architect's Handbook of Formulas, Tables, and Mathematical Calculations.* New Jersey: Prentice-Hall, 1988.

___ (8) Ballast, David Kent. *Architect's Handbook of Construction Detailing.* New Jersey: Prentice-Hall, 1990.

___ (9) Better Homes and Gardens, *Decorating Book,* 1975.

___ (10) Bovill, Carl. *Architectural Design, Integration of Structural and Environmental Systems.* New York: Van Nostrand Reinhold Co., 1991.

___ (10A) Building Green, Inc., 2001. *Green Spec Directory: Product Directory with Guideline Specifications, 1999–2001,* Environmental Building News, eds.

___ (11) Building News. *Facilities Manager's 2002 Cost Book.* William D. Mahoney, editor-in-chief.

___ (11A) Butler, Robert Brown. *Standard Handbook of Architectural Engineering.* New York: McGraw-Hill, 1989.

___ (12) Ching, Francis. *Building Construction Illustrated.* New York: Van Nostrand Reinhold Co., 1975.

___ (13) Ching, Francis. *Drawing—A Creative Process.* New York: Van Nostrand Reinhold Co., 1990.

— (14) Ching, Francis. *Interior Design Illustrated.* New York: Van Nostrand Reinhold Co., 1987.

— (15) Construction Specifications Institute, 1995. *Master-Format: Master List of Section Titles and Numbers.* The numbers and titles used in this product are from *MasterFormat*™ (1995 edition) and *UniFormat*™ (1998 edition) and are published by the Construction Specifications Institute (CSI) and Construction Specifications Canada (CSC), and are used with permission from CSI, 2002.

For those interested in a more in-depth explanation of *MasterFormat* and *UniFormat* and their use in the construction industry, contact:

The Construction Specifications Institute (CSI)
99 Canal Center Plaza, Suite 300
Alexandria, VA 22314
800-689-2900; 703-684-0300
CSINet: http://www.csinet.org

— (16) Craftsman Book Co. *2000 National Construction Estimator.* Edited by Martin D. Kiley and William M. Moselle. 2000.

— (17) Craftsman Book Co. *2000 Building Cost Manual.* Edited by Martin D. Kiley and Michael L. Kiley. 2000.

— (18) De Chiara, Panero, and Zelnik. *Time Saver Standards for Interior Design and Space Planning.* New York: McGraw-Hill, 1991.

— (19) De Chiara and Callender. *Time Saver Standards for Building Types.* 3rd ed. New York: McGraw-Hill, 1990.

— (19A) Department of the Air Force AFM 88-54. *Air Force Civil Engineer Handbook.*

— (20) Federal Register. *Americans with Disabilities Act* (ADA). 1991.

— (21) Foote, Rosslynn F. *Running an Office for Fun and Profit: Business Techniques for Small Design Firms.* Pennsylvania: Dowden, Hutchinson & Ross, 1978.

— (22) Flynn, Kremers, Segil, and Steffy. *Architectural Interior Systems.* 3rd Edition.

— (23) Friedmann, Arnold. *Commonsense Design.* New York: Scribners, 1976.

— (23A) Harmon, Sharon and Katherine Kennon. *The Codes Guidebook for Interiors,* 2nd ed., New York: John Wiley & Sons, Inc., 2001.

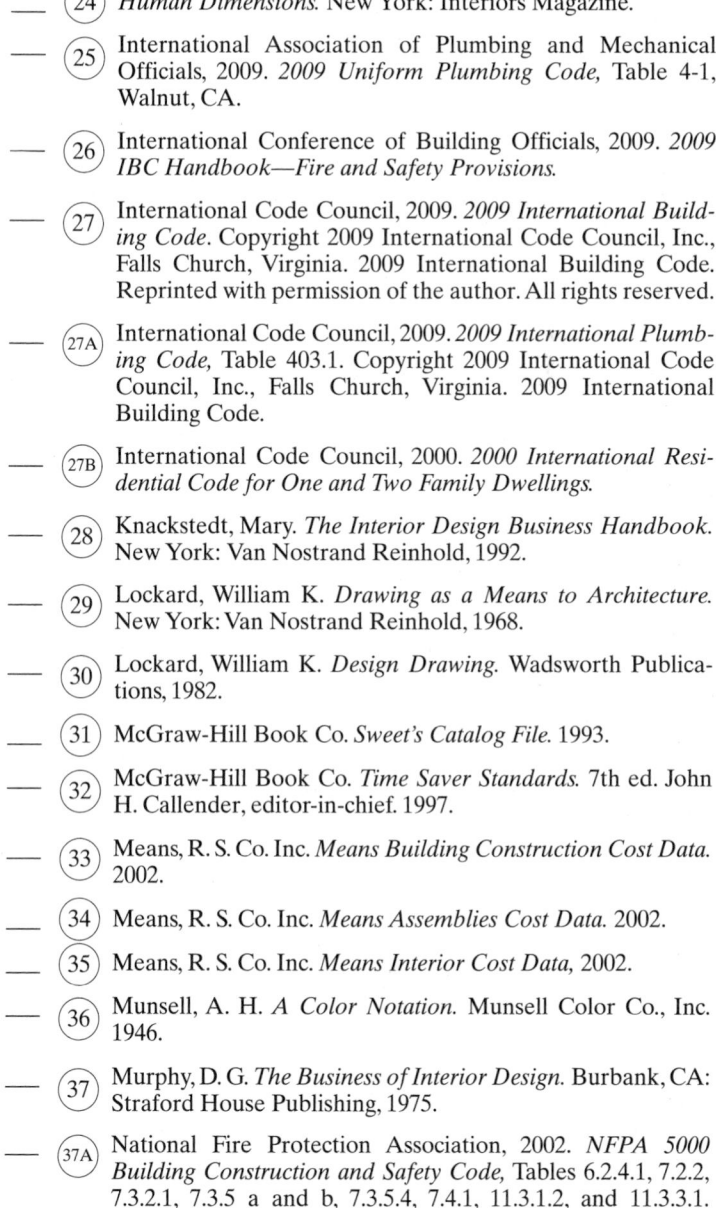

___ (24) *Human Dimensions.* New York: Interiors Magazine.

___ (25) International Association of Plumbing and Mechanical Officials, 2009. *2009 Uniform Plumbing Code,* Table 4-1, Walnut, CA.

___ (26) International Conference of Building Officials, 2009. *2009 IBC Handbook—Fire and Safety Provisions.*

___ (27) International Code Council, 2009. *2009 International Building Code.* Copyright 2009 International Code Council, Inc., Falls Church, Virginia. 2009 International Building Code. Reprinted with permission of the author. All rights reserved.

___ (27A) International Code Council, 2009. *2009 International Plumbing Code,* Table 403.1. Copyright 2009 International Code Council, Inc., Falls Church, Virginia. 2009 International Building Code.

___ (27B) International Code Council, 2000. *2000 International Residential Code for One and Two Family Dwellings.*

___ (28) Knackstedt, Mary. *The Interior Design Business Handbook.* New York: Van Nostrand Reinhold, 1992.

___ (29) Lockard, William K. *Drawing as a Means to Architecture.* New York: Van Nostrand Reinhold, 1968.

___ (30) Lockard, William K. *Design Drawing.* Wadsworth Publications, 1982.

___ (31) McGraw-Hill Book Co. *Sweet's Catalog File.* 1993.

___ (32) McGraw-Hill Book Co. *Time Saver Standards.* 7th ed. John H. Callender, editor-in-chief. 1997.

___ (33) Means, R. S. Co. Inc. *Means Building Construction Cost Data.* 2002.

___ (34) Means, R. S. Co. Inc. *Means Assemblies Cost Data.* 2002.

___ (35) Means, R. S. Co. Inc. *Means Interior Cost Data,* 2002.

___ (36) Munsell, A. H. *A Color Notation.* Munsell Color Co., Inc. 1946.

___ (37) Murphy, D. G. *The Business of Interior Design.* Burbank, CA: Straford House Publishing, 1975.

___ (37A) National Fire Protection Association, 2002. *NFPA 5000 Building Construction and Safety Code,* Tables 6.2.4.1, 7.2.2, 7.3.2.1, 7.3.5 a and b, 7.3.5.4, 7.4.1, 11.3.1.2, and 11.3.3.1.

Reprinted with permission from NFPA 5000™ Building Construction and Safety Code. Copyright © 2002, National Fire Protection Association, Quincy, MA 02269. This reprinted material is not the complete and official position of the National Fire Protection Association, on the referenced subject which is represented only by the standard in its entirety.

—— (38) Peña, William, with Parshall and Kelley. *Problem Seeking.* Washington, DC: A.I.A. Press, 1987.

—— (39) Piotrowski. *Professional Practice for Interior Designers.* New York: Van Nostrand Reinhold, 1989.

—— (40) Reznikoff, S. C. *Interior Graphic and Design Standards.* New York: Whitney Library of Design, 1986.

—— (41) Reznikoff, S. C. *Specifications for Commercial Interiors.* New York: Whitney Library of Design, 1979.

—— (41A) Running Press. *Cyclopedia: The Portable Visual Encyclopedia.* Philadelphia: Running Press, 1995.

—— (42) Saylor Publications, Inc. *1997 Current Construction Costs.* 1997.

—— (43) Schiler, Marc. *Simplified Design of Building Lighting.* New York: John Wiley & Sons, 1992.

—— (44) Sleeper, Harold. *Building Planning and Design Standards.* New York: John Wiley & Sons Inc., 1955.

—— (45) Stasiowski, Frank A. *Staying Small Successfully.* New York: John Wiley & Sons, 1991.

—— (45A) Guthrie, Pat. *The Architect's Portable Handbook: First-Step Rules of Thumb for Building Design*, 4th ed. New York: McGraw-Hill, 2010.

—— (46) U.S. Navy. *Basic Construction Techniques for Houses and Small Buildings.* New York: Dover Publications, 1972.

—— (47) Watson, Donald, and Kenneth Labs. *Climatic Design.* New York: McGraw-Hill Book Co., 1983.

—— (48) Watson, Crosbie, and Callender. *Time Saver Standards for Architectural Design Data.* 7th ed. New York: McGraw-Hill, 1997.

—— (49) Wing, Charlie. *The Visual Handbook of Building and Remodeling.* Pennsylvania: Rodale Press, 1990.

—— (50) Wood, R. S. and Co. *The Pocket Size Carpenter's Helper.* 1985.

NOTES

NOTES

Index

A

Abrasive, 385
Absorption, sound, 232
Access flooring, 360
Accessibility requirements [see
    ADA (Americans with Dis-
    abilities Act) accessibility
    requirements]
Accessories (bath), 370
Accounting reports, 4
Acoustical treatments, 340
Acoustics, 231
Acrylic (*see* Carpet)
Active building design, 224
ADA (Americans with Disabili-
    ties Act) accessibility
    requirements, 215
  alterations, 215
  area of rescue, 185
  controls, 497
  doors, 315
  elevators, 423
  general, 215
  hardware, 329
  lighting, 486
  ramps, 187
  residential, 165–166
  route, 216
  seating requirements, 48, 62

ADA (*Cont.*):
  signage requirements, 363
  for telephones, 497
  toilets/baths, 95–98
  wheelchair reaches (*see* Seat-
      ing requirements)
Airborne sound (*see* Acoustics)
Air delivery systems, 461
Air handlers, 457–471
Air patterns, 9
Air/water delivery systems, 469
Algebra, simple, 129
Alkyd paint, 349
Aluminum, 277
  doors, 319
  windows, 327
American Institute of Architects,
    fee breakdown, 3
Amps, 495
Analogous color schemes, 246
Angles, 136
Apartments/condominiums,
    costs and design fees, 515
Architects, 159, 169
Area:
  of circles, 141, 143
  gross and net, 29
  unassigned, 30
Areas, table of, 146
Artwork and accessories, 386

NOTES